Evolutionary Large-Scale Multi-Objective Optimization and Applications

IEEE Press
445 Hoes Lane
Piscataway, NJ 08854

Evolutionary Large-Scale Multi-Objective Optimization and Applications

Integrating Evolutionary Computation, Machine Learning, and Data Science

Xingyi Zhang
Anhui University
Hefei, China

Ran Cheng
Southern University of Science and Technology
Shenzhen, China

Ye Tian
Anhui University
Hefei, China

Yaochu Jin
Westlake University
Hangzhou, China

IEEE Press

WILEY

Published by John Wiley & Sons, Inc., Hoboken, New Jersey.
Published simultaneously in Canada.

Library of Congress Cataloging-in-Publication Data

Names: Zhang, Xingyi (Professor), author. | Cheng, Ran (Associate Professor), author. | Tian, Ye (Associate Professor), author. | Jin, Yaochu (Professor), author.
Title: Evolutionary large-scale multi-objective optimization and applications : integrating evolutionary computation, machine learning, and data science / Xingyi Zhang, Ran Cheng, Ye Tian, Yaochu Jin.
Description: Hoboken, New Jersey : Wiley, [2024] | Includes bibliographical references and index.
Identifiers: LCCN 2024018004 (print) | LCCN 2024018005 (ebook) | ISBN 9781394178414 (hardback) | ISBN 9781394178421 (adobe pdf) | ISBN 9781394178438 (epub)
Subjects: LCSH: Mathematical optimization. | Evolutionary computation. | Machine learning.
Classification: LCC QA402.5 .Z4193 2024 (print) | LCC QA402.5 (ebook) | DDC 519.6--dc23/eng/20240510
LC record available at https://lccn.loc.gov/2024018004
LC ebook record available at https://lccn.loc.gov/2024018005

Cover Design: Wiley
Cover Image: © Flavio Coelho/Getty Images, © Simfo/Getty Images

Set in 9.5/12.5pt STIXTwoText by Straive, Chennai, India

Contents

About the Authors

Xingyi Zhang, PhD, is a Professor in the School of Computer Science and Technology at Anhui University, Hefei, China. He serves as an Associate Editor of the *IEEE Transactions on Evolutionary Computation*, and a member of the editorial board for *Complex and Intelligent Systems*.

Ran Cheng, PhD, is an Associate Professor in the Department of Computer Science and Engineering at the Southern University of Science and Technology, China. He is an Associate Editor for the *IEEE Transactions on Evolutionary Computation, IEEE Transactions on Artificial Intelligence, IEEE Transactions on Emerging Topics in Computational Intelligence, IEEE Transactions on Cognitive and Developmental Systems, and ACM Transactions on Evolutionary Learning and Optimization.*

Ye Tian, PhD, is an Associate Professor in School of Computer Science and Technology at Anhui University, Hefei, China. He serves as an Associate Editor of the *IEEE Transactions on Evolutionary Computation.*

Yaochu Jin, PhD, is a Chair Professor of Artificial Intelligence, Head of the Trustworthy and General Artificial Intelligence Laboratory, Westlake University, China. He was an Alexander von Humboldt Professor of Artificial Intelligence at the Bielefeld University, Germany, and Distinguished Chair in Computational Intelligence at the University of Surrey, United Kingdom.

Foreword

The use of evolutionary algorithms for solving multi-objective problems originated with David Schaffer's PhD thesis (from 1984) and the two conference papers that he published out of it, in 1985. However, it was until the mid-1990s that this area, which is now called evolutionary multi-objective optimization (EMO), started to attract the attention of an important number of researchers.

During the past 30 years, multi-objective evolutionary algorithms (MOEAs) have become more sophisticated and powerful, evolving from the original non-elitist Pareto-based approaches proposed in the mid-1990s to the elitist decomposition-based and indicator-based schemes that most researchers use today. In this process, and associated to the evolution experienced by MOEAs, new (and more challenging) test problems and performance indicators have been proposed. Additionally, the research in the area has become more and more specialized.

However, interestingly enough, one topic that had been somehow disregarded for many years is large-scale multi-objective optimization. The topic, analogous to many-objective optimization, is a bit elusive to define, but most researchers consider as "large-scale" any multi-objective optimization problem having at least 100 decision variables. Although there had been studies of large-scale multi-objective optimization and its effect on MOEAs, and several applications reported in the literature at least in the past 15 years, large-scale multi-objective optimization was not considered a specific research topic (appearing as such in the title of papers) until relatively recently. Particularly, in the past 10 years, there has been a remarkable growth in the research conducted in this area.

This book contains a comprehensive and much-needed review of evolutionary large-scale multi-objective optimization. The book is divided into two parts. The first four chapters provide some basic concepts, a gentle introduction to evolutionary large-scale multi-objective optimization including techniques and algorithms representative of the state of the art in the area and the study of sparse large-scale multi-objective optimization problems. The second part, which consists of five chapters, covers applications in different domains, including community detection, capacitated vehicle routing and shelter location, ratio error estimation for voltage transformers, intensity-modulated radiotherapy, and deep neural network training. I am convinced that this book is a fundamental text for anyone interested in conducting research on evolutionary large-scale multi-objective optimization, and as such, it is meant to become a classic in the field that nobody should miss.

December 2023

Carlos A. Coello Coello
CINV ESTAV-IPN
Department of Computer Science
Mexico City, Mexico

Preface

My initial encounter with evolutionary computation occurred during my doctoral studies at Huazhong University of Science and Technology, China, in 2006. Our research group was engaged in designing reliable DNA sequences for DNA computing, with the aim of leveraging biochemical operations to efficiently resolve complex problems. Naturally, we turned to evolutionary computation as a potential solution for this NP-hard problem – a multi-objective optimization problem laden with several thermodynamic and combinatorial constraints. During this work, two difficulties in implementing evolutionary computation left a lasting impression. The first was the challenge of employing a unified constraint handling method for the constraints, as distinct strategies had to be adopted for each constraint. The second was the sheer volume of decision variables involved in DNA sequence design. We regularly encountered perplexing situations where a well-designed evolutionary algorithm performed admirably on small-sized DNA sequences, but its performance substantially deteriorated when tasked with designing DNA sequences involving hundreds of decision variables. My understanding of evolutionary computation became more structured following my visit to the University of Surrey in 2013. It was there that I recognized the unique importance and challenge of large-scale optimization, especially in multi-objective optimization scenarios. Since then, one of my primary research interests has centered on the application of evolutionary algorithms to solve varied types of large-scale multi-objective optimization problems in the real world. This rapidly evolving field within evolutionary computation is now commonly referred to as evolutionary large-scale multi-objective optimization.

In contrast to large-scale optimization for a single objective, large-scale multi-objective optimization has emerged as a vibrant research focus within the domain of evolutionary computation. The majority of significant advancements in this area have been achieved in the past decade, a development that can partly be attributed to the increasing accessibility of high-performance computational devices for general users. At present, there is no established definition as to how many decision variables constitute a "large scale" for multi-objective optimization, a determination that may well depend on the complexity of the problem being addressed. For a typical multi-objective optimization problem, it is widely accepted that 100 decision variables present a formidable challenge for evolutionary algorithms, and thus problems featuring at least 100 decision variables are often classified as large-scale multi-objective optimization problems. The challenges inherent in evolutionary large-scale multi-objective optimization primarily stem from two factors. First, the problems under consideration are usually NP-hard, implying that the search space will increase exponentially as the problem size grows. This "curse of dimensionality" presents a significant hurdle for evolutionary algorithms, which are tasked with identifying optimal or quasi-optimal solutions using an acceptable number of function evaluations. Second, the problems are likely to encompass challenging characteristics, such as sparse optimal solutions, discrete decision variables, or stringent constraints, particularly in real-world scenarios. Therefore, the adaptation of new search strategies is essential to address these characteristics; failing to do so could mean that evolutionary algorithms are unable to produce any feasible or valid solutions. As a result of these complexities, large-scale multi-objective optimization problems have necessitated the development of a range of advanced evolutionary algorithms, which are superior to conventional evolutionary algorithms in navigating high-dimensional search spaces.

This book offers an exhaustive overview of the latest developments in large-scale multi-objective evolutionary algorithms, catering to diverse readers including undergraduate and graduate students, researchers, and industry professionals in the fields of computer science, artificial intelligence, cybernetics, and automation. The book comprises nine chapters, each designed to deepen readers' understanding of this complex subject. For the self-containedness of the book, Chapter 1 gives an introduction to evolutionary multi-objective optimization and presents the details of several classical multi-objective evolutionary algorithms. Chapter 2 extends this foundation, introducing the definitions of widely used large-scale multi-objective benchmark problems as well as indicators for evaluating the performance of evolutionary algorithms. Chapters 3 and 4 delve into the details of representative evolutionary algorithms. Chapter 3 covers four evolutionary algorithms for large-scale multi-objective optimization, while Chapter 4 discusses three evolutionary algorithms for sparse large-scale multi-objective optimization. From Chapter 5 onward, this book explores several emerging applications of evolutionary large-scale multi-objective optimization. These include community detection in Chapter 5, capacitated vehicle routing and shelter location in Chapter 6, ratio error estimation for voltage transformers in Chapter 7, intensity-modulated radiotherapy in Chapter 8, and deep neural network training in Chapter 9. The contributions to this book are divided among four authors. Chapters 2, 5, and 6 are written by me, Chapters 1 and 7 are written by Ran Cheng, Chapters 3, 4, and 8 are written by Ye Tian, and Chapter 9 is written by Yaochu Jin. Ye Tian was my graduate student from 2012 to 2018 and became my colleague since 2018, while Ran Cheng and Yaochu Jin became my collaborators since I visited the University of Surrey in 2013.

For all the evolutionary algorithms, benchmark problems, and performance indicators introduced in Chapters 2–4, their MATLAB codes are encompassed within PlatEMO and can be obtained from https://github.com/BIMK/PlatEMO. With PlatEMO, readers can easily reproduce the experimental results reported in this book, or use the evolutionary algorithms to address specific large-scale multi-objective optimization problems.

This scholarly endeavor would not have materialized without the invaluable contribution and support from a multitude of colleagues, collaborators, and doctoral and master's students of mine. Foremost, my gratitude extends to Prof. Yaochu Jin, with whom I got insightful comments and suggestions on evolutionary computation from 2013 in University of Surrey and still maintained close collaboration when I completed my visit. I thank Prof. Kay Chen Tan from The Hong Kong Polytechnic University, with whom I also collaborate with on evolutionary large-scale multi-objective optimization and who gave me a number of valuable suggestions. I offer my sincere thanks to Prof. Carlos A. Coello Coello, who wrote the valuable foreword for this book and gave me some insightful suggestions for the future topics of evolutionary large-scale multi-objective optimization and evolutionary computation. The following collaborators and previous or current PhD/master's students of mine have contributed to part of the work presented in this book: Prof. Fan Cheng, Prof. Yansen Su, Prof. Handing Wang, Prof. Jianhua Xiao, Prof. Chunhou Zheng, Dr. Ruifen Cao, Dr. Ran Cheng, Dr. Cheng He, Dr. Hao Jiang, Dr. Miqing Li, Dr. Haiping Ma, Dr. Xi Pei, Dr. Ye Tian, Dr. Chao Wang, Dr. Xiaoshu Xiang, Dr. Shangshang Yang, Dr. Lei Zhang, Dr. Yajie Zhang, Ms. Haowen Chen, Mr. Fuchen Duan, Mr. Yuandong Feng, Mr. Yaopei Guang, Ms. Zhongxiang Jin, Mr. Chunlong Liu, Mr. Ruchen Liu, Mr. Zhenjie Liu, Mr. Chang Lu, Mr. Hebin Pan, Ms. Jingwen Pan, Mr. Shichen Peng, Mr. Langchun Si, Mr. Fei Zhan, Mr. Tao Zhang, Ms. Xiutao Zheng, and Mr. Kefei Zhou. I take this opportunity to thank Prof. Xiaodong Li, Prof. Hisao Ishibuchi, Prof. Yew-Soon Ong, Prof. Ke Tang, Prof. Yao Xin, Prof. Shengxiang Yang, Prof. Gary Yen, Prof. Qingfu Zhang, and Prof. Mengjie Zhang. Financial support from the National Key Research and Development Program of China (No. 2018AAA0100100), the National Natural Science Foundation of China (No. U21A20512, No. 62276001, No. 62136008), the Anhui Provincial

Natural Science Foundation (No. 2308085J03), the Excellent Youth Foundation of Anhui Provincial Colleges (No. 2022AH030013), and the Anhui Provincial Key Research Institute for Artificial Intelligence Teaching Materials are gratefully acknowledged. Finally, I thank my wife and two daughters (Ziran Zhang and Zixin Zhang), who have given great support in doing my research and completing this book.

September 2023 *Xingyi Zhang*
Hefei, China

Acronyms

AE	auto-encoder
APD	angle-penalized distance
AR-MOEA+SA	clustering-based surrogate-assisted multi-objective evolutionary algorithm with reference point adaptation
CC	cooperative coevolution
CCGDE3	cooperative coevolution generalized differential evolution 3
CCR	correct classification rate
CE	cross-entropy loss
CNNs	convolutional neural networks
CSD	convergence–sparsity–diversity
CSO	competitive swarm optimizer
CTs	current transformers
CVRP	capacitated vehicle routing problem
DAE	denoising auto-encoder
DE	differential evolution
DG2	improved differential grouping
DNNs	deep neural networks
EAs	evolutionary algorithms
EMRG-HA	evolutionary multi-objective route grouping-based heuristic algorithm
FC	fully connected
FEs	function evaluations
FLP	facility location problem
GAs	genetic algorithms
GD	generational distance
GDG	global differential grouping
GrEA	grid-based evolutionary algorithm
G-MOEA	guided multi-objective evolutionary algorithm
HSOs	hypervolume by slicing objectives

HV	hypervolume
HypE	Hypervolume estimation algorithm
IBEA	binary indicator ($I_{\epsilon+}$)-based evolutionary algorithm
IGD	inverted generation distance
IGD-NS	inverted generation distance with noncontributing solution detection
IMRT	intensity-modulated radiotherapy
KKM	kernel k-means
KnEA	knee point-driven evolutionary algorithm
LMEA	large-scale many-objective evolutionary algorithm
LMOCSO	large-scale multi-objective competitive swarm optimizer
LSCVRP	large-scale capacitated vehicle routing problem
LSMaOPs	many-objective optimization problems with a large number of decision variables
LSMOF	large-scale multi-objective framework
LSMOPs	multi-objective optimization problems with a large number of decision variables
LSSOPs	single-objective optimization problems with a large number of decision variables
LSTM	long short-term memory
LSUSLP	large-scale uncertain shelter location problem
MaOP	many-objective optimization problem
MLC	multileaf collimator
MOEAs	multi-objective evolutionary algorithms
MOEA/D	multi-objective evolutionary algorithm based on decomposition
MOEA/IMRT	multi-objective evolutionary algorithm for intensity-modulated radiotherapy planning
MOEA/PSL	multi-objective evolutionary algorithm based on Pareto optimal subspace learning
MOPs	multi-objective optimization problems
MS	maximum spread
MSE	mean squared error
MVU	maximum variance unfolding
NMI	normalized mutual information
NNs	neural networks
NSGA-II	elitist non-dominated sorting genetic algorithm
OAR	organ at risk
PAMs	partitioning around medoids

PCA	principal component analysis
PCSEA	Pareto corner search evolutionary algorithm
PESA	Pareto envelope-based selection algorithm
PF	Pareto front
PM	polynomial mutation
PMOEAs	parallel framework for multi-objective evolutionary algorithms
PM-MOEA	pattern mining-based multi-objective evolutionary algorithm
PS	Pareto set
PSO	particle swarm optimization
PTV	planning tumor volume
PV	primary voltage
Q	modularity
RBF	Radial basis function
RBM	restricted Boltzmann machine
RC	ratio cut
RCGAs	real-coded genetic algorithms
RE	ratio error
REs	relative errors
RMOEA	network reduction-based multi-objective evolutionary algorithm
RNIs	ratio of non-dominated individuals
RVEA	reference vector-guided evolutionary algorithm
SAE	stacked auto-encoder
SBX	simulated binary crossover
SDE	shift-based density estimation
SET	sparse evolutionary training
SGD	stochastic gradient descent
SMS-EMOA	S-metric selection evolutionary multi-objective algorithm
SOPs	single-objective optimization problems
SP	spacing
SparseEA	sparse evolutionary algorithm
Sparse LSMOPs	sparse large-scale multi-objective optimization problems
SPEA2	strength Pareto evolutionary algorithms 2
SSGD	sparse stochastic gradient descent
SV	secondary voltage
TREE	time-varying ratio error estimation
TS-LS	tabu search-based local search

USLP	uncertain shelter location problem
VRP	vehicle routing problem
VTs	voltage transformers
VUF	voltage unbalance factor
XDG	extended differential grouping
ϵ R-EMO	ϵ-ranking multi-objective optimizer

Symbols

$\mathbf{f}$	objective vector
x	decision variable or input variable (feature)
$\mathbf{x}$	decision vector
$\mathbb{R}$	real number space
$f(\mathbf{x})$	objective function
X	decision space
Y	objective space
P	population
N	population size
M	number of objectives
D	number of decision variables
$O(N)$	time complexity
$maxgen$	maximum generations of algorithm
z^*	reference point
$\mathbf{z}^{max}$	nadir point
$\mathbf{z}^{min}$	ideal point
ϵ	threshold
pbest	the personal best of a particle
gbest	the global best of the swarm
$\mathbf{g}$	distance function
$\mathbf{h}$	shape function
$\mathbf{x}^p$	position-related part of a decision vector
$\mathbf{x}^d$	distance-related part of a decision vector
u	upper bound of a decision variable
l	lower bound of a decision variable
θ	sparsity of Pareto optimal solutions
$ErrorRate$	error rate of a classifier
xb	binary vector of a decision vector
xr	real vector of a decision vector

G	complex network
c_p	probability of crossover
m_p	probability of mutation
T	neighborhood size
V	set of vertices
E	set of edges

1

Multi-Objective Evolutionary Algorithms and Evolutionary Large-Scale Optimization

1.1 Introduction

The word *optimization* initially appeared in the mid-19th century, meaning "to make the best or most of." Today, optimization has deeply penetrated into our thoughts and actions: engineers aim to achieve the best performance of their designs; businessmen seek to maximize the profits of their deals; manufacturers expect the highest efficiency of their production processes; investors try to minimize the risk of investment while maximizing rate of return, etc. With sophisticated formulations and modelings, such optimization-driven targets become *optimization problems*.

With problems always come solvers, as known as *algorithms* in the context of optimization. In the early age, the conventional optimization algorithms were often gradient based, e.g., Newton's method and the hill climbing method, assuming that the problems to be solved were convex (or at least differentiable) – and indeed – problems were supposed to be well defined by mathematicians. Nonetheless, the information age commencing from the mid-20th century has led modernized science, engineering, industry, and modernized optimization as well. While enjoying the benefits of modernization, we are facing emerging challenges – optimization problems are increasingly complex, far beyond the scope of conventional optimization algorithms.

Scientists have always been dedicated to looking into nature to find answers to research questions, so as in the case of solving optimization problems. The *nature-inspired optimization algorithms* are exactly a family of modern algorithms for global optimization. Facing the challenging features of complex optimization problems (e.g., nonlinearity, non-differentiability, or multi-modality), nature-inspired optimization algorithms benefit from unique advantage of robustness – treating optimization problems as *black-boxes*. Particularly, inspired by the mechanisms of biological evolution, the evolutionary

Evolutionary Large-Scale Multi-Objective Optimization and Applications: Integrating Evolutionary Computation, Machine Learning, and Data Science,
First Edition. Xingyi Zhang, Ran Cheng, Ye Tian, and Yaochu Jin.

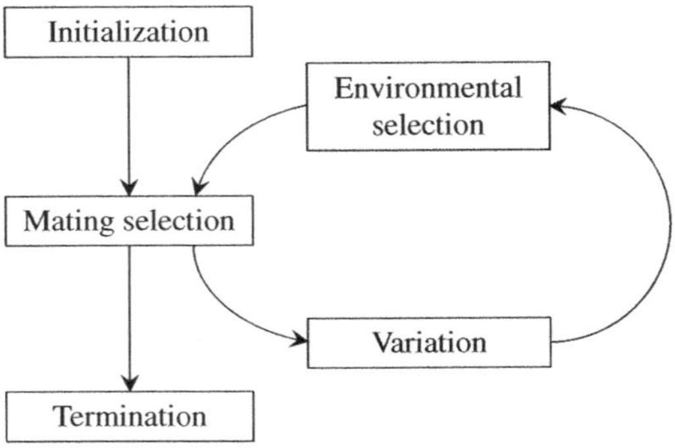

Figure 1.1 The generic flowchart of an *Evolutionary Algorithm* (*EA*).

algorithms (EAs) [1] are among the most representative niches of the family. Generally speaking, EAs encompass genetic algorithms (GAs), evolutionary strategies, evolutionary programming, and genetic programming.

As illustrated in Fig. 1.1, a generic flowchart of EAs comprises five main components: initialization, mating selection, variation, environmental selection, and termination. Normally, an EA starts with a randomly initialization population of candidate solutions. Then, the population of candidate solutions is iteratively updated via the main loop consisting of mating selection, variation, and environmental selection. Specifically, mating selection randomly picks up pairwise (parent) candidate solutions for generating new (offspring) candidate solutions via variation, and environmental selections decide which candidate solutions can survive to the next iteration (generation). Eventually, when a certain condition (e.g., the maximum iteration number) is satisfied, the EA will terminate and output the candidate solutions in the final population as the approximate optimal solution(s) to the optimization problem being solved.

During the past three decades, EAs have achieved remarkable success in solving complex optimization problems; recently, however, they are facing a new challenge brought by the surging development of information sciences – the *scalability*. Just as many other computer algorithms, the nature-inspired optimization algorithms suffer from a serious *curse of dimensionality*, i.e., as the scale of the optimization problems increases, the performance of the algorithms substantially deteriorates. Now, we are stepping forward into new adventures – scaling up EAs for solving *large-scale optimization problems*.

Before moving toward the main course of large-scale optimization problems, we might start with an appetizer – general optimization problems. Generally, an optimization problem often consists of three essential components: objective(s), decision variable(s), and constraint(s): an objective depicts the performance of the system under study, and a decision variable is a certain characteristic in correlation with the objective(s); the spaces where the objectives and decision variables are defined are often known as objective space and decision space, respectively, and a constraint specifies the limitation(s) in either the objective space or decision space. Correspondingly, an optimization problem can be mathematically formulated as[1]:

$$\begin{aligned} \min_{\mathbf{x}} \ & \mathbf{f}(\mathbf{x}) = (f_1(\mathbf{x}), f_2(\mathbf{x}), \dots, f_m(\mathbf{x})), \\ \text{s.t.} \ & \mathbf{x} \in X, \quad \mathbf{f} \in Y, \end{aligned} \tag{1.1}$$

where $X \subset \mathbb{R}^n$ is the decision space and $\mathbf{x} = (x_1, x_2, \dots, x_n) \in X$ is the decision vector; $Y \subset \mathbb{R}^m$ is the objective space and $\mathbf{f} \in Y$ is the objective vector. The decision vector $\mathbf{x}$ is composed of n decision variables while the objective vector $\mathbf{f}$ is composed of m objective functions that map $\mathbf{x}$ from X to Y.

Considering the number of objectives, optimization problems can be intrinsically categorized into two types: single-objective optimization problems (SOPs) (i.e., $m = 1$) and multi-objective optimization problems (MOPs) (i.e., $m > 1$). For an SOP, the optimal solution is often a single decision vector minimizing (or maximizing) the objective. For an MOP, since the multiple objectives are often conflicting with each other, it is impossible to find any single solution minimizing (or maximizing) all objectives simultaneously; instead, the optima to an MOP is often a set of solutions trading-off between different objectives, known as the Pareto optimal solutions. Specially, an MOP with more than three objectives is also known as a many-objective optimization problem (MaOP) due to its particular challenges [2].

As illustrated in Fig. 1.2, optimization problems can be large scale in terms of either objectives or decision variables, intrinsically comprising four types: (i) SOPs with a large number of decision variables (LSSOPs), (ii) MOPs with a large number of decision variables (LSMOPs), (iii) MaOPs with a normal scale of decision variables, and (iv) MaOPs with a large number of decision variables (LSMaOPs).

For SOPs, the optimization target is quite straightforward – searching for the global optimum solution $\mathbf{x}^*$ which minimizes the objective function $f(\mathbf{x}^*)$. By contrast, for MOPs/MaOPs, due to the possible conflicts between different objectives, no single solution is able to minimize all objectives simultaneously. As a consequence, there exist a set of optimal solutions trading-off between different objectives, namely, the *Pareto optimality* – defining a situation where no solution can be better off without making at least another solution worse off. In other words, the optimality of MOPs/MaOPs is defined upon the pairwise relationship (the *dominance relationship*) among candidate solutions considering the trade-offs between different optimization objectives. To be specific, a solution

1 This book only considers minimization problems, while maximization problems can be equivalently transformed into minimization problems by negating the objective functions.

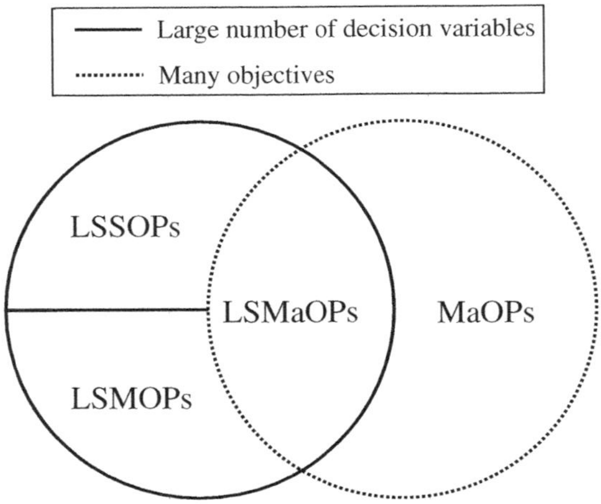

Figure 1.2 This figure illustrates the classification of large optimization problems involved in this chapter. Generally, the problems can have either a large number of decision variables or many objectives, which can be further categorized into four classes: single-objective problems with a large number of decision variables (LSSOPs), multi-objective optimization problems with a large number of decision variables (LSMOPs), many-objective optimization problems (MaOPs), and many-objective optimization problems with a large number of decision variables (LSMaOPs).

$\mathbf{x}_1$ is said to *dominate* another solution $\mathbf{x}_2$ (denoted as $\mathbf{x}_1 \succ \mathbf{x}_2$) iff

$$\begin{cases} \forall i \in 1,2,\dots,m : f_i(\mathbf{x}_1) \leq f_i(\mathbf{x}_2), \\ \exists j \in 1,2,\dots,m : f_j(\mathbf{x}_1) < f_j(\mathbf{x}_2). \end{cases} \tag{1.2}$$

If a solution $\mathbf{x}^*$ cannot be dominated by any other feasible solutions, $\mathbf{x}^*$ is said to be *Pareto optimal* and the union of all $\mathbf{x}^*$ is called *Pareto set (PS)*, while the union of $\mathbf{f}(\mathbf{x}^*)$ is termed *Pareto front (PF)*. An illustrative example of Pareto optimality is given in Fig. 1.3.

For most continuous MOPs/MaOPs, the PS (PF) often consists of an infinite number of Pareto optimal solutions. Therefore, in practice, only a representative subset of the PS (PF) can be approximated using a limited number of *non-dominated solutions*, which are often the candidate solutions not dominated by any other in the final solution set obtained by an algorithm.

In this chapter, we mainly focus on problems of type (ii) and type (iv), i.e., LSMOPs and LSMaOPs. Since MaOPs are essentially MOPs of a special case, for simplicity, we unified the abbreviations of both LSMOPs and LSMaoPs to be LSMOPs. Specifically, we refer to research adopting EAs to solve large-scale optimization problems as *evolutionary large-scale optimization*. The rest of this chapter is organized as follows: Section 1.2 briefly introduces multi-objective evolutionary algorithms (MOEAs), Section 1.3 overviews evolutionary large-scale optimization, and Section 1.4 summarizes this chapter.

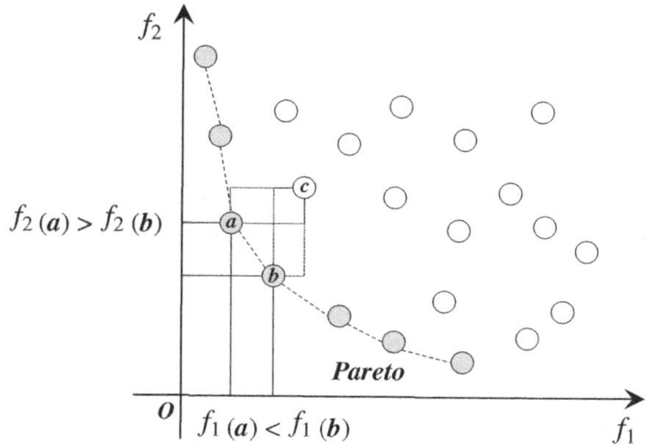

Figure 1.3 An illustrative example to *Pareto optimality*. The images of a number of candidate solutions (**a**, **b**, **c**, etc.) are spreading in the objective space. **a** and **b**, non-dominated to each other, are among the Pareto optimal solutions on the PF (PS). **c** is dominated by both **a** and **b**.

1.2 Multi-Objective Evolutionary Algorithms (MOEAs)

Dating back to the GA proposed by John Holland in 1975 [3], the EAs, generally referring to the family of soft computing paradigms inspired by natural evolution mechanisms, have witnessed booming development during the past half-century. Given an optimization problem, an EA iteratively maintains a population of candidate solutions undergoing evolutionary operators such as crossover, mutation, and selection. As a consequence, the population is expected to move toward optimum to the optimization problem, just as species adapting to the natural environments.

1.2.1 MOEAs for Solving MOPs

Considering the population-based character, EAs are intrinsically suited for solving MOPs – to obtain a set of candidate solutions trading-off between multiple optimization objectives. Since Schaffer proposed the vector-evaluated GA [4] in 1984, a large number of MOEAs have been proposed. In general, MOEAs proposed for solving MOPs can be categorized into three groups: dominance-based ones, performance-indicator-based ones, and decomposition[2] -based ones.

Dominance-based MOEAs have been prevalent over the past two decades. Early dominance-based non-elitism MOEAs include the non-dominated sorting GA [5], the niched-Pareto GA [6], and the multi-objective genetic algorithm [7]. It was found out later on that the use of elitism strategies can substantially improve the convergence of MOEAs. As a result, several popular elitism MOEAs have been developed, including the elitist non-dominated sorting genetic algorithm (NSGA-II) [8], the strength Pareto evolutionary algorithm 2

2 Decomposition in multi- or many-objective optimization specifically refers to the decomposition in the objective space.

(SPEA2) [9], and the Pareto envelope-based selection algorithm (PESA) [10, 11]. Although dominance-based MOEAs are very powerful for solving bi-objective or three-objective MOPs, their convergence performance dramatically degrades when the number of objectives is larger than three, mainly due to the loss of selection pressure.

Decomposition-based approaches divide a complex MOP into a number of sub-problems and solve them in a collaborative manner. There are mainly two types of decomposition-based approaches. In the first type of decomposition-based approaches, an MOP is decomposed into a group of single-objective problems (SOPs), including the weighted aggregation-based approaches in the early days [12, 13], and the more recent multi-objective evolutionary algorithm based on decomposition (MOEA/D), where more explicit collaboration strategies between the solutions of the subproblems were introduced. A number of variants of MOEA/D have been proposed for enhancing the selection strategy for each subproblem to strike a better balance between convergence and diversity [14–16]. In the second type of decomposition-based approaches, an MOP is decomposed into a group of sub-MOPs. For instance, MOEA/D-M2M [17, 18] divides the whole PF into a group of segments, and each segment can be regarded as a subproblem.

Since performance indicators are used to measure the solution qualities, it is quite intuitive to also use them as selection criteria to guide the search of the algorithms to obtain high-quality solutions. Among various performance indicators, the hypervolume indicator (or the S metric) is most widely applied in various MOEAs, due to the fact that it is the only metric known to be strictly monotonic to the Pareto dominance [19]. The first well-known hypervolume indicator-based algorithm was proposed by Emmerich *et al.* in 2005, known as the s-metric selection evolutionary multi-objective algorithm (SMS-EMOA) [20]. In SMS-EMOA, a $(\mu + 1)$ evolutionary strategy is adopted to generate one offspring candidate solution from a number of μ parents. To select a number of μ new candidate solutions for the next generation, the hypervolume contribution of each candidate solution on the last non-dominated front is calculated, and the one with the smallest hypervolume contribution is removed. Apart from the hypervolume indicator, some other performance indicators have also been adopted to guide the search of MOEAs. Zitzler and Künzli have proposed a binary indicator ($I_{\epsilon+}$)-based evolutionary algorithm (IBEA) [21], where $I_{\epsilon+}$ is considered to be compliant with the Pareto dominance. Tian *et al.* have proposed an IGD-NS-based multi-objective evolutionary algorithm (AR-MOEA) [22], where a reference point adaptation method is developed to adjust the reference points for the calculation of IGD-NS [23] at each generation.

Despite the various MOEAs as introduced above, all of them have more or less inherited the merits from NSGA-II – either the framework or the main algorithm components. On the one hand, the simple and flexible framework of NSGA-II allows users to replace the main algorithm components with their tailored ones; on the other hand, the main algorithm components in NSGA-II can be easily modified as required. More recently, NSGA-II has also shed light on the literature of evolutionary large-scale multi-objective optimization, and it is highly related to the works to be presented in this book. Hence, before moving to the rest of chapters of this book, the readers are encouraged to go deep into the details of NSGA-II as given in the following.

As presented in Algorithm 1.1, the main loop of NSGA-II follows the workflow of a typical EA (Fig. 1.1), involving three main components: mating selection, variation, and environmental selection. More specifically, the environmental selection in NSGA-II consists of another two subcomponents: the non-dominated sorting (Line 2) and the crowding distance sorting (Line 3), which are two tailored sorting methods considering convergence criterion and diversity criterion when performing environmental selection, respectively. In the following, we will detail each component in NSGA-II one by one.

Algorithm 1.1: Procedure of NSGA-II

Input: N (population size)
Output: P (final population)

```
1 P ← Randomly generate N solutions;
2 Front ← NondominatedSorting(P); //Algorithm 1.2
3 Crowd ← CrowdingDistance(P, Front); //Algorithm 1.3
4 while termination criterion not fulfilled do
5     P' ← MatingSelection(P, N, Front, Crowd); //Algorithm 1.4
6     O ← Variation(P'); //Algorithm 1.5
7     [P, Front, Crowd] ← EnvironmentalSelection(P ∪ O, N);
          //Algorithm 1.6
8 return P;
```

In evolutionary multi-objective optimization, the candidate solutions in the final population of an MOEA are expected to be an approximation to the PF of the MOP to be solved, and the motivation of non-dominated sorting in NSGA-II is exactly to meet the basic requirement (i.e., Pareto dominance, the *convergence criterion*) of the candidate solutions.

Basically, the *non-dominated sorting* method (Algorithm 1.2) is motivated to partition the candidate solutions into a number of *non-dominated fronts* (fronts for short) according to the dominance relationship as defined by Eq. (1.2). Consequently, as illustrated in Fig. 1.4, the candidate solutions inside each front

Algorithm 1.2: *NondominatedSorting(P)*

Input: P (population)

Output: *Front* (non-dominated front number of each solution)

```
//Efficient non-dominated sort with sequential search
P ← Sort the solutions in P in an ascending order of the first objective values;
Front ← A 1 × |P| vector of zeros;
maxF ← 0; //Maximum non-dominated front number
while not all Front_i > 0 do
    maxF ← maxF + 1;
    for i = 1, ..., |P| do
        if Front_i == 0 then
            Dominated ← false; //Whether solution P_i is dominated
            for j = i − 1, ..., 1 do
                if Solution P_j dominates solution P_i then
                    Dominated ← true;
                    break;
            if Dominated == false then
                Front_i ← maxF;
    return Front;
```

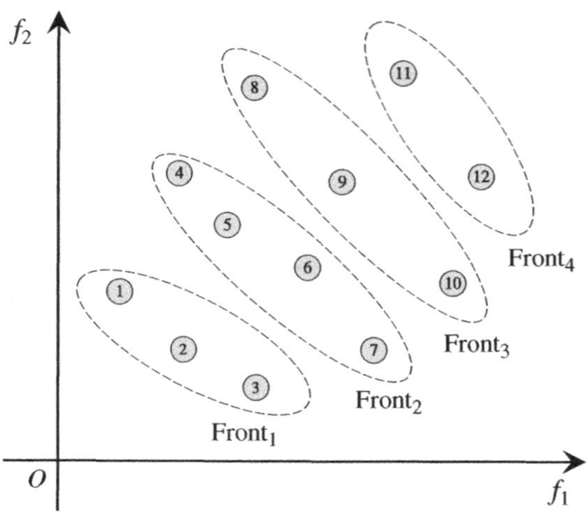

Figure 1.4 An illustrative example to *non-dominated sorting*. The 12 candidate solutions are partitioned into four fronts by non-dominated sorting. The candidate solutions inside each front are indexed with the same front number. The candidate solutions inside $Front_1$ dominate those on the other three fronts.

are non-dominated to each other and share the same ranking index, while those in the last front (i.e., the one with the minimum front number) dominate all the rest.

In order to obtain a population of candidate solutions as an approximation to the PF of a given MOP, apart from the convergence criterion as considered in non-dominated sorting, it is equally important to take the *diversity criterion* into consideration. Specifically, in evolutionary multi-objective optimization, the diversity criterion often refers to degree of how uniformly/evenly the candidate solutions are spreading over the PF. It is due to the fact that, without any preference, the decision-maker can be interested in diverse candidate solutions diversely trading-off between different objectives.

The crowding distance sorting (Algorithm 1.3) in NSGA-II is exactly motivated to maintain diversity of the candidate solutions. As indicated by the name of the sorting method, crowding distance sorting aims to rank candidate solutions according to their crowding degrees (i.e., how much each candidate

---

Algorithm 1.3: *CrowdingDistance(Front)*

Input: P (population), *Front* (non-dominated front number of each solution)
Output: *Crowd* (crowding distance of each solution)

$Crowd \leftarrow$ A $1 \times |P|$ vector of zeros;
$maxF \leftarrow 0$; `//Current non-dominated front number`
while *not all* $Crowd_i > 0$ **do**
  $maxF \leftarrow maxF + 1$;
  $I \leftarrow \{i = 1, \dots, |P| \,|\, Front_i = maxF\}$;
  for $m = 1, \dots, M$ **do**
    $fmax = max_{i \in I} f_m(P_i)$; `//`$f_m(P_i)$ `is the` m`th objective value of the` i`th solution in` P
    $fmin = min_{i \in I} f_m(P_i)$;
    $I' \leftarrow$ Sort the solutions in I in an ascending order of the mth objective values;
    for $i = 1, \dots, |I'|$ **do**
      if $i == 1$ *or* $i == |I'|$ **then**
        $Crowd_{I'_i} = \inf$;
      else
        $Crowd_{I'_i} = Crowd_{I'_i} + \frac{f_m(P_{I'_{i+1}}) - f_m(P_{I'_{i-1}})}{fmax - fmin}$;
  return *Crowd*;

---

is crowded with respect of its closest neighbors). Eventually, the candidate solutions primarily with the best convergence (i.e., those on or as close to the last non-dominated front as possible) and secondarily best diversity (those having the least crowding degrees) will be selected.

The *mating selection* in NSGA-II is very similar to the conventional tournament selection (Algorithm 1.4) as adopted in single-objective GAs [24], except that it performs ranking according to two (instead of one) criteria – non-dominate front indexing ($Fit1$, ranking results by Algorithm 1.2) and crowding indexing ($Fit1$, ranking results by Algorithm 1.3). As indicated by the name itself, tournament selection is designed to select candidate solutions in a population via random "tournaments." Consequently, the winner of each tournament (i.e., the one with the best fitness value) is selected to go through *variation* (i.e., crossover/mutation).

Algorithm 1.4: $MatingSelection(P, N, Fit1, Fit2)$

Input: P (population), N (number of selected parents), $Fit1$ (primary fitness of solutions), $Fit2$ (secondary fitness of solutions)

Output: P' (a set of parents)

$P' \leftarrow \emptyset$;
for $i = 1, \dots, N$ **do**
    $[i, j] \leftarrow$ Randomly select two indexes from $\{1, \dots, |P|\}$;
    if $Fit1_i < Fit1_j$ **then**
        $P' \leftarrow P' \cup \{P_i\}$;
    else if $Fit1_i > Fit1_j$ **then**
        $P' \leftarrow P' \cup \{P_j\}$;
    else if $Fit2_i < Fit2_j$ **then**
        $P' \leftarrow P' \cup \{P_i\}$;
    else
        $P' \leftarrow P' \cup \{P_j\}$;
return P';

It is worth noting that tournament selection (as well as any other mating selection method) is not necessarily an indispensable component of an MOEA, while it is expected to be more diverse, yet high-quality (in terms of fitness) candidate solutions when adopting it. The key mechanism is the random pairing of the candidate solutions according to probabilistic (but not deterministic) measurement of them. From a more general perspective, such kind of randomization mechanism is widely seen and shown to be effective when coupled with MOEAs.

The *variation* operation (Algorithm 1.5) in NSGA-II adopts the classic *simulated binary crossover* and *polynomial mutation*, which are also widely adopted in real-coded genetic algorithms (RCGAs) [25]. Essentially, a variation operation is expected to generate high-quality new candidate solutions toward future evolution. As in the case of EA, crossover and mutation are the two most commonly seen variation operations. On the one hand, the crossover operation randomly recombines a pair of candidate solutions by exchange part of each; on the other hand, the mutation operation randomly modifies variable(s) of a candidate solution.

Algorithm 1.5: *Variation*(P')

Input: P' (a set of parents)
Output: O (offspring population)

$O \leftarrow \emptyset$;
while $P' \neq \emptyset$ **do**
  $[\mathbf{x}, \mathbf{y}] \leftarrow$ Randomly select two parents from P';
  $P' \leftarrow P' \setminus \{\mathbf{x}, \mathbf{y}\}$;
  $[\mathbf{p}, \mathbf{q}] \leftarrow$ Perform simulated binary crossover and polynomial mutation based on $\mathbf{x}$ and $\mathbf{y}$;
  $O \leftarrow O \cup \{\mathbf{p}, \mathbf{q}\}$;
return O;

Since NSGA-II was initially proposed for solving MOPs with continuous search space, the variation operation using classic simulated binary crossover and polynomial mutation works pretty efficiently in generating candidate solutions for real-coded small-scale MOPs. From the probabilistic point of view, a variation operation can be seen as a process of sampling new data toward optimum in the decision space. Hence, a variation operation in high-dimensional space may also face challenges of *curse of dimensionality*, thus losing efficiency when dealing with LSMOPs. To this end, this book (Chapters 3 and 4) will introduce some novel variation operators tailored for solving LSMOPs.

Since *environmental selection* substantially determines where the candidate solutions evolve toward, it plays a vital role in an EA. Particularly, in an MOEA, environmental selection must be delicately designed to balance convergence criterion and diversity criterion, aiming to obtain a population of candidate solutions

Algorithm 1.6: *EnvironmentalSelection*(P, N)

Input: P (combined population), N (population size)

Output: P (population for next generation), *Front* (non-dominated front number of each solution), *Crowd* (crowding distance of each solution)

1 $Front \leftarrow NondominatedSorting(P)$; `//Algorithm 1.2`
2 $Crowd \leftarrow CrowdingDistance(P, Front)$; `//Algorithm 1.3`
3 $i \leftarrow 1$;
4 **while** $\{P_j | Front_j = i\} \neq \emptyset$ **do**
5     $F_i \leftarrow \{P_j | Front_j = i\}$; `//Set of solutions in the ith front`
6     $i \leftarrow i + 1$;
7 $k \leftarrow \text{argmin}_i |F_1 \bigcup \cdots \bigcup F_i| \geq N$;
8 $F_k \leftarrow$ Delete $|F_1 \bigcup \cdots \bigcup F_k| - N$ solutions from F_k with the smallest *Crowd*;
9 $P \leftarrow F_1 \bigcup \cdots \bigcup F_k$;
10 *Front* $\leftarrow$ Retain the elements in *Front* whose corresponding solutions are in P;
11 *Crowd* $\leftarrow$ Retain the elements in *Crowd* whose corresponding solutions are in P;
12 **return** P, *Front*, and *Crowd*;

uniformly spreading over the approximated PF. To this end, the environmental selection in NSGA-II (Algorithm 1.6) consists of two subcomponents: the non-dominated sorting (Line 1) and the crowding distance sorting (Line 2), as introduced above, respectively.

Essentially, the reason for using two sorting methods together is to simultaneously take into consideration the convergence criterion (Pareto dominance) and the diversity criterion (crowding distance). Thus, the environmental selection can be seen as *multi-index sorting* procedure. As the *primary sorting*, the non-dominated sorting divides the candidate solutions into a number of fronts by considering the convergence criterion of Pareto optimality. Ideally, if the number of candidate solutions in the last front exactly equals the predefined number of candidate solutions to be selected (e.g., N out of $2N$), the environmental selection will end up here. However, it is very likely that the number of candidate solutions in the last front may exceed or not reach the total number of candidates to be selected and passed to the next generation. Hence, the environmental selection will continue to perform crowding distance sorting as the *secondary sorting* until the predefined number of selected candidate solutions is met (Lines 4–9).

1.2.2 MOEAs for Solving MaOPs

MOEAs have been shown to perform well on a wide range of MOPs with two or three objectives; however, MOEAs have experienced substantial difficulties when they are adopted to tackle MOPs with more than three objectives, i.e., the MaOPs [26–29]. As a result, MaOPs have attracted increasing attention in evolutionary optimization.

One major reason behind the failure of most conventional MOEAs in solving MaOPs can be attributed to the loss of selection pressure, i.e., the pressure for the population to converge toward the PF, when dominance is adopted as a criterion for selecting individuals with a limited population size [30]. For example, the elitist NSGA-II [8] and the SPEA2 [9], both of which use dominance-based selection, will fail to work properly when applied to the optimization of MaOPs. It is due to the fact that most candidate solutions generated in a population of a limited size are non-dominated, thus making the dominance-based selection criterion hardly possible to distinguish the candidate solutions, even in a very early stage of the search.

Another important reason for the degraded performance of MOEAs on MaOPs is the difficulty in maintaining a good population diversity in a high-dimensional objective space. Generally speaking, since the PFs of most continuous MOPs are piecewise continuous [31, 32], it is practically unlikely to approximate all Pareto optimal solutions. Instead, most MOEAs aim to find a set of evenly distributed representative solutions to approximate the PF. When the number of objectives is two or three, where the PF is typically a one-dimensional curve or two-dimensional surface, maintaining a good diversity of the solutions is relatively straightforward. As the dimension of the objective space increases, it becomes increasingly challenging to maintain a good population diversity, as the candidate solutions distribute very sparsely in the high-dimensional objective space. This consequently causes immense difficulties to the diversity management strategies widely used in MOEAs, e.g., the crowding distance-based diversity method in NSGA-II [33–35].

To enhance the performance of most traditional MOEAs in solving MaOPs, a number of approaches have been proposed [2, 36, 37], which can be roughly divided into three classes.

The first class is the *modified dominance*-based approaches. The basic idea is to enhance the convergence pressure by modifying the Pareto dominance to enlarge the dominating areas, such that some of the original non-dominated solutions become dominated by others. An early work of this class is the ϵ-dominance proposed by Laummans *et al.* [38], where the original dominance relation can be relaxed to a certain extent by tuning the parameter $\epsilon \in (0, 1)$. The bigger the ϵ is set, the more relaxed the dominance relation is, and vice versa. And, based on ϵ-dominance, Deb *et al.* have proposed a steady-state algorithm called ϵ-MOEA

[39]. In addition, some extended variants of the ϵ-dominance have also been proposed. In [40], Aguirre and Tanaka have proposed an ϵ-ranking multi-objective optimizer (ϵ R-EMO), where an adaptive ϵ-ranking procedure is performed on top of a space partitioning strategy. Another representative modified dominance is known as the α-domination, proposed by Ikeda *et al.* in [26]. Given two solutions, the α-domination allows a solution to be dominated by the other if it is slightly superior in one objective but substantially inferior in others. The α-domination has been successfully integrated into some MOEAs for solving MOPs, such as the guided multi-objective evolutionary algorithm (G-MOEA) [41]. As a combination of ϵ-dominance and α-dominance, Batista *et al.* have proposed another modified dominance, namely, the cone ϵ-dominance [42]. It has been reported that the cone ϵ-dominance is promising for solving MaOPs with up to 50 objectives [29]. Recently, Yang *et al.* have proposed a grid based evolutionary algorithm (GrEA) [43], where a grid dominance is designed on top of some adaptively constructed grid coordinates in the objective space. The grid dominance has an attractive feature that allows one solution to dominate another if it is slightly inferior in some objectives but substantially superior in some others, which is useful in distinguishing the non-dominated solutions in many-objective optimization.

The second class is the *diversity management enhancement*-based approaches. Since in many-objective optimization, the distribution of the candidate solutions becomes very sparse in the high-dimensional objective space, diversity management becomes particularly meaningful to alleviate the negative impact of the excessive diversity on the population convergence. An early work of this class is from Adra and Fleming, who have proposed a diversity management strategy to quantitatively evaluate the degree of the population diversity according to the distribution of the candidate solutions [34], and if the population is found excessively diverse, the diversity promotion mechanism in the selection procedure will be deactivated. Another more recent work of this class is the shift-based density estimation (SDE) proposed by Li *et al.* [44], where the basic idea is to penalize poorly converged solutions by assigning them a high-density value, such that only candidate solutions with both good convergence and distribution can survive to the next generation. In [45], Wang *et al.* have proposed to coevolve preferences with a population of candidate solutions for diversity management in many-objective optimization. Most recently, another representative algorithm falling into this class is the NSGA-III [46], where the diversity is managed by a set of predefined reference points. In addition, the recently proposed knee point-driven evolutionary algorithm (KnEA) also belongs to this category, where a knee point-based diversity secondary selection strategy is adopted on top of traditional non-dominated sorting [47].

While the first two classes aim to improve the performance of MOEAs on MaOPs, the third class, known as the *objective reduction* approaches, tries to transform MaOPs to MOPs by reducing the number of redundant or irrelevant objectives, such that the problems can be directly solved with traditional MOEAs [48, 49]. In general, there are two types of objective reduction approaches: offline approaches and online approaches. Offline objective reduction approaches can be seen as preprocessing techniques, which work independently of specific MOEAs, but require a set of approximate Pareto optimal solutions as *a priori* knowledge. In [50], Brockhoff and Zitzler have presented both deterministic and stochastic algorithms for objective reduction based on the objective conflicts measured by ϵ-dominance. Sexane *et al.* have proposed to use principal component analysis (PCA) [51] and maximum variance unfolding (MVU) [52] for linear and nonlinear objective reduction, respectively [49]. Singh *et al.* have proposed to use corner solutions for objective reduction, which are obtained using a Pareto corner search evolutionary algorithm (PCSEA) [48]. Online objective reduction approaches are usually embedded in an existing MOEA and perform as part of the evolutionary procedure, where the idea is to update the reduced objectives iteratively using the solutions obtained by an MOEA during the optimization procedure. In [50], Brockhoff and Zitzler have proposed to embed the PCA approach into NSGA-II to build an online objective reduction approach, known as the PCA-NSGA-II. In [53], Jaimes *et al.* have proposed an online objective reduction algorithm to find a k-sized objective subset based on correlations between the objectives. Recently, Guo *et al.* have proposed to adopt the partitioning around medoids (PAM) [54] clustering algorithm to detect the correlations between objectives and divide them into different clusters [55].

Apart from the three classes of approaches, the decomposition-based approaches and the performance indicator-based approaches, which were originally proposed for solving MOPs, are also applicable to many-objective optimization. However, although neither of these two types of approaches are dominance based, they still suffer from some other issues.

For decomposition-based approaches, it is always required to predefine a set of weight (or reference) vectors before running the algorithms. In high-dimensional objective space, it becomes particularly challenging to specify a proper vector set to well approximate the PF. To remedy this issue, Li *et al.* have proposed an MOEA based on both dominance and decomposition, known as the MOEA/DD [56].

For performance indicator-based approaches, the major issue is the significant *curse of dimensionality* regarding the computational efficiency [57]. For example, the runtime of one generation in SMS-EMOA is as high as $O(N^{M/2+1})$, where N is the population size and M is the number of objectives. To tackle this issue, Bader and Zitzler have proposed a hypervolume estimation algorithm (HypE) [19], where the Monte Carlo sampling method is adopted to obtain the approximate hypervolume values for many-objective cases. Similar to SMS-MOEA, the selection in HypE is still based on the hypervolume contribution of the candidate

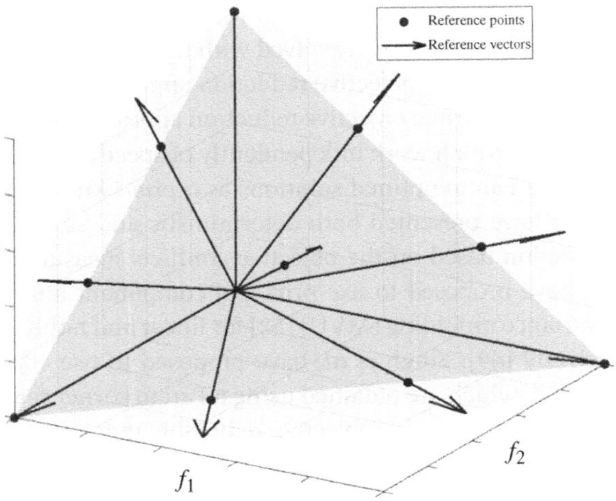

Figure 1.5 An illustration of how to generate the uniformly distributed reference vectors in a three-objective space. In this case, 10 uniformly distributed reference points are first generated on a hyperplane, and then they are mapped to a hypersphere to generate the 10 reference vectors.

solutions, while a minor difference is that the contribution is calculated based on a randomly chosen subset of the non-dominated solution set, where the size of the subset equals the number of solutions to be removed. In addition, some faster hypervolume calculation methods have been proposed recently, e.g., the hypervolume by slicing objectives (HSO) [58] and the walking fish group (WFG) hypervolume algorithm [59] (Fig. 1.5).

In summary, the approaches for solving MaOPs are either tailored on the basis of conventional MOEAs (originally proposed for solving MOPs) or adopting some strategies transforming MaOPs to MOPs. Moving toward LSMaOPs, an MOEA may suffer from the curse of dimensionality in both decision space and objective space, thus posing stiff challenges in convergence/diversity management of the candidate solutions. Recently, Cheng *et al.* have proposed a reference vector guided evolutionary algorithm (RVEA) [60] for solving MaOPs. RVEA inherited the generic elitism-selection-based framework (similar to that of NSGA-II) while adopting delicately tailored environmental selection inspired by both decomposition- and indicator-based MEOAs. Thanks to the generic framework and powerful convergence/diversity management capability, RVEA has proven to have certain potentials in scaling up to LSMaOPs, as will be elaborated in the rest of this book. Hence, as preliminary knowledge, we will introduce the details of RVEA in the following.

As summarized in Algorithm 1.7, RVEA adopts the framework of traditional elitism selection-based EAs – very similar to that of NSGA-II (Algorithm 1.1).

Algorithm 1.7: Procedure of RVEA

Input: N (population size)
Output: P (final population)

1 $P \leftarrow$ Randomly generate N solutions;
2 $V^0 \leftarrow$ *UniformSampling*(N, M); `//Algorithm 1.8`
3 $V \leftarrow V^0$;
4 **while** *termination criterion not fulfilled* **do**
5     $P' \leftarrow$ Randomly select N solutions from P;
6     $O \leftarrow$ *Variation*(P'); `//Algorithm 1.5`
7     $[P, \mathit{Fitness}] \leftarrow$ *EnvironmentalSelection*$(P \cup O, V, N)$; `//Algorithm 1.9`
8     $V \leftarrow$ *VectorAdaptation*(P, V, V^0); `//Algorithm 1.10`
9 **return** P;

To begin with, RVEA samples a set of reference vectors (Line 2). As illustrated in Fig. 1.5, the method of generating reference vectors in RVEA is inspired by the method of generating reference points for decomposition-based methods, using the *simplex-lattice design* in the case of requiring that the reference vectors uniformly distributed (Algorithm 1.8). Despite the deep connections between reference points and reference vectors, their working mechanisms are substantially different: a reference point specifies a region in the objectives space where

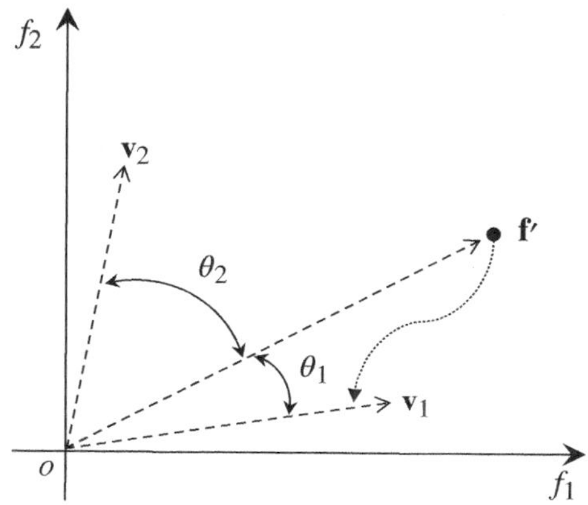

Figure 1.6 An example showing how to associate an individual with a reference vector. In this example, $\mathbf{f}'$ is a translated objective vector, $\mathbf{v}_1$ and $\mathbf{v}_2$ are two unit reference vectors. θ_1 and θ_2 are the angles between $\mathbf{f}'$ and $\mathbf{v}_1$, $\mathbf{v}_2$, respectively. Since $\theta_1 < \theta_2$, the candidate solution related to $\mathbf{f}'$ is associated with reference vector $\mathbf{v}_1$.

Algorithm 1.8: *UniformSampling*(N, M)

Input: N (number of vectors), M (number of objectives)
Output: V (uniform vector set)

$H \leftarrow \text{argmax}_h C_{h+M-1}^{M-1} \leq N$; `//Number of divisions on each objective`
$X \leftarrow$ All the $(M-1)$-combinations of $\{\frac{0}{H}, \frac{1}{H}, \ldots, \frac{H+M-2}{H}\}$;
for *each* $x_{ij} \in X$ **do**
    $x_{ij} \leftarrow x_{ij} - \frac{j-1}{H}$; `//`$x_{ij}$ `is the` j`th element of the` i`th combination`
$V \leftarrow$ A $C_{H+M-1}^{M-1} \times M$ matrix of zeros;
for *each* $v_{ij} \in V$ **do**
    if $j == 1$ **then**
        $v_{ij} = x_{ij}$;
    else if $j == M$ **then**
        $v_{ij} = 1 - x_{i(j-1)}$;
    else
        $v_{ij} = x_{ij} - x_{i(j-1)}$;
return V;

---

a decision-maker can be interested; by contrast, a reference vector specifies a direction that a candidate solution is expected to converge along, probably toward a unique Pareto optimal solution on the PF, as illustrated in Fig. 1.6.

In the main loop of RVEA (Lines 4–8), the variation component is also the same as that in NSGA-II. The main differences between RVEA and NSGA-II lie in the other two key components:*environmental selection* (Line 7) and *vector adaptation* (Line 8).

The *environmental selection* (Algorithm 1.9) in RVEA is designed to perform single-objective selections inside the subpopulations comprising the candidate solutions associated with their closest reference vectors (Line 4). As the selection criterion, a scalarization function known as the angle-penalized distance (APD) is proposed to aggregate the convergence criterion and the diversity criterion:

$$d_{t,i,j} = (1 + P(\theta_{t,i,j})) \cdot \|\mathbf{f}_{t,i} - \mathbf{z}_t^{min}\|, \tag{1.3}$$

where $d_{t,i,j}$ denotes the APD value of solution $p_{t,i}$ with respect to reference vector $\mathbf{v}_{t,j}$ in the tth generation; $\|\mathbf{f}_{t,i} - \mathbf{z}_t^{min}\|$, as the convergence criterion, is the

Algorithm 1.9: *EnvironmentalSelection(P, V, N)*

Input: P (combined population), V(uniform vector set), N (population size)
Output: P (population for next generation)

//Objective value translation
$\mathbf{z}^{ideal} \leftarrow$ Minimum objective values in P;
for $i = 1, \dots, |P|$ **do**
  $\mathbf{f}'(P_i) \leftarrow \mathbf{f}(P_i) - \mathbf{z}^{ideal}$;

//Population partition
$Associate \leftarrow$ A $1 \times |P|$ vector of zeros; //The index of vector that each solution belongs to
for $i = 1, \dots, |P|$ **do**
  for $j = 1, \dots, |V|$ **do**
    $\theta_{ij} \leftarrow \langle \mathbf{f}'(P_i), V_j \rangle$; //Angle between the solution and the vector
  $Associate_i \leftarrow \text{argmin}_j \theta_{ij}$;

//Angle-penalized distance based selection
$P_{next} \leftarrow \emptyset$;
for $j = 1, \dots, |V|$ **do**
  $I \leftarrow \{i = 1, \dots, |P| | Associate_i = j\}$;
  for *each* $i \in I$ **do**
    Calculate the angle-penalized distances using Eqs. (1.3)–(1.5);
  $k \leftarrow \text{argmin}_{i \in I} d_{ij}$;
  $P_{next} \leftarrow P_{next} \cup \{P_k\}$;
$P \leftarrow P_{next}$;
return P;

Euclidean distance from the candidate solution $p_{t,i}$ to the ideal vector $\mathbf{z}_t^{min}$; and $P(\theta_{t,i,j})$, as the diversity criterion, is a penalty function related to the angle $\theta_{t,i,j}$ between candidate solution $p_{t,i}$ and reference vector $\mathbf{v}_{t,j}$:

$$P(\theta_{t,i,j}) = M \cdot \left(\frac{t}{t_{max}}\right)^{\alpha} \cdot \frac{\theta_{t,i,j}}{\gamma_{\mathbf{v}_{t,j}}}, \tag{1.4}$$

with

$$\gamma_{\mathbf{v}_{t,j}} = \min_{i\in\{1,\ldots,N\}, i\neq j} \langle \mathbf{v}_{t,i}, \mathbf{v}_{t,j} \rangle, \tag{1.5}$$

where M is the number of objectives, t_{max} is the predefined maximum number of generations, $\gamma_{\mathbf{v}_{t,j}}$ is the smallest angle value between reference vector $\mathbf{v}_{t,j}$, and α is a user-defined parameter controlling the changing rate of the diversity criterion with respect to the number of generations. It is worth noting that the design of penalty function $P(\theta_{t,i,j})$ is out of several empirical observations. First, since there are only a limited number of candidate solutions sparsely distributed in the high-dimensional objective space, it is particularly important to enhance the convergence and diversity of the population at the early stage and late stages of the evolutionary search, respectively; hence, $P(\theta_{t,i,j})$ is designed to be linked to the generation number t. Second, due to the sparse distribution of the candidate solutions, the angles between the candidate solutions and reference vectors can vary a lot; hence, the angles are normalized into $[0, 1]$ with $\frac{\theta_{t,i,j}}{\gamma_{\mathbf{v}_{t,j}}}$. Third, the more objectives there are, the more sparsely the candidate solutions are distributed; hence, $P(\theta_{t,i,j})$ is designed to be related with the number of objectives M.

The *vector adaption* (Algorithm 1.10) in RVEA is designed to adjust the distribution of the reference vectors according to the different ranges of objectives on the PFs. For dominance-based algorithms such as the NSGA-II, normalization of objective functions does not change the partial orders of dominance relations, and thus has no influence on the selection process; by contrast, in RVEA, since the selection criterion is related to the specific objective function values of the candidate solutions, dynamically normalizing the objective values during the

---

Algorithm 1.10: *VectorAdaptation*(P, V, V^0)

Input: P (population), V (uniform vector set), V^0 (original vector set)
Output: V (adapted uniform vector set)

if *t is a multiple of* $0.1t_{max}$ **then**
  `//t is the current generation number and` t_{max} `is the maximum generation number`
  $\mathbf{z}^{ideal} \leftarrow$ Minimum objective values in P;
  $\mathbf{z}^{nadir} \leftarrow$ Maximum objective values in P;
  for $i = 1, \ldots, |V|$ **do**
    Normalize the reference vectors in P using Eq. (1.6);
return V;

---

evolutionary search will substantially perturb the convergence of the algorithm. Therefore, instead of normalizing the objective functions, RVEA adapts the reference vectors in the following manner:

$$\mathbf{v}_{t+1,i} = \frac{\mathbf{v}_{0,i} \circ \left(\mathbf{z}_{t+1}^{max} - \mathbf{z}_{t+1}^{min}\right)}{\|\mathbf{v}_{0,i} \circ \left(\mathbf{z}_{t+1}^{max} - \mathbf{z}_{t+1}^{min}\right)\|}, \tag{1.6}$$

where $\circ$ operator denotes the Hadamard product that element-wisely multiplies two vectors (or matrices) of the same size, $i = 1, \dots, N$, $\mathbf{v}_{0,i}$ denotes the ith uniformly distributed reference vector, which is generated in the initialization stage (in Step 1 in Algorithm 1.7) and $\mathbf{v}_{t+1,i}$ denotes the ith adapted reference vector for the next generation $t+1$, $\mathbf{z}_{t+1}^{max} = (z_{t+1,1}^{max}, z_{t+1,2}^{max}, \dots, z_{t+1,m}^{max})$ and $\mathbf{z}_{t+1}^{min} = (z_{t+1,1}^{min}, z_{t+1,2}^{min}, \dots, z_{t+1,m}^{min})$ denote the nadir point and the ideal point estimated with the candidate solutions in population P_{t+1}, respectively. Besides, in order to control the frequency of the reference vector adaptation, a predefined parameter f_r is introduced.

1.3 Evolutionary Large-Scale Optimization

Adapting to the environment is an important trait of natural evolution; hence, it is intuitive to take such adaptability as a unique advantage when applying EAs – treating the problems to be optimized as *black-box*. Indeed, EAs work in a way requiring no specific information of the problems at hand; however, when facing large-scale problems, leaving all labors to EAs would probably lead to prohibitively expensive computational costs due to the enormous search space. Therefore, the literature of evolutionary large-scale optimization [61, 62] has been mainly dedicated to accelerating the search speed of EAs by transforming black-box problems to be *gray-box*, mainly on the basis of *a priori* knowledge or *a posteriori* information as mined from the problems. For such purposes, the *cooperative coevolution (CC)* provides a general approach coupled with *decision variable grouping* techniques.

1.3.1 Decision Variable Grouping

The intuitive idea of decision variable grouping is borrowed from classical operation research – *divide-and-conquer*, aiming to decompose a large-scale optimization problems into a set of subproblems of smaller scales [63]. Specifically, an optimization problem $f(\mathbf{x})$ can be decomposed into subproblems by considering the *separability* of the decision vector (i.e., vector comprising the decision variables) $\mathbf{x}$, mainly in two cases.

Ideally, an optimization problem $f(\mathbf{x})$ can be decomposed into subproblems to be optimized fully independently *iff* the original objective function $f(\mathbf{x})$ is *fully separable* with respect to any decision variable x_k related to each subproblem (Case 1):

$$\underset{\mathbf{x}}{\text{argmin}}\ f(\mathbf{x}) = \left(\underset{x_k}{\text{argmin}}\ f(\mathbf{x}), \underset{\forall x_j, j \neq k}{\text{argmin}}\ f(\mathbf{x}) \right). \tag{1.7}$$

However, the subproblems can only be optimized component-wise independently *iff* $f(\mathbf{x})$ is *partially separable* with respect to any decision variable x_k related to each subproblem (Case 2):

$$\underset{\mathbf{x}}{\text{argmin}}\ f(\mathbf{x}) = \left(\underset{\mathbf{x}_1}{\text{argmin}}\ f(\mathbf{x}_1, \dots), \dots, \underset{\mathbf{x}_m}{\text{argmin}}\ f(\dots, \mathbf{x}_m) \right), \tag{1.8}$$

where $\mathbf{x}_1, \dots, \mathbf{x}_m$ are disjoint sub-vectors of $\mathbf{x}$, and the interactions only exist among decision variables inside each sub-vector, but the decision variables in different sub-vectors do not interact with each other.

The *separability* properties as defined above provide a theoretical basis for the development of decision variable grouping techniques. Practically, in order to check the separability of a given function, it requires *variable interaction detection* among each pair of decision variables in the decision vector:

$$\begin{aligned} &f(\mathbf{x})|_{x_i=a_2, x_j=b_1} < f(\mathbf{x})|_{x_i=a_1, x_j=b_1} \land \\ &f(\mathbf{x})|_{x_i=a_2, x_j=b_2} > f(\mathbf{x})|_{x_i=a_1, x_j=b_2}, \text{ with} \end{aligned} \tag{1.9}$$

$$f(\mathbf{x})|_{x_i=a_2, x_j=b_1} \triangleq f(x_1, \dots, x_{i-1}, a_2, \dots, x_{j-1}, b_1, \dots, x_d),$$

where x_i and x_j are known to be *interacted* with each other, and thus not *separable*.

On the basis of separability properties coupled with variable interaction detection, a number of decision variable grouping techniques have been developed. As the most representative work along this line, the differential grouping [64] proposed to permutate each pair of decision variables according to Eq. (1.9) and checked the variable interactions with respect to a threshold ϵ; consequently, the interacted decision variables are divided into the same group to be a subproblem. Considering to the sensitivity to parameter ϵ and poor accuracy in detecting interacted variables on functions with overlapping components in differential grouping, the global differential grouping (GDG) [65] and the eXtended Differential Grouping (XDG) [66] were proposed. More recently, the improved DG (DG2) was further proposed to reduce the number of function evaluations (FEs) required for pairwise variable interaction detection [67].

Despite the theoretical soundness of differential grouping, it intrinsically suffers from the burdens of requiring a large number of FEs for pairwise variable interaction detection. By contrast, some other variable grouping techniques attempted to circumvent explicit variable interaction detection. For instance, the random

Algorithm 1.11: $(x^{\star}, f^{\star}) = \texttt{CC}(f)$

/*Main Framework of CC*/
$\mathbf{P} \leftarrow$ randomized initial population;
$\mathbf{c} \leftarrow$ randomized initial context vector;
//grouping stage
$\mathcal{G} = \texttt{VariableGrouping}(f)$;
//optimization stage
while *Termination Condition is Not Satisfied* **do**
    for $\kappa = 1$ *to* $|\mathcal{G}|$
    do
        $(\mathbf{P}, \mathbf{c}) = \texttt{Optimizer}(\mathbf{P}, \mathbf{c}, \mathcal{G}_{\kappa})$;
$\mathbf{x}^{\star} = \mathbf{c}$; $f^{\star} = f(\mathbf{x}^{\star})$;
return $(\mathbf{x}^{\star}, f^{\star})$;

static/dynamic grouping techniques proposed to randomly divide the decision variables into k even groups dynamically [68, 69].

1.3.2 Cooperative Coevolution

The cooperative coevolutionary (CC) framework was initially proposed by Potter and De Jong [70] as a general framework to optimize the subcomponents of a given problem in a collaborative manner. As introduced in the Section 1.3.1, large-scale optimization problems can be decomposed into a set of subproblems of smaller scales via decision variable grouping; hence, it is intuitively reasonable to adopt CC framework for the optimization of the subproblems as decomposed [71, 72].

As shown in Algorithm 1.11, the main framework of CC consists of two main stages: the grouping stage and the optimization stage. At the grouping stage, the original optimization problem is decomposed into a set of subproblems using the methods as introduced in Section 1.3.1

At the optimization stage, the decision variables in each nonseparable group (i.e., $\mathcal{G}_{\kappa}$) are iteratively optimized in a coevolutionary manner, where the optimizer can be any derivative-free optimization algorithm supporting *variable-wise* update of the decision vectors. In other words, the CC framework requires that an optimization algorithm can flexibly determine which decision variable(s) to be independent updated. As a representative example, the particle swarm optimization (PSO) has proven to be well suitable as an embedded optimizer in a CC framework [72] due to its *pseudo-gradient* characteristic.

PSO is one powerful and widely used swarm intelligence paradigm introduced by Kennedy and Eberhart in 1995 [73], originally intended for simulating social behaviors such as bird flocking, and later simplified to be applicable for solving optimization problems. PSO iteratively maintains a swarm of particles, each of which has a pair of position and velocity. In PSO, it is assumed that each particle is able to memorize the best position found in history, i.e., the best position that has ever been found by the whole swarm, termed *global best*, and the best position that has ever been found by each particle, known as *personal best*. To find the global optimum of the optimization problem, the particles learn from the personal best and global best positions. To be specific, the learning mechanisms in the canonical PSO can be summarized as follows:

$$\begin{aligned}\mathbf{v}_i^{t+1} &= \omega \mathbf{v}_i^t + c_1 \mathbf{r}_1^t \circ (\mathbf{pbest}_i^t - \mathbf{x}_i^t) \\ &\quad + c_2 \mathbf{r}_2^t \circ (\mathbf{gbest}^t - \mathbf{x}_i^t),\end{aligned} \tag{1.10}$$

$$\mathbf{x}_i^{t+1} = \mathbf{x}_i^t + \mathbf{v}_i^{t+1}, \tag{1.11}$$

where t is the generation number, $\mathbf{v}_i^t$ and $\mathbf{x}_i^t$ represent the velocity and position of the ith particle, respectively; ω is termed inertia weight [74], c_1 and c_2 are the acceleration coefficients [75], $\mathbf{r}_1^t$ and $\mathbf{r}_2^t$ are two vectors randomly generated within $[0, 1]^n$, with n being the dimension of the search space; $\mathbf{pbest}_i^t$ and $\mathbf{gbest}^t$ denote the personal best of the ith particle and the global best of the swarm, respectively. The operator $\circ$ is the Hadamard product which element-wisely multiplies two vectors of the same size.

1.4 Summary

This chapter begins with a brief overview of how optimization problems changed along with the modernization of science, engineering, and industry. As the optimization problems became increasingly complex, researchers attempted to look into nature for possible solutions – to develop the nature-inspired optimization algorithms, e.g., EAs, a representative family.

Then, this chapter summarized different types of large-scale optimization problems, in terms of either a large number of decision variables or objectives. This book is mainly focused on the MOPs with a large number of decision variables, i.e., large-scale multi-/many-objective optimization problems (LSMOPs).

On the one hand, to tailor EAs for solving LSMOPs, the related works are highly related to MOEAs. While conventional MOEAs were developed to solve problems with relatively small-scale decision variables, the core mechanisms behind them have founded further development toward evolutionary large-scale multi-objective optimization. On the other hand, when dealing with large-scale decision variables, ideas are also borrowed from the single-objective optimization literature. Particularly, the divide-and-conquer strategy is among the most representative and effective ones to deal with large-scale decision variables. When embedded with EAs, the divide-and-conquer strategy is often implemented inside the CC framework on top of the decision variable grouping methods.

The rest of this book is organized as follows: Chapter 2 presents some background knowledge of evolutionary large-scale multi-objective optimization; Chapters 3 and 4 present several representative EAs for general and sparse large-scale multi-objective optimization, respectively; Chapters 5–9 present several representative applications of evolutionary large-scale multi-objective optimization in areas of community detection in complex networks, logistic scheduling, power systems, radiotherapy planning, and deep learning.

References

1 F.-A. Fortin, F.-M. De Rainville, M.-A. G. Gardner, M. Parizeau, and C. Gagné, "DEAP: Evolutionary algorithms made easy," *The Journal of Machine Learning Research*, vol. 13, no. 1, pp. 2171–2175, 2012.

2 B. Li, J. Li, K. Tang, and X. Yao, "Many-objective evolutionary algorithms: a survey," *ACM Computing Surveys*, vol. 48, no. 1, pp. 1–35, 2015.

3 J. H. Holland, *Adaptation in natural and artificial systems: an introductory analysis with applications to biology, control, and artificial intelligence*. MIT Press, 1992.

4 J. D. Schaffer, "Multiple objective optimization with vector evaluated genetic algorithms," in *Proceedings of the 1st International Conference on Genetic Algorithms and Their Applications*, 2014, pp. 93–100.

5 N. Srinivas and K. Deb, "Muiltiobjective optimization using nondominated sorting in genetic algorithms," *Evolutionary Computation*, vol. 2, no. 3, pp. 221–248, 1994.

6 J. Horn, N. Nafpliotis, and D. E. Goldberg, "A niched Pareto genetic algorithm for multi-objective optimization," in *Proceedings of the 1st IEEE Conference on Evolutionary Computation*, 1994, pp. 82–87.

7 C. M. Fonseca and P. J. Fleming, "Genetic algorithms for multiobjective optimization: formulationdiscussion and generalization," in *Proceedings of the 5th International Conference on Genetic Algorithms*, vol. 93, July, 1993, pp. 416–423.

8 K. Deb, A. Pratap, S. Agarwal, and T. Meyarivan, "A fast and elitist multi-objective genetic algorithm: NSGA-II," *IEEE Transactions on Evolutionary Computation*, vol. 6, no. 2, pp. 182–197, 2002.

9 E. Zitzler, M. Laumanns, and L. Thiele, "SPEA2: Improving the strength Pareto evolutionary algorithm," *TIK-report*, vol. 103, 2001.

10 D. W. Corne, J. D. Knowles, and M. J. Oates, "The Pareto envelope-based selection algorithm for multi-objective optimization," in *Proceedings of International Conference on Parallel Problem Solving from Nature*, 2000, pp. 839–848.

11 D. W. Corne, N. R. Jerram, J. D. Knowles, and M. J. Oates, "PESA-II: Region-based selection in evolutionary multiobjective optimization," in *Proceedings of the 3rd Annual Conference on Genetic and Evolutionary Computation*, 2001, pp. 283–290.

12 Y. Jin, M. Olhofer, and B. Sendhoff, "Dynamic weighted aggregation for evolutionary multi-objective optimization: why does it work and how?" in *Proceedings of the Genetic and Evolutionary Computation Conference*, 2001, pp. 1042–1049.

13 T. Murata, H. Ishibuchi, and M. Gen, "Specification of genetic search directions in cellular multi-objective genetic algorithms," in *Proceedings of the 1st International Conference on Evolutionary Multi-Criterion Optimization*, 2001, pp. 82–95.

14 Z. Wang, Q. Zhang, M. Gong, and A. Zhou, "A replacement strategy for balancing convergence and diversity in MOEA/D," in *Proceedings of the 2014 IEEE Congress on Evolutionary Computation*, 2014, pp. 2132–2139.

15 K. Li, Q. Zhang, S. Kwong, M. Li, and R. Wang, "Stable matching-based selection in evolutionary multi-objective optimization," *IEEE Transactions on Evolutionary Computation*, vol. 18, no. 6, pp. 909–923, 2013.

16 K. Li, S. Kwong, Q. Zhang, and K. Deb, "Interrelationship-based selection for decomposition multi-objective optimization," *IEEE Transactions on Cybernetics*, vol. 45, no. 10, pp. 2076–2088, 2014.

17 H.-L. Liu, F. Gu, and Q. Zhang, "Decomposition of a multi-objective optimization problem into a number of simple multi-objective subproblems," *IEEE Transactions on Evolutionary Computation*, vol. 18, no. 3, pp. 450–455, 2013.

18 L. Chen, H.-L. Liu, C. Lu, Y.-m. Cheung, and J. Zhang, "A novel evolutionary multi-objective algorithm based on S metric selection and M2M population decomposition," in *Proceedings of the Asia Pacific Symposium on Intelligent and Evolutionary Systems*, 2015, pp. 441–452.

19 J. Bader and E. Zitzler, "HypE: An algorithm for fast hypervolume-based many-objective optimization," *Evolutionary Computation*, vol. 19, no. 1, pp. 45–76, 2011.

20 M. Emmerich, N. Beume, and B. Naujoks, "An EMO algorithm using the hypervolume measure as selection criterion," in *Proceedings of International Conference on Evolutionary Multi-Criterion Optimization*, 2005, pp. 62–76.

21 E. Zitzler and S. Künzli, "Indicator-based selection in multi-objective search," in *Proceedings of International Conference on Parallel Problem Solving from Nature*, 2004, pp. 832–842.

22 Y. Tian, R. Cheng, X. Zhang, F. Cheng, and Y. Jin, "An indicator-based multiobjective evolutionary algorithm with reference point adaptation for better versatility," *IEEE Transactions on Evolutionary Computation*, vol. 22, no. 4, pp. 609–622, 2018.

23 Y. Tian, X. Zhang, R. Cheng, and Y. Jin, "A multi-objective evolutionary algorithm based on an enhanced inverted generational distance metric," in *2016 IEEE Congress on Evolutionary Computation (CEC)*, 2016, pp. 5222–5229.

24 B. L. Miller and D. E. Goldberg, "Genetic algorithms, tournament selection, and the effects of noise," *Complex Systems*, vol. 9, no. 3, pp. 193–212, 1995.

25 R. B. Agrawal, K. Deb, and R. B. Agrawal, "Simulated binary crossover for continuous search space," *Complex Systems*, vol. 9, no. 3, pp. 115–148, 1994.

26 K. Ikeda, H. Kita, and S. Kobayashi, "Failure of Pareto-based MOEAs: does non-dominated really mean near to optimal?" in *Proceedings of the 2001 IEEE Congress on Evolutionary Computation*, vol. 2, 2001, pp. 957–962.

27 D. Brockhoff, T. Friedrich, N. Hebbinghaus, C. Klein, F. Neumann, and E. Zitzler, "On the effects of adding objectives to plateau functions," *IEEE Transactions on Evolutionary Computation*, vol. 13, no. 3, pp. 591–603, 2009.

28 O. Schutze, A. Lara, and C. A. Coello Coello, "On the influence of the number of objectives on the hardness of a multiobjective optimization problem," *IEEE Transactions on Evolutionary Computation*, vol. 15, no. 4, pp. 444–455, 2010.

29 L. S. Batista, F. Campelo, F. G. Guimar aes, and J. Ramírez, "A comparison of dominance criteria in many-objective optimization problems," in *Proceedings of the IEEE Congress on Evolutionary Computation*, 2011, pp. 2359–2366.

30 D. W. Corne and J. D. Knowles, "Techniques for highly multiobjective optimisation: some nondominated points are better than others," in *Proceedings of the 9th Annual Conference on Genetic and Evolutionary Computation*, 2007, pp. 773–780.

31 Y. Jin and B. Sendhoff, "Connectedness, regularity and the success of local search in evolutionary multi-objective optimization," in *Proceedings of the IEEE Congress on Evolutionary Computation*, 2003, pp. 1910–1917.

32 Q. Zhang, A. Zhou, and Y. Jin, "RM-MEDA: A regularity model-based multi-objective estimation of distribution algorithm," *IEEE Transactions on Evolutionary Computation*, vol. 12, no. 1, pp. 41–63, 2008.

33 R. C. Purshouse and P. J. Fleming, "On the evolutionary optimization of many conflicting objectives," *IEEE Transactions on Evolutionary Computation*, vol. 11, no. 6, pp. 770–784, 2007.

34 S. F. Adra and P. J. Fleming, "Diversity management in evolutionary many-objective optimization," *IEEE Transactions on Evolutionary Computation*, vol. 15, no. 2, pp. 183–195, 2010.

35 M. Li, S. Yang, and X. Liu, "Diversity comparison of Pareto front approximations in many-objective optimization," *IEEE Transactions on Cybernetics*, vol. 44, no. 12, pp. 2568–2584, 2014.

36 H. Ishibuchi, N. Tsukamoto, and Y. Nojima, "Evolutionary many-objective optimization: a short review," in *Proceedings of the 2008 IEEE Congress on Evolutionary Computation*, 2008, pp. 2419–2426.

37 Z. He and G. G. Yen, "Ranking many-objective evolutionary algorithms using performance metrics ensemble," in *Proceedings of 2013 IEEE Congress on Evolutionary Computation*, 2013, pp. 2480–2487.

38 M. Laumanns, L. Thiele, K. Deb, and E. Zitzler, "Combining convergence and diversity in evolutionary multiobjective optimization," *Evolutionary Computation*, vol. 10, no. 3, pp. 263–282, 2002.

39 K. Deb, M. Mohan, and S. Mishra, "Evaluating the ε-domination based multi-objective evolutionary algorithm for a quick computation of pareto-optimal solutions," *Evolutionary Computation*, vol. 13, no. 4, pp. 501–525, 2005.

40 H. Aguirre and K. Tanaka, "Space partitioning with adaptive ε-ranking and substitute distance assignments: a comparative study on many-objective mnk-landscapes," in *Proceedings of the 11th Annual Conference on Genetic and Evolutionary Computation*, 2009, pp. 547–554.

41 J. Branke, T. Kaußler, and H. Schmeck, "Guidance in evolutionary multi-objective optimization," *Advances in Engineering Software*, vol. 32, no. 6, pp. 499–507, 2001.

42 L. S. Batista, F. Campelo, F. G. Guimaraes, and J. A. Ramírez, "Pareto cone ε-dominance: improving convergence and diversity in multiobjective evolutionary algorithms," in *Proceedings of the 6th International Conference Evolutionary Multi-Criterion Optimization*, 2011, pp. 76–90.

43 S. Yang, M. Li, X. Liu, and J. Zheng, "A grid-based evolutionary algorithm for many-objective optimization," *IEEE Transactions on Evolutionary Computation*, vol. 17, no. 5, pp. 721–736, 2013.

44 M. Li, S. Yang, and X. Liu, "Shift-based density estimation for Pareto-based algorithms in many-objective optimization," *IEEE Transactions on Evolutionary Computation*, vol. 18, no. 3, pp. 348–365, 2013.

45 R. Wang, R. C. Purshouse, and P. J. Fleming, "Preference-inspired coevolutionary algorithms for many-objective optimization," *IEEE Transactions on Evolutionary Computation*, vol. 17, no. 4, pp. 474–494, 2012.

46 K. Deb and H. Jain, "An evolutionary many-objective optimization algorithm using reference-point-based nondominated sorting approach, Part I: solving problems with box constraints," *IEEE Transactions on Evolutionary Computation*, vol. 18, no. 4, pp. 577–601, 2013.

47 X. Zhang, Y. Tian, and Y. Jin, "A knee point driven evolutionary algorithm for many-objective optimization," *IEEE Transactions on Evolutionary Computation*, vol. 19, no. 6, pp. 761–776, 2015.

48 H. K. Singh, A. Isaacs, and T. Ray, "A Pareto corner search evolutionary algorithm and dimensionality reduction in many-objective optimization problems," *IEEE Transactions on Evolutionary Computation*, vol. 15, no. 4, pp. 539–556, 2011.

49 D. K. Saxena, J. A. Duro, A. Tiwari, K. Deb, and Q. Zhang, "Objective reduction in many-objective optimization: linear and nonlinear algorithms," *IEEE Transactions on Evolutionary Computation*, vol. 17, no. 1, pp. 77–99, 2012.

50 D. Brockhoff and E. Zitzler, "Objective reduction in evolutionary multi-objective optimization: theory and applications," *Evolutionary Computation*, vol. 17, no. 2, pp. 135–166, 2009.

51 I. T. Jolliffe, "Principal component analysis," *Technometrics*, vol. 45, no. 3, p. 276, 2003.

52 K. Q. Weinberger and L. K. Saul, "An introduction to nonlinear dimensionality reduction by maximum variance unfolding," in *Proceedings of the 21st National Conference on Artificial Intelligence*, 2006, pp. 1683–1686.

53 A. L. Jaimes, C. A. Coello Coello, and J. E. U. Barrientos, "Online objective reduction to deal with many-objective problems," in *Proceedings of IEEE International Conference on Evolutionary Multi-criterion Optimization*, 2009, pp. 423–437.

54 L. Kaufman and P. J. Rousseeuw, *Finding groups in data: an introduction to cluster analysis*. Wiley, 1990.

55 X. Guo, X. Wang, M. Wang, and Y. Wang, "A new objective reduction algorithm for many-objective problems: employing mutual information and clustering algorithm," in *Proceedings of International Conference on Computational Intelligence and Security*, 2012, pp. 11–16.

56 K. Li, K. Deb, Q. Zhang, and S. Kwong, "An evolutionary many-objective optimization algorithm based on dominance and decomposition," *IEEE Transactions on Evolutionary Computation*, vol. 19, no. 5, pp. 694–716, 2014.

57 K. Bringmann, T. Friedrich, F. Neumann, and M. Wagner, "Approximation-guided evolutionary multi-objective optimization," in *Proceedings of International Joint Conference on Artificial Intelligence*, 2011, p. 1198.

58 L. While, P. Hingston, L. Barone, and S. Huband, "A faster algorithm for calculating hypervolume," *IEEE Transactions on Evolutionary Computation*, vol. 10, no. 1, pp. 29–38, 2006.

59 L. While, L. Bradstreet, and L. Barone, "A fast way of calculating exact hypervolumes," *IEEE Transactions on Evolutionary Computation*, vol. 16, no. 1, pp. 86–95, 2012.

60 R. Cheng, Y. Jin, M. Olhofer, and B. Sendhoff, "A reference vector guided evolutionary algorithm for many-objective optimization," *IEEE Transactions on Evolutionary Computation*, vol. 20, no. 5, pp. 773–791, 2016.

61 M. N. Omidvar, X. Li, and X. Yao, "A review of population-based metaheuristics for large-scale black-box global optimizationPart I," *IEEE Transactions on Evolutionary Computation*, vol. 26, no. 5, pp. 802–822, 2021.

62 M. N. Omidvar, X. Li, and X. Yao, "A review of population-based metaheuristics for large-scale black-box global optimizationPart II," *IEEE Transactions on Evolutionary Computation*, vol. 26, no. 5, pp. 823–843, 2021.

63 M. N. Omidvar, X. Li, and K. Tang, "Designing benchmark problems for large-scale continuous optimization," *Information Sciences*, vol. 316, pp. 419–436, 2015.

64 Y. Ling, H. Li, and B. Cao, "Cooperative co-evolution with graph-based differential grouping for large scale global optimization," in *Proceedings of the 12th International Conference on Natural Computation, Fuzzy Systems and Knowledge Discovery*, 2016, pp. 95–102.

65 Y. Mei, M. N. Omidvar, X. Li, and X. Yao, "A competitive divide-and-conquer algorithm for unconstrained large-scale black-box optimization," *ACM Transactions on Mathematical Software*, vol. 42, no. 2, pp. 1–24, 2016.

66 Y. Sun, M. Kirley, and S. K. Halgamuge, "Extended differential grouping for large scale global optimization with direct and indirect variable interactions," in *Proceedings of the 2015 Annual Conference on Genetic and Evolutionary Computation*, 2015, pp. 313–320.

67 M. N. Omidvar, M. Yang, Y. Mei, X. Li, and X. Yao, "DG2: A faster and more accurate differential grouping for large-scale black-box optimization," *IEEE Transactions on Evolutionary Computation*, vol. 21, no. 6, pp. 929–942, 2017.

68 M. N. Omidvar, X. Li, Z. Yang, and X. Yao, "Cooperative co-evolution for large scale optimization through more frequent random grouping," in *Proceedings of the 2010 IEEE Congress on Evolutionary Computation*, 2010, pp. 1–8.

69 R. Liu, J. Liu, Y. Li, and J. Liu, "A random dynamic grouping based weight optimization framework for large-scale multi-objective optimization problems," *Swarm and Evolutionary Computation*, vol. 55, p. 100684, 2020.

70 M. A. Potter and K. A. De Jong, "A cooperative coevolutionary approach to function optimization," in *Proceedings of IEEE International Conference on Parallel Problem Solving from Nature*, 1994, pp. 249–257.

71 Z. Yang, K. Tang, and X. Yao, "Large scale evolutionary optimization using cooperative coevolution," *Information Sciences*, vol. 178, no. 15, pp. 2985–2999, 2008.

72 X. Li and X. Yao, "Cooperatively coevolving particle swarms for large scale optimization," *IEEE Transactions on Evolutionary Computation*, vol. 16, no. 2, pp. 210–224, 2011.

73 J. Kennedy and R. Eberhart, "Particle swarm optimization," in *Proceedings of IEEE International Conference on Neural Networks*, 1995, pp. 1942–1948.

74 Y. Shi and R. Eberhart, "A modified particle swarm optimizer," in *Proceedings of IEEE International Conference on Evolutionary Computation*, 1998, pp. 69–73.

75 R. Eberhart and J. Kennedy, "A new optimizer using particle swarm theory," in *Proceedings of International Symposium on Micro Machine and Human Science*, 1995, pp. 39–43.

2

Evolutionary Large-Scale Multi-Objective Optimization

2.1 Introduction

Large-scale multi-objective optimization problems (LSMOPs) refer to a subset of optimization problems which have both a large number of decision variables and multiple conflicting objectives. This class of problems is extensively present in many applications. For example, in the community detection which aims to minimize the inter-link density and maximize the intra-link density, the number of network nodes is usually in the order of tens of thousands [1, 2]. In the deep neural network training which aims to minimize the network complexity and maximize the prediction accuracy, there are often millions of weights to be trained [3]. Similarly, problems in logistic scheduling [4, 5], power systems [6], radiotherapy planning [7], and many other real-world applications have also shown the characteristics of LSMOPs.

Up to now, there still does not exist a unified opinion on the minimum number of decision variables for LSMOPs, which may differ for different types of problems. In this book, LSMOPs are defined as the optimization problems which have at least two conflicting objectives and one hundred decision variables. Generally, LSMOPs are more difficult to be handled than the problems with a few decision variables, since the search space is exponentially related to the number of decision variables. On the one hand, with the increasing of the number of decision variables, the number of local optima increases considerably even exponentially. This will easily lead to multi-objective evolutionary algorithms (MOEAs) converging prematurely to a local optimum [8]. On the other hand, with the scaling up of the number of decision variables, the probability that decision variables interact with each other increases, which makes the landscape of search space more complicated and the conflicts between multiple objectives more serious [9].

In the past decades, a lot of efforts have been concentrated on studying large-scale single-objective optimization problems (LSOPs). A large number of test

Evolutionary Large-Scale Multi-Objective Optimization and Applications: Integrating Evolutionary Computation, Machine Learning, and Data Science, First Edition. Xingyi Zhang, Ran Cheng, Ye Tian, and Yaochu Jin.

problems have been designed for large-scale single-objective optimization [10–14]. The methods for solving LSOPs have also achieved remarkable progress, mainly including the decomposition-based and non-decomposition-based methods [15–17]. Recently, with the improvement of computing power, increasing attention has been paid to using MOEAs to address LSMOPs. While different types of benchmark test problems are designed for large-scale multi-objective optimization, a handful of promising MOEAs have also been customized for solving LSMOPs. In addition, some LSMOPs with specific properties have been extensively investigated, especially those with sparse Pareto optimal solutions. The sparse large-scale multi-objective optimization problems, namely, sparse LSMOPs, refer to a class of LSMOPs whose decision variables in Pareto optimal solutions are mostly zero-valued. As an emerging research topic in LSMOPs, some benchmark test problems and a number of MOEAs have also been developed for sparse LSMOPs [18].

The remainder of this chapter is organized as follows. Section 2.2 gives the benchmark test problems for large-scale multi-objective optimization, and Section 2.3 presents some commonly used indicators for evaluating the performance of MOEAs in solving LSMOPs. Section 2.4 introduces the test problems tailored for sparse large-scale multi-objective optimization, followed by a performance indicator in Section 2.5 which can assess the convergence, sparsity, and diversity of solution sets simultaneously. Finally, Section 2.6 finishes this chapter with a brief summary.

2.2 Test Problems for Large-Scale Multi-Objective Optimization

In this section, some primary concepts related to multi-objective optimization are first given. Then, we introduce some test problems that are commonly used for large-scale multi-objective optimization.

2.2.1 Primary Concepts

To better mimic the problems in real world, a benchmark test suite should capture as much problem characteristics as possible, including fitness landscape, variable interactions, Pareto front geometries, separability of objective functions, and among many others. In the following, some primary concepts related to large-scale multi-objective test problems are introduced in detail.

Fitness landscape. The goal of solving a multi-objective optimization problem is to obtain a set of trade-off solutions in the decision space, and the quality of

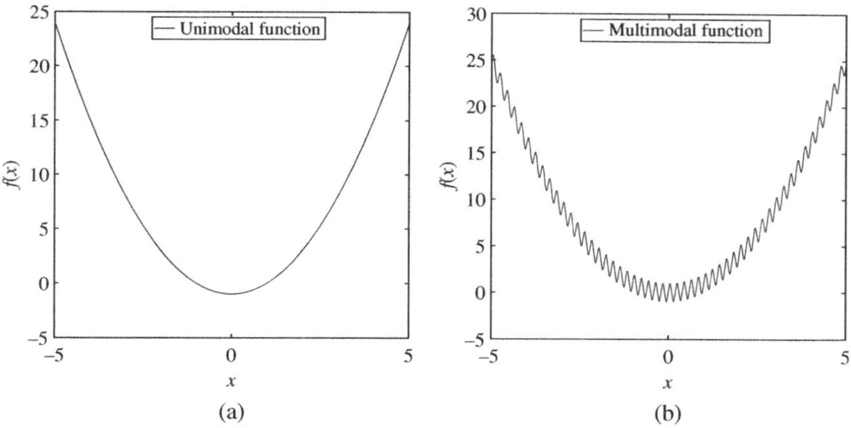

Figure 2.1 Examples of a unimodal function and a multimodal function. (a) Minimize $f(x) = x^2 - 1, x \in [-5, 5]$. (b) Minimize $f(x) = x^2 + \sin(10\pi x), x \in [-5, 5]$.

these solutions is usually assessed in the objective space. The mapping from the decision space to the objective space defines the fitness landscape.

Modality. The modality is related to the number of local optima of an objective function. Specifically, an objective function is said to be *multimodal* if it has multiple local optima, and an objective function is *unimodal* if it has only a single optimum. Figure 2.1 gives the examples of a unimodal function and a multimodal function. Typically, multimodal functions are more difficult to be optimized than unimodal functions, since solutions are prone to trap into the local optima in multimodal functions.

Deceptive. For multimodal functions, there exists a special case, namely, the deceptive function, which has at least two optima, i.e., a true optimum and a deceptive optimum. Besides, the majority of the search space must favor the deceptive optimum. Figure 2.2(a) gives an example of deceptive function. The true optimum lies in the position of $x = 0$ which is very prone to be missed by an optimization algorithm.

Bias. The bias attribute is used to describe whether an evenly distributed sample of solutions in the decision space maps to an evenly distributed set of solutions in the objective space. If not, the mapping from the Pareto set to the Pareto front is said to be *biased*. Figure 2.2(b) plots a large number of objective vectors of a biased multi-objective optimization problem, where these objective vectors are obtained by randomly sampling 60,000 decision vectors. As can be seen from the figure, these objective vectors get denser toward the Pareto front.

Degenerate. The Pareto front geometries of different multi-objective optimization problems are usually diverse (e.g., linear, concave, convex, disconnected, and

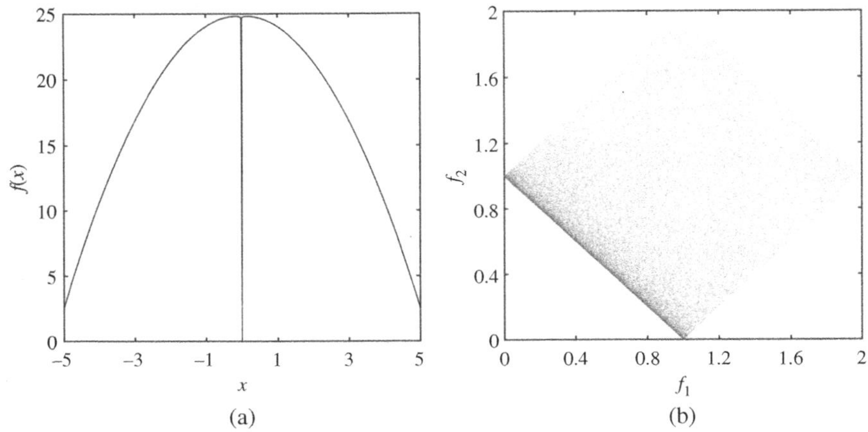

Figure 2.2 Examples of a deceptive function and a biased multi-objective optimization problem. (a) Minimize $f(x) = -0.9x^2 + (5|x|^{0.001}/5^{0.001})^2, x \in [-5, 5]$. (b) Minimize $f_1(x_1, x_2) = (x_1)^5 + x_2, f_2(x_1, x_2) = (x_1)^5 + 1 - x_2, x_1, x_2 \in [0, 1]$.

mixed). Among these geometries, there exists a special kind of Pareto front, namely, the degenerate Pareto front, which is of lower dimension than the objective space where it is embedded. For example, in the bi-objective optimization problems, a front consisting of a number of points is degenerate, while a front that is a curve is not degenerate. In the three-objective optimization problems, a front that is a curve is degenerate, while a front that is a two-dimensional surface is not degenerate. Figure 2.3 provides examples of Pareto fronts with different geometries.

Nonseparability. The nonseparability describes the dependency relationship between decision variables. A decision variable x_k is said to be *separable* on an objective function $f(\mathbf{x})$ if the following condition is satisfied:

$$f(\mathbf{x}) < f(\mathbf{x}') \Rightarrow f(\mathbf{y}) < f(\mathbf{y}'), \tag{2.1}$$

where $\mathbf{x} = [x_1, \ldots, x_k, \ldots, x_n]$, $\mathbf{x}' = [x_1, \ldots, x_k', \ldots, x_n]$, $\mathbf{y} = [y_1, \ldots, y_k, \ldots, y_n]$, and $\mathbf{y}' = [y_1, \ldots, y_k', \ldots, y_n]$ are four different decision vectors. Otherwise, the decision variable x_k is *nonseparable* on $f(\mathbf{x})$. If each decision variable is separable, the corresponding objective function is a *separable objective*. Conversely, if each decision variable is nonseparable, the corresponding objective function is a *nonseparable objective*. A special case is that the nonseparable decision variables only interact with some of the other decision variables, and the corresponding objective function is a *partially separable objective*.

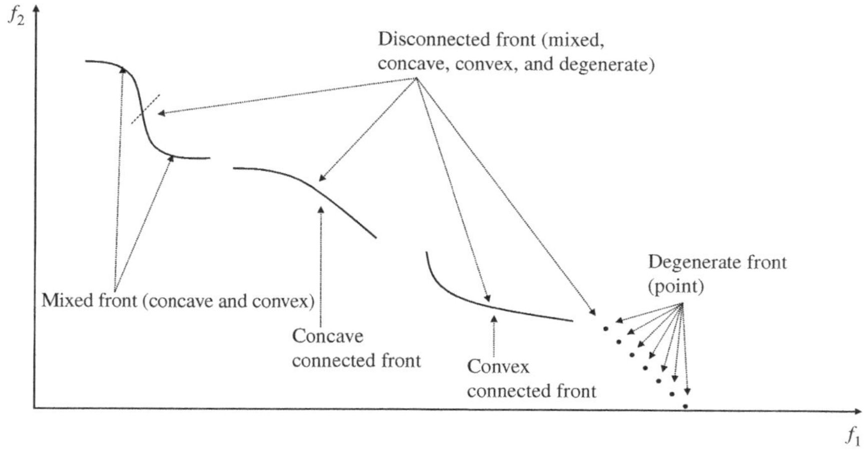

Figure 2.3 Examples of Pareto fronts with different geometries.

2.2.2 Multi-Objective Optimization Test Problems

In the earlier years, the performance of MOEAs for large-scale multi-objective optimization was usually tested on the problems that were initially designed for multi-objective optimization, such as the Zitzler, Deb, and Thiele (ZDT) benchmark suite [19], Deb, Thiele, Laumanns, and Zitzler (DTLZ) benchmark suite [20], and Walking Fish Group (WFG) benchmark suite [21].

As one of the first benchmark suites designed for multi-objective optimization, the ZDT benchmark suite contains six bi-objective test problems [19]. Each of them has the same structure and consists of three functions, namely, the distribution function $\mathbf{f}_1$, the distance function $\mathbf{g}$ and the shape function $\mathbf{h}$:

$$\begin{cases} \text{Minimize } \mathbf{F}(\mathbf{x}) = (\mathbf{f}_1(x_1), f_2(\mathbf{x})), \\ \text{subject to } f_2(\mathbf{x}) = \mathbf{g}(x_2, \dots, x_n)\mathbf{h}(\mathbf{f}_1(x_1), \mathbf{g}(x_2, \dots, x_n)), \\ \qquad \text{where } \mathbf{x} = (x_1, \dots, x_n). \end{cases} \tag{2.2}$$

The distribution function $\mathbf{f}_1$ is used for assessing the diversity maintenance ability of an MOEA, which is only related to the first decision variable of $\mathbf{F}(\mathbf{x})$. The distance function $\mathbf{g}$ is related to the remaining $n-1$ decision variables of $\mathbf{F}(\mathbf{x})$, which is used to assess the convergence ability of an MOEA. The shape function $\mathbf{h}$ determines the geometries of Pareto fronts, and its inputs are the function values of $\mathbf{f}_1$ and $\mathbf{g}$. Table 2.1 gives detailed information of the six ZDT test problems, which contains the features of Pareto front, the features of landscape, the

Table 2.1 Detailed Information of the Six ZDT Test Problems.

Test problem	Features of Pareto front	Features of landscape	Separability of objective functions	Problem formula
ZDT1	Convex	Unimodal	Separable	$\begin{cases} \min & f_1(x_1) = x_1 \\ \min & f_2(\mathbf{x}) = g(x_2, \ldots, x_n)h(f_1(x_1), g(x_2, \ldots, x_n)) \\ \text{where} & g(x_2, \ldots, x_n) = 1 + 9\sum_{i=2}^{n} x_i/(n-1) \\ & h(f_1, g) = 1 - \sqrt{f_1/g} \\ & n = 30, x_i \in [0, 1] \end{cases}$
ZDT2	Concave	Unimodal	Separable	$\begin{cases} \min & f_1(x_1) = x_1 \\ \min & f_2(\mathbf{x}) = g(x_2, \ldots, x_n)h(f_1(x_1), g(x_2, \ldots, x_n)) \\ \text{where} & g(x_2, \ldots, x_n) = 1 + 9\sum_{i=2}^{n} x_i/(n-1) \\ & h(f_1, g) = 1 - (f_1/g)^2 \\ & n = 30, x_i \in [0, 1] \end{cases}$
ZDT3	Convex, disconnected	Multimodal	Separable	$\begin{cases} \min & f_1(x_1) = x_1 \\ \min & f_2(\mathbf{x}) = g(x_2, \ldots, x_n)h(f_1(x_1), g(x_2, \ldots, x_n)) \\ \text{where} & g(x_2, \ldots, x_n) = 1 + 9\sum_{i=2}^{n} x_i/(n-1) \\ & h(f_1, g) = 1 - \sqrt{f_1/g} - (f_1/g)\sin(10\pi f_1) \\ & n = 30, x_i \in [0, 1] \end{cases}$

ZDT4	Convex	Multimodal	Separable	$\begin{cases} \min & f_1(x_1) = x_1 \\ \min & f_2(\mathbf{x}) = g(x_2, \ldots, x_n)h(f_1(x_1), g(x_2, \ldots, x_n)) \\ \text{where} & g(x_2, \ldots, x_n) = 1 + 10(n-1) + \sum_{i=2}^{n}(x_i^2 - 10\cos(4\pi x_i)) \\ & h(f_1, g) = 1 - \sqrt{f_1/g} \\ & n = 10, x_1 \in [0,1], x_2, \ldots, x_n \in [-5,5] \end{cases}$
ZDT5	Convex	Multimodal	Separable	$\begin{cases} \min & f_1(x_1) = 1 + u(x_1) \\ \min & f_2(\mathbf{x}) = g(x_2, \ldots, x_n)h(f_1(x_1), g(x_2, \ldots, x_n)) \\ \text{where} & g(x_2, \ldots, x_n) = \sum_{i=2}^{n} v(u(x_i)) \\ & h(f_1, g) = 1/f_1 \\ & v(u(x_i)) = \begin{cases} 2 + u(x_i) & \text{if } u(x_i) < 5 \\ 1 & \text{if } u(x_i) = 5 \end{cases} \\ & n = 11, x_1 \in \{0,1\}^{30}, x_2, \ldots, x_n \in \{0,1\}^5 \end{cases}$
ZDT6	Concave	Multimodal	Separable	$\begin{cases} \min & f_1(x_1) = 1 - \exp(-4x_1)\sin^6(6\pi x_1) \\ \min & f_2(\mathbf{x}) = g(x_2, \ldots, x_n)h(f_1(x_1), g(x_2, \ldots, x_n)) \\ \text{where} & g(x_2, \ldots, x_n) = 1 + 9((\sum_{i=2}^{n} x_i)/(n-1))^{0.25} \\ & h(f_1, g) = 1 - (f_1/g)^2 \\ & n = 10, x_i \in [0,1] \end{cases}$

separability of objective functions, and problem formulas. As can be seen from the table, the six ZDT test problems are diverse in terms of the features of Pareto front and landscape, but they are all separable in terms of the separability of objective functions.

While the ZDT benchmark suite is widely used for assessing the performance of MOEAs, one limitation of the six ZDT test problems is that they are not scalable in the objective space. Considering that many problems in the real world have three or more objectives and MOEAs need to demonstrate their efficacy in solving these problems, Deb *et al.* proposed another benchmark test suite, i.e., the DTLZ benchmark test suite [20]. All the problems in the DTLZ benchmark suite follow a generic design principle, where the first $M-1$ decision variables define the Pareto front and the remaining $n-M+1$ decision variables control the convergence property. The n and M represent the number of decision variables and the number of objectives, respectively. The detailed information of the seven DTLZ test problems is presented in Table 2.2, it is worth noting that in the problem formulas, the last $n-M+1$ decision variables are represented as $\mathbf{x}_M$. As can be seen from the table, the DTLZ benchmark suite captures more diverse problem characteristics than the ZDT benchmark suite. For example, while the DTLZ test problems have degenerate Pareto fronts and biased landscapes, the ZDT test problems do not capture these problem characteristics. Besides, from the problem formulas, it can be seen that each DTLZ test problem is scalable to have any number of objectives. But, all the DTLZ test problems are still separable in terms of the separability of objective functions.

In addition to the ZDT and DTLZ benchmark test suites, some other benchmark test suites have also been designed and used for performance assessment of large-scale multi-objective optimization. For example, in the WFG benchmark suite [21], some problems introduce decision variable interactions, which makes them more consistent with the problems in real world. The BT benchmark suite [22] is specially designed to test the performance of MOEAs for solving multi-objective optimization problems with bias landscape. In the nine BT test problems, two kinds of biases, namely, the position-related bias and distance-related bias, are considered. The MaF benchmark suite [23] is designed for assessing the performance of MOEAs on problems with more than three objectives. In the 15 MaF test problems, a variety of problem characteristics that are widely presented in real-world applications are captured.

Although the benchmark suites as introduced above have effectively promoted the research of evolutionary multi-objective optimization and some of them are used as the test problems for large-scale multi-objective optimization, the majority of them are designed by introducing difficulties in the objective space. Few of them pay attention to the challenges brought by the huge decision space. Specifically, on the one hand, some of them are not scalable in the decision

Table 2.2 Detailed Information of the Seven DTLZ Test Problems.

Test problem	Features of Pareto front	Features of landscape	Separability of objective functions	Problem formula
DTLZ1	Linear	Multimodal	Separable	$\begin{cases} \min & f_1(\mathbf{x}) = \frac{1}{2}x_1x_2\cdots x_{M-1}(1+g(\mathbf{x}_M)) \\ \min & f_2(\mathbf{x}) = \frac{1}{2}x_1x_2\cdots(1-x_{M-1})(1+g(\mathbf{x}_M)) \\ & \vdots\vdots \\ \min & f_{M-1}(\mathbf{x}) = \frac{1}{2}x_1(1-x_2)(1+g(\mathbf{x}_M)) \\ \min & f_M(\mathbf{x}) = \frac{1}{2}(1-x_1)(1+g(\mathbf{x}_M)) \\ \text{where} & g(\mathbf{x}_M) = 100[\lvert\mathbf{x}_M\rvert + \sum_{x_i\in\mathbf{x}_M}(x_i-0.5)^2 - \cos(20\pi(x_i-0.5))] \\ & x_i \in [0,1],\ \text{for } i=1,2,\ldots,n \end{cases}$;
DTLZ2	Concave	Unimodal	Separable	$\begin{cases} \min & f_1(\mathbf{x}) = (1+g(\mathbf{x}_M))\cos(x_1\pi/2)\cdots\cos(x_{M-2}\pi/2)\cos(x_{M-1}\pi/2) \\ \min & f_2(\mathbf{x}) = (1+g(\mathbf{x}_M))\cos(x_1\pi/2)\cdots\cos(x_{M-2}\pi/2)\sin(x_{M-1}\pi/2) \\ \min & f_3(\mathbf{x}) = (1+g(\mathbf{x}_M))\cos(x_1\pi/2)\cdots\sin(x_{M-2}\pi/2) \\ & \vdots\vdots \\ \min & f_M(\mathbf{x}) = (1+g(\mathbf{x}_M))\sin(x_1\pi/2) \\ \text{where} & g(\mathbf{x}_M) = \sum_{x_i\in\mathbf{x}_M}(x_i-0.5)^2 \\ & x_i \in [0,1],\ \text{for } i=1,2,\ldots,n \end{cases}$

(Continued)

Table 2.2 (Continued)

Test problem	Features of Pareto front	Features of landscape	Separability of objective functions	Problem formula
DTLZ3	Concave	Multimodal	Separable	$\begin{cases} \min & f_1(\mathbf{x}) = (1+g(\mathbf{x}_M))\cos(x_1\pi/2)\cdots\cos(x_{M-2}\pi/2)\cos(x_{M-1}\pi/2) \\ \min & f_2(\mathbf{x}) = (1+g(\mathbf{x}_M))\cos(x_1\pi/2)\cdots\cos(x_{M-2}\pi/2)\sin(x_{M-1}\pi/2) \\ \min & f_3(\mathbf{x}) = (1+g(\mathbf{x}_M))\cos(x_1\pi/2)\cdots\sin(x_{M-2}\pi/2) \\ & \vdots\vdots \\ \min & f_M(\mathbf{x}) = (1+g(\mathbf{x}_M))\sin(x_1\pi/2) \\ \text{where} & g(\mathbf{x}_M) = 100[\lvert\mathbf{x}_M\rvert + \sum_{x_i\in\mathbf{x}_M}(x_i-0.5)^2 - \cos(20\pi(x_i-0.5))] \\ & x_i \in [0,1],\ \text{for } i = 1,2,\ldots,n \end{cases}$
DTLZ4	Concave	Unimodal, biased	Separable	$\begin{cases} \min & f_1(\mathbf{x}) = (1+g(\mathbf{x}_M))\cos(x_1{}^{\alpha}\pi/2)\cdots\cos(x_{M-2}{}^{\alpha}\pi/2)\cos(x_{M-1}{}^{\alpha}\pi/2) \\ \min & f_2(\mathbf{x}) = (1+g(\mathbf{x}_M))\cos(x_1{}^{\alpha}\pi/2)\cdots\cos(x_{M-2}{}^{\alpha}\pi/2)\sin(x_{M-1}{}^{\alpha}\pi/2) \\ \min & f_3(\mathbf{x}) = (1+g(\mathbf{x}_M))\cos(x_1{}^{\alpha}\pi/2)\cdots\sin(x_{M-2}{}^{\alpha}\pi/2) \\ & \vdots\vdots \\ \min & f_M(\mathbf{x}) = (1+g(\mathbf{x}_M))\sin(x_1{}^{\alpha}\pi/2) \\ \text{where} & g(\mathbf{x}_M) = \sum_{x_i\in\mathbf{x}_M}(x_i-0.5)^2 \\ & x_i \in [0,1],\ \text{for } i = 1,2,\ldots,n \\ & \alpha = 100 \end{cases}$

DTLZ5	Concave, degenerate	Unimodal	Separable	$\left\{\begin{array}{ll} \min & f_1(\mathbf{x}) = (1+g(\mathbf{x}_M))\cos(\theta_1\pi/2)\cdots\cos(\theta_{M-2}\pi/2)\cos(\theta_{M-1}\pi/2) \\ \min & f_2(\mathbf{x}) = (1+g(\mathbf{x}_M))\cos(\theta_1\pi/2)\cdots\cos(\theta_{M-2}\pi/2)\sin(\theta_{M-1}\pi/2) \\ \min & f_3(\mathbf{x}) = (1+g(\mathbf{x}_M))\cos(\theta_1\pi/2)\cdots\sin(\theta_{M-2}\pi/2) \\ & \vdots\vdots \\ \min & f_M(\mathbf{x}) = (1+g(\mathbf{x}_M))\sin(\theta_1\pi/2) \\ \text{where} & g(\mathbf{x}_M) = \sum_{x_i\in\mathbf{x}_M}(x_i-0.5)^2 \\ & \theta_i = \frac{\pi}{4(1+g(\mathbf{x}_M))}(1+2g(\mathbf{x}_M)x_i),\ for\ i=2,3,\ldots,(M-1) \\ & x_i\in[0,1],\ \text{for } i=1,2,\ldots,n \end{array}\right.$
DTLZ6	Concave, degenerate	Unimodal, biased	Separable	$\left\{\begin{array}{ll} \min & f_1(\mathbf{x}) = (1+g(\mathbf{x}_M))\cos(\theta_1\pi/2)\cdots\cos(\theta_{M-2}\pi/2)\cos(\theta_{M-1}\pi/2) \\ \min & f_2(\mathbf{x}) = (1+g(\mathbf{x}_M))\cos(\theta_1\pi/2)\cdots\cos(\theta_{M-2}\pi/2)\sin(\theta_{M-1}\pi/2) \\ \min & f_3(\mathbf{x}) = (1+g(\mathbf{x}_M))\cos(\theta_1\pi/2)\cdots\sin(\theta_{M-2}\pi/2) \\ & \vdots\vdots \\ \min & f_M(\mathbf{x}) = (1+g(\mathbf{x}_M))\sin(\theta_1\pi/2) \\ \text{where} & g(\mathbf{x}_M) = \sum_{x_i\in\mathbf{x}_M}(x_i)^{0.1} \\ & \theta_i = \frac{\pi}{4(1+g(\mathbf{x}_M))}(1+2g(\mathbf{x}_M)x_i),\ \text{for } i=2,3,\ldots,(M-1) \\ & x_i\in[0,1],\ \text{for } i=1,2,\ldots,n \end{array}\right.$
DTLZ7	Disconnected	Multimodal	Separable	$\left\{\begin{array}{ll} \min & f_1(\mathbf{x}_1) = x_1 \\ \min & f_2(\mathbf{x}_2) = x_2 \\ & \vdots\vdots \\ \min & f_{M-1}(\mathbf{x}_{M-1}) = x_{M-1} \\ \min & f_M(\mathbf{x}) = (1+g(\mathbf{x}_M))h(f_1,f_2,\ldots,f_{M-1},g) \\ \text{where} & g(\mathbf{x}_M) = 1+\frac{9}{\lvert\mathbf{x}_M\rvert}\sum_{x_i\in\mathbf{x}_M}x_i \\ & h(f_1,f_2,\ldots,f_{M-1},g) = M-\sum_{i=1}^{M-1}\left[\frac{f_i}{1+g}(1+\sin(3\pi f_i))\right] \\ & x_i\in[0,1],\ \text{for } i=1,2,\ldots,n \end{array}\right.$

space. On the other hand, while some of them are scalable to have any number of decision variables, the decision variables are separable without complex variable interactions. Hence, they are not suitable for evaluating the performance of MOEAs in solving LSMOPs, but these problems just exist widely in many practical applications. To this end, some benchmark test problems have been specially designed for large-scale multi-objective optimization, which will be elaborated in Section 2.2.3.

2.2.3 Large-Scale Multi-Objective Optimization Test Problems

As the first benchmark suite that is customized for large-scale multi-objective optimization, the LSMOP benchmark suite [24] is designed by using the following uniform design formulation:

$$\begin{cases} f_1(\mathbf{x}) &= h_1(\mathbf{x}^f)(1+g_1(\mathbf{x}^s)), \\ f_2(\mathbf{x}) &= h_2(\mathbf{x}^f)(1+g_2(\mathbf{x}^s)), \\ & \cdots \\ f_M(\mathbf{x}) &= h_M(\mathbf{x}^f)(1+g_M(\mathbf{x}^s)), \end{cases} \tag{2.3}$$

where f_1 to f_M are the objective functions, $\mathbf{x}^f = (x_1, \dots, x_{M-1})$ and $\mathbf{x}^s = (x_M, \dots, x_n)$ are the position-related part and distance-related part of the decision vector $\mathbf{x}$, respectively. The functions $h_1, \dots, h_M$ together define the geometry of Pareto front, and they are used to test the diversity maintenance ability of MOEAs. The functions $g_1, \dots, g_M$ together define the fitness landscape, and they are used to test the convergence ability of MOEAs. Based on the above design formulation, the LSMOP test problems are scalable to have any number of decision variables and objectives. Besides, by strengthening the design of fitness landscape functions, the LSMOP test problems capture some important characteristics that are related to decision variables, which are listed as follows:

(1) The decision variables are nonuniformly divided into multiple groups.
(2) Different groups of decision variables are correlated with different objectives.
(3) The decision variables have mixed separability.
(4) The decision variables have linkages on the Pareto optimal sets.

By introducing the above four characteristics, the LSMOP test problems mimic the problems in real world better, which enables them to comprehensively test the ability of MOEAs in solving LSMOPs. For example, when the ZDT, DTLZ, and WFG test problems are used for performance testing in large-scale multi-objective optimization, once a single solution on the Pareto front is obtained, it can be spread along the whole Pareto front by perturbing the variables that are related to the shape functions. By contrast, the LSMOP test problems introduce linkages between position variables in $\mathbf{x}^f$ and distance variables in $\mathbf{x}^s$, perturbing the position variables of solutions on the Pareto front will affect the distance variables, thus worsening the convergence of solutions.

Table 2.3 presents detailed information of the LSMOP test problems. As can be seen from the table, despite that the nine LSMOP test problems have simple Pareto front geometries, a wide variety of difficulties brought by decision variables are introduced, such as the mixed landscape and mixed separability of objective functions. It is worth noting that, in the problem formulas, $c_{i,j}$ is used to define the correlation between decision variables and objective functions. The u_i and l_i are the upper bound and lower bound of the ith decision variable, respectively. The $\eta_1, \eta_2, \ldots, \eta_6$ are six basic single-objective optimization problems, and their detailed information is provided in Table 2.4.

Large-scale multi-objective function (LMF) [25] is another benchmark test suite that is specially designed for large-scale multi-objective optimization. It first proposes to construct large-scale multi-objective optimization test problems by a hybrid formulation model, which is shown as follows:

$$\begin{cases} f_{i=1:2:m}(\mathbf{x}) = h_i(\mathbf{x}^p) \times (1 + g_i(\mathbf{x}^d)), \\ f_{i=2:2:m}(\mathbf{x}) = h_i(\mathbf{x}^p) \times g_i(\mathbf{x}^d), \end{cases} \tag{2.4}$$

where $\mathbf{x}^p$ and $\mathbf{x}^d$ are the position-related part and distance-related part of the decision vector $\mathbf{x}$, respectively. The h_i and g_i are the ith shape function and landscape function, respectively. As can be seen from the formula, both the multiplication-based formulation model and addition-based formulation model are used in the LMF benchmark test suite, which is obviously different from the benchmarks as introduced above that only use a single formulation model.

To better mimic and reflect the characteristics of LSMOPs in real-world applications, LMF inherits the common features in the LSMOP benchmark suite, e.g., the nonuniform grouping of decision variables, the mixed separability of objective functions, and variable linkages. Besides, LMF makes the following improvements.

(1) Hybrid linkages between position-related variables and distance-related variables are introduced.
(2) More flexible grouping of decision variables is adopted.
(3) The shape functions are strengthened by adding the correlation between position-related variables.
(4) The imbalanced contribution of decision variables to objective functions is created.

Table 2.3 Detailed Information of the Nine LSMOP Test Problems.

Test problem	Features of Pareto front	Features of landscape	Separability of objective functions	Problem formula
LSMOP1	Linear	Unimodal	Separable	Variable linkage: $\begin{cases} \mathbf{x}^s \leftarrow \left(1+\frac{i}{\lvert\mathbf{x}^s\rvert}\right)\times(x_i^s-l_i)-x_1^f\times(u_i-l_i) \\ i=1,\ldots,\lvert\mathbf{x}^s\rvert \end{cases}$ Objective functions: $\begin{cases} f_1(\mathbf{x})=x_1^f\cdots x_{M-1}^f\left(1+\sum_{j=1}^{M}c_{1,j}\times\overline{g}_1(x_j^s)\right) \\ f_2(\mathbf{x})=x_1^f\cdots\left(1-x_{M-1}^f\right)\left(1+\sum_{j=1}^{M}c_{2,j}\times\overline{g}_2(x_j^s)\right) \\ \cdots \\ f_{M-1}(\mathbf{x})=x_1^f\left(1-x_2^f\right)\left(1+\sum_{j=1}^{M}c_{M-1,j}\times\overline{g}_{M-1}(x_j^s)\right) \\ f_M(\mathbf{x})=\left(1-x_1^f\right)\left(1+\sum_{j=1}^{M}c_{M,j}\times\overline{g}_M(x_j^s)\right) \end{cases}$ where $\begin{cases} c_{i,j}=1,\ if\ i=j \\ c_{i,j}=0,\ \text{otherwise} \end{cases}$ $\begin{cases} \overline{g}_{2k-1}(\mathbf{x}_i^s)=\frac{1}{n_k}\sum_{j=1}^{n_k}\frac{\eta_1(\mathbf{x}_{i,j}^s)}{\lvert\mathbf{x}_{i,j}^s\rvert} \\ \overline{g}_{2k}(\mathbf{x}_i^s)=\frac{1}{n_k}\sum_{j=1}^{n_k}\frac{\eta_1(\mathbf{x}_{i,j}^s)}{\lvert\mathbf{x}_{i,j}^s\rvert} \\ k=1,\ldots,\lceil\frac{M}{2}\rceil \end{cases}$
LSMOP2	Linear	Mixed	Partially separable	Variable linkage: it is the same as that of LSMOP1 Objective functions: they are the same as those of LSMOP1 $\begin{cases} c_{i,j}=1,\ if\ i=j \\ c_{i,j}=0,\ \text{otherwise} \end{cases}$ $\begin{cases} \overline{g}_{2k-1}(\mathbf{x}_i^s)=\frac{1}{n_k}\sum_{j=1}^{n_k}\frac{\eta_5(\mathbf{x}_{i,j}^s)}{\lvert\mathbf{x}_{i,j}^s\rvert} \\ \overline{g}_{2k}(\mathbf{x}_i^s)=\frac{1}{n_k}\sum_{j=1}^{n_k}\frac{\eta_2(\mathbf{x}_{i,j}^s)}{\lvert\mathbf{x}_{i,j}^s\rvert} \\ k=1,\ldots,\lceil\frac{M}{2}\rceil \end{cases}$
LSMOP3	Linear	Multimodal	Mixed	Variable linkage: it is the same as that of LSMOP1 Objective functions: they are the same as those of LSMOP1 $\begin{cases} c_{i,j}=1,\ if\ i=j \\ c_{i,j}=0,\ \text{otherwise} \end{cases}$ $\begin{cases} \overline{g}_{2k-1}(\mathbf{x}_i^s)=\frac{1}{n_k}\sum_{j=1}^{n_k}\frac{\eta_4(\mathbf{x}_{i,j}^s)}{\lvert\mathbf{x}_{i,j}^s\rvert} \\ \overline{g}_{2k}(\mathbf{x}_i^s)=\frac{1}{n_k}\sum_{j=1}^{n_k}\frac{\eta_3(\mathbf{x}_{i,j}^s)}{\lvert\mathbf{x}_{i,j}^s\rvert} \\ k=1,\ldots,\lceil\frac{M}{2}\rceil \end{cases}$

LSMOP4	Linear	Mixed	Mixed	Variable linkage: it is the same as that of LSMOP1 Objective functions: they are the same as those of LSMOP1 $\begin{cases} c_{i,j} = 1, \ if \ i = j \\ c_{i,j} = 0, \ \text{otherwise} \end{cases} \quad \begin{cases} \overline{g}_{2k-1}(\mathbf{x}_i^s) = \frac{1}{n_k} \sum_{j=1}^{n_k} \frac{\eta_6(\mathbf{x}_{i,j}^s)}{\lvert\mathbf{x}_{i,j}^s\rvert} \\ \overline{g}_{2k}(\mathbf{x}_i^s) = \frac{1}{n_k} \sum_{j=1}^{n_k} \frac{\eta_5(\mathbf{x}_{i,j}^s)}{\lvert\mathbf{x}_{i,j}^s\rvert} \\ k = 1, \ldots, \lceil \frac{M}{2} \rceil \end{cases}$
LSMOP5	Convex	Unimodal	Separable	Variable linkage: $\begin{cases} \mathbf{x}^s \leftarrow (1 + \cos(0.5\pi \frac{i}{\lvert\mathbf{x}^s\rvert})) \times (x_i^s - l_i) - x_1^f \times (u_i - l_i) \\ i = 1, \ldots, \lvert\mathbf{x}^s\rvert \end{cases}$ Objective functions: $\begin{cases} f_1(\mathbf{x}) = \cos(\frac{\pi}{2}x_1^f) \cdots \cos(\frac{\pi}{2}x_{M-2}^f)\cos(\frac{\pi}{2}x_{M-1}^f) \times (1 + \sum_{j=1}^{M} c_{1,j} \times \overline{g}_1(x_j^s)) \\ f_2(\mathbf{x}) = \cos(\frac{\pi}{2}x_1^f) \cdots \cos(\frac{\pi}{2}x_{M-2}^f)\sin(\frac{\pi}{2}x_{M-1}^f) \times (1 + \sum_{j=1}^{M} c_{2,j} \times \overline{g}_2(x_j^s)) \\ \cdots \\ f_{M-1}(\mathbf{x}) = \cos(\frac{\pi}{2}x_1^f)\sin(\frac{\pi}{2}x_2^f) \times (1 + \sum_{j=1}^{M} c_{M-1,j} \times \overline{g}_{M-1}(x_j^s)) \\ f_M(\mathbf{x}) = \sin(\frac{\pi}{2}x_1^f) \times (1 + \sum_{j=1}^{M} c_{M,j} \times \overline{g}_M(x_j^s)) \end{cases}$ where $\begin{cases} c_{i,j} = 1, \ if \ j = i \ or \ j = i+1 \\ c_{i,j} = 0, \ \text{otherwise} \end{cases} \quad \begin{cases} \overline{g}_{2k-1}(\mathbf{x}_i^s) = \frac{1}{n_k} \sum_{j=1}^{n_k} \frac{\eta_1(\mathbf{x}_{i,j}^s)}{\lvert\mathbf{x}_{i,j}^s\rvert} \\ \overline{g}_{2k}(\mathbf{x}_i^s) = \frac{1}{n_k} \sum_{j=1}^{n_k} \frac{\eta_1(\mathbf{x}_{i,j}^s)}{\lvert\mathbf{x}_{i,j}^s\rvert} \\ k = 1, \ldots, \lceil \frac{M}{2} \rceil \end{cases}$
LSMOP6	Convex	Mixed	Partially separable	Variable linkage: it is the same as that of LSMOP5 Objective functions: they are the same as those of LSMOP5 $\begin{cases} c_{i,j} = 1, \ if \ j = i \ or \ j = i+1 \\ c_{i,j} = 0, \ \text{otherwise} \end{cases} \quad \begin{cases} \overline{g}_{2k-1}(\mathbf{x}_i^s) = \frac{1}{n_k} \sum_{j=1}^{n_k} \frac{\eta_3(\mathbf{x}_{i,j}^s)}{\lvert\mathbf{x}_{i,j}^s\rvert} \\ \overline{g}_{2k}(\mathbf{x}_i^s) = \frac{1}{n_k} \sum_{j=1}^{n_k} \frac{\eta_2(\mathbf{x}_{i,j}^s)}{\lvert\mathbf{x}_{i,j}^s\rvert} \\ k = 1, \ldots, \lceil \frac{M}{2} \rceil \end{cases}$
LSMOP7	Convex	Multimodal	Mixed	Variable linkage: it is the same as that of LSMOP5 Objective functions: they are the same as those of LSMOP5 $\begin{cases} c_{i,j} = 1, \ if \ j = i \ or \ j = i+1 \\ c_{i,j} = 0, \ \text{otherwise} \end{cases} \quad \begin{cases} \overline{g}_{2k-1}(\mathbf{x}_i^s) = \frac{1}{n_k} \sum_{j=1}^{n_k} \frac{\eta_6(\mathbf{x}_{i,j}^s)}{\lvert\mathbf{x}_{i,j}^s\rvert} \\ \overline{g}_{2k}(\mathbf{x}_i^s) = \frac{1}{n_k} \sum_{j=1}^{n_k} \frac{\eta_3(\mathbf{x}_{i,j}^s)}{\lvert\mathbf{x}_{i,j}^s\rvert} \\ k = 1, \ldots, \lceil \frac{M}{2} \rceil \end{cases}$

(Continued)

Table 2.3 (Continued)

Test problem	Features of Pareto front	Features of landscape	Separability of objective functions	Problem formula
LSMOP8	Convex	Mixed	Mixed	Variable linkage: it is the same as that of LSMOP5 Objective functions: they are the same as those of LSMOP5 $\begin{cases} c_{i,j} = 1, \ if \ j = i \ or \ j = i+1 \\ c_{i,j} = 0, \ \text{otherwise} \end{cases}$ $\begin{cases} \overline{g}_{2k-1}(\mathbf{x}_i^s) = \frac{1}{n_k} \sum_{j=1}^{n_k} \frac{\eta_5(\mathbf{x}_{i,j}^s)}{\|\mathbf{x}_{i,j}^s\|} \\ \overline{g}_{2k}(\mathbf{x}_i^s) = \frac{1}{n_k} \sum_{j=1}^{n_k} \frac{\eta_1(\mathbf{x}_{i,j}^s)}{\|\mathbf{x}_{i,j}^s\|} \\ k = 1, \dots, \lceil \frac{M}{2} \rceil \end{cases}$
LSMOP9	Disconnected	Mixed	Separable	Variable linkage: it is the same as that of LSMOP5 Objective functions: $\begin{cases} f_1(\mathbf{x}) = x_1^f \\ f_2(\mathbf{x}) = x_2^f \\ \cdots \\ f_{M-1}(\mathbf{x}) = x_{M-1}^f \\ f_M(\mathbf{x}) = \left(M - \sum_{i=1}^{M-1} \frac{x_i^f (1+\sin(3\pi x_i^f))}{2+\sum_{j=1}^{M} c_{M,j} \times \overline{g}_M(x_j^s)} \right) \times (2 + \sum_{j=1}^{M} c_{M,j} \times \overline{g}_M(x_j^s)) \end{cases}$ where $\begin{cases} \overline{g}_{2k-1}(\mathbf{x}_i^s) = \frac{1}{n_k} \sum_{j=1}^{n_k} \frac{\eta_1(\mathbf{x}_{i,j}^s)}{\|\mathbf{x}_{i,j}^s\|} \\ \overline{g}_{2k}(\mathbf{x}_i^s) = \frac{1}{n_k} \sum_{j=1}^{n_k} \frac{\eta_6(\mathbf{x}_{i,j}^s)}{\|\mathbf{x}_{i,j}^s\|} \end{cases}$ $\begin{cases} c_{i,j} = 1 \\ k = 1, \cdots, \lceil \frac{M}{2} \rceil \end{cases}$

Table 2.4 Detailed Information of the Six Basic Single-Objective Optimization Problems Which Are Used in the LSMOP Benchmark Test Suite. The Function Value of All the Six Optimization Problems Falls Between $[0, 10]^{|\mathbf{x}|}$.

Problem	Modality	Separability	Problem formula
Sphere function	Unimodal	Separable	$\eta_1(\mathbf{x}) = \sum_{i=1}^{\lvert\mathbf{x}\rvert} (x_i)^2$
Schwefel's problem 2.21	Unimodal	Nonseparable	$\eta_2(\mathbf{x}) = \max_i \{\lvert x_i\rvert, 1 \le i \le \lvert\mathbf{x}\rvert\}$
Rosenbrock's function	Multimodal	Nonseparable	$\eta_3(\mathbf{x}) = \sum_{i=1}^{\lvert\mathbf{x}\rvert-1} [100(x_i^2 - x_{i+1})^2 + (x_i - 1)^2]$
Rastrigin's function	Multimodal	Separable	$\eta_4(\mathbf{x}) = \sum_{i=1}^{\lvert\mathbf{x}\rvert} (x_i^2 - 10\cos(2\pi x_i) + 10)$
Griewank's function	Multimodal	Nonseparable	$\eta_5(\mathbf{x}) = \sum_{i=1}^{\lvert\mathbf{x}\rvert} \frac{x_i^2}{4000} - \prod_{i=1}^{\lvert\mathbf{x}\rvert} \cos\left(\frac{x_i}{\sqrt{i}}\right) + 1$
Ackley's function	Multimodal	Separable	$\eta_6(\mathbf{x}) = -20\exp\left(-0.2\sqrt{\frac{1}{\lvert\mathbf{x}\rvert}\sum_{i=1}^{\lvert\mathbf{x}\rvert} x_i^2}\right) - \exp\left(\frac{1}{\lvert\mathbf{x}\rvert}\sum_{i=1}^{\lvert\mathbf{x}\rvert} \cos(2\pi x_i)\right) + 20 + e$

Based on the above design principle and improvements, there are totally 12 problems that have been defined in the LMF benchmark test suite. Tables 2.5 and 2.6 present detailed information of the 12 LMF test problems. It can be seen that the 12 test problems are diverse in terms of Pareto front geometries. While the commonly seen concave, convex, and linear Pareto fronts are included, the inverted concave and inverted linear Pareto fronts are also designed. In terms of the features of landscape, the unimodal functions, multimodal functions, and mixed functions are used. Besides, the most objectives of the 12 test problems are partially separable, considering that the partially separable objectives have wider existence in practical applications compared with the two extreme cases, fully separable and nonseparable.

It is worth noting that, in the problem formulas, the position-related variables in $\mathbf{x^p}$ are first divided into $m - 1$ groups $\mathbf{x}^{\mathbf{p},1}, \ldots, \mathbf{x}^{\mathbf{p},m-1}$, where m is the number of objectives. Then, the variables in each $\mathbf{x}^{\mathbf{p},i}$ are transformed to the intermediate variable y_i as follows:

$$y_i = \left\| \frac{1}{\lvert\mathbf{x}^{\mathbf{p},i}\rvert} \sum_{j=1}^{\lvert\mathbf{x}^{\mathbf{p},i}\rvert} x_j^{\mathbf{p},i} \right\|, \tag{2.5}$$

Table 2.5 Detailed Information of the LMF1–LMF6 Test Problems.

Test problem	Features of Pareto front	Features of landscape	Separability of objective functions	Problem formula
LMF1	Concave	Unimodal	Partially separable	Shape functions: $H_1(\mathbf{x^p}) \begin{cases} h_1(\mathbf{x^p}) & = \cos(0.5\pi y_1)\cdots\cos(0.5\pi y_{m-2})\cos(0.5\pi y_{m-1}) \\ h_2(\mathbf{x^p}) & = \cos(0.5\pi y_1)\cdots\cos(0.5\pi y_{m-2})\sin(0.5\pi y_{m-1}) \\ & \cdots \\ h_{m-1}(\mathbf{x^p}) & = \cos(0.5\pi y_1)\sin(0.5\pi y_2) \\ h_m(\mathbf{x^p}) & = \sin(0.5\pi y_1) \end{cases}$ Landscape functions: $g_i(\mathbf{x^{d,i}}) = \begin{cases} \sum_{j=1}^{T_i} w_j b_1(\mathbf{x^{d,i,j}}) & \text{if } i \bmod 2 = 0 \\ \sum_{j=1}^{T_i} w_j b_2(\mathbf{x^{d,i,j}}) & \text{otherwise} \end{cases}$ $b_1(\mathbf{x^{d,i,j}}) = \begin{cases} bf_1(\mathbf{x^{d,i,j}}) & \text{if } j \bmod 2 = 0 \\ bf_2(\mathbf{x^{d,i,j}}) & \text{otherwise} \end{cases}$ $b_2(\mathbf{x^{d,i,j}}) = \begin{cases} bf_1(\mathbf{x^{d,i,j}}) & \text{if } j \bmod 2 = 0 \\ bf_3(\mathbf{x^{d,i,j}}) & \text{otherwise} \end{cases}$
LMF2	Convex	Multimodal	Separable	Shape functions: $H_2(\mathbf{x^p}) \begin{cases} h_1(\mathbf{x^p}) & = [\cos(0.5\pi y_1)\cdots\cos(0.5\pi y_{m-2})\cos(0.5\pi y_{m-1})]^4 \\ h_2(\mathbf{x^p}) & = [\cos(0.5\pi y_1)\cdots\cos(0.5\pi y_{m-2})\sin(0.5\pi y_{m-1})]^4 \\ & \cdots \\ h_{m-1}(\mathbf{x^p}) & = [\cos(0.5\pi y_1)\sin(0.5\pi y_2)]^4 \\ h_m(\mathbf{x^p}) & = [\sin(0.5\pi y_1)]^2 \end{cases}$ Landscape functions: $g_i(\mathbf{x^{d,i}}) = \begin{cases} \sum_{j=1}^{T_i} w_j b_1(\mathbf{x^{d,i,j}}) & \text{if } i \bmod 2 = 0 \\ \sum_{j=1}^{T_i} w_j b_2(\mathbf{x^{d,i,j}}) & \text{otherwise} \end{cases}$ $b_1(\mathbf{x^{d,i,j}}) = \begin{cases} bf_1(\mathbf{x^{d,i,j}}) & \text{if } j \bmod 2 = 0 \\ bf_3(\mathbf{x^{d,i,j}}) & \text{otherwise} \end{cases}$ $b_2(\mathbf{x^{d,i,j}}) = \begin{cases} bf_1(\mathbf{x^{d,i,j}}) & \text{if } j \bmod 2 = 0 \\ bf_4(\mathbf{x^{d,i,j}}) & \text{otherwise} \end{cases}$

LMF3	Linear	Multimodal	Partially separable	Shape functions: $H_3(\mathbf{x^p})$ $\begin{cases} h_1(\mathbf{x^p}) &= y_1 y_2 \cdots y_{m-1} \\ h_2(\mathbf{x^p}) &= y_1 y_2 \cdots (1-y_{m-1}) \\ & \cdots \\ h_{m-1}(\mathbf{x^p}) &= y_1(1-y_2) \\ h_m(\mathbf{x^p}) &= 1-y_1 \end{cases}$ Landscape functions: $g_i(\mathbf{x^{d,i}}) = \begin{cases} \sum_{j=1}^{T_i} w_j b_1(\mathbf{x^{d,i,j}}) & \text{if } i \bmod 2 = 0 \\ \sum_{j=1}^{T_i} w_j b_2(\mathbf{x^{d,i,j}}) & \text{otherwise} \end{cases}$ $b_1(\mathbf{x^{d,i,j}}) = \begin{cases} bf_3(\mathbf{x^{d,i,j}}) & \text{if } j \bmod 2 = 0 \\ bf_4(\mathbf{x^{d,i,j}}) & \text{otherwise} \end{cases}$ $b_2(\mathbf{x^{d,i,j}}) = \begin{cases} bf_2(\mathbf{x^{d,i,j}}) & \text{if } j \bmod 2 = 0 \\ bf_5(\mathbf{x^{d,i,j}}) & \text{otherwise} \end{cases}$
LMF4	Concave	Mixed	Partially separable	Shape functions: $H_1(\mathbf{x^p})$ as described in LMF1 Landscape functions: $g_i(\mathbf{x^{d,i}}) = \begin{cases} \sum_{j=1}^{T_i} w_j b_1(\mathbf{x^{d,i,j}}) & \text{if } i \bmod 2 = 0 \\ \sum_{j=1}^{T_i} w_j b_2(\mathbf{x^{d,i,j}}) & \text{otherwise} \end{cases}$ $b_1(\mathbf{x^{d,i,j}}) = \begin{cases} bf_2(\mathbf{x^{d,i,j}}) & \text{if } j \bmod 2 = 0 \\ bf_1(\mathbf{x^{d,i,j}}) & \text{otherwise} \end{cases}$ $b_2(\mathbf{x^{d,i,j}}) = \begin{cases} bf_1(\mathbf{x^{d,i,j}}) & \text{if } j \bmod 2 = 0 \\ bf_6(\mathbf{x^{d,i,j}}) & \text{otherwise} \end{cases}$

(Continued)

Table 2.5 (Continued)

Test problem	Features of Pareto front	Features of landscape	Separability of objective functions	Problem formula
LMF5	Convex	Mixed	Mixed	Shape functions: $H_2(\mathbf{x^p})$ as described in LMF2 Landscape functions: $g_i(\mathbf{x^{d,i}}) = \begin{cases} \sum_{j=1}^{T_i} w_j b_1(\mathbf{x^{d,i,j}}) & \text{if } i \bmod 2 = 0 \\ \sum_{j=1}^{T_i} w_j b_2(\mathbf{x^{d,i,j}}) & \text{otherwise} \end{cases}$ $b_1(\mathbf{x^{d,i,j}}) = \begin{cases} bf_1(\mathbf{x^{d,i,j}}) & \text{if } j \bmod 3 = 0 \\ bf_2(\mathbf{x^{d,i,j}}) & \text{if } j \bmod 3 = 1 \\ bf_3(\mathbf{x^{d,i,j}}) & \text{otherwise} \end{cases}$ $b_2(\mathbf{x^{d,i,j}}) = \begin{cases} bf_1(\mathbf{x^{d,i,j}}) & \text{if } j \bmod 3 = 0 \\ bf_4(\mathbf{x^{d,i,j}}) & \text{if } j \bmod 3 = 1 \\ bf_5(\mathbf{x^{d,i,j}}) & \text{otherwise} \end{cases}$
LMF6	Linear	Multimodal	Partially separable	Shape functions: $H_3(\mathbf{x^p})$ as described in LMF3 Landscape functions: $g_i(\mathbf{x^{d,i}}) = \begin{cases} \sum_{j=1}^{T_i} w_j b_1(\mathbf{x^{d,i,j}}) & \text{if } i \bmod 2 = 0 \\ \sum_{j=1}^{T_i} w_j b_2(\mathbf{x^{d,i,j}}) & \text{otherwise} \end{cases}$ $b_1(\mathbf{x^{d,i,j}}) = \begin{cases} bf_1(\mathbf{x^{d,i,j}}) & \text{if } j \bmod 3 = 0 \\ bf_2(\mathbf{x^{d,i,j}}) & \text{if } j \bmod 3 = 1 \\ bf_4(\mathbf{x^{d,i,j}}) & \text{otherwise} \end{cases}$ $b_2(\mathbf{x^{d,i,j}}) = \begin{cases} bf_1(\mathbf{x^{d,i,j}}) & \text{if } j \bmod 3 = 0 \\ bf_3(\mathbf{x^{d,i,j}}) & \text{if } j \bmod 3 = 1 \\ bf_5(\mathbf{x^{d,i,j}}) & \text{otherwise} \end{cases}$

Table 2.6 Detailed Information of the LMF7–LMF12 Test Problems.

Test problem	Features of Pareto front	Features of landscape	Separability of objective functions	Problem formula
LMF7	Concave	Multimodal	Partially separable	Shape functions: $H_1(\mathbf{x^p})$ as described in LMF1 Landscape functions: $g_i(\mathbf{x}^{\mathbf{d},i}) = \begin{cases} \sum_{j=1}^{T_i} w_j b_1(\mathbf{x}^{\mathbf{d},i,j}) & \text{if } i \bmod 2 = 0 \\ \sum_{j=1}^{T_i} w_j b_2(\mathbf{x}^{\mathbf{d},i,j}) & \text{otherwise} \end{cases}$ $b_1(\mathbf{x}^{\mathbf{d},i,j}) = \begin{cases} bf_1(\mathbf{x}^{\mathbf{d},i,j}) & \text{if } j \bmod 3 = 0 \\ bf_2(\mathbf{x}^{\mathbf{d},i,j}) & \text{if } j \bmod 3 = 1 \\ bf_6(\mathbf{x}^{\mathbf{d},i,j}) & \text{otherwise} \end{cases}$ $b_2(\mathbf{x}^{\mathbf{d},i,j}) = \begin{cases} bf_2(\mathbf{x}^{\mathbf{d},i,j}) & \text{if } j \bmod 3 = 0 \\ bf_3(\mathbf{x}^{\mathbf{d},i,j}) & \text{if } j \bmod 3 = 1 \\ bf_6(\mathbf{x}^{\mathbf{d},i,j}) & \text{otherwise} \end{cases}$
LMF8	Linear	Multimodal	Partially separable	Shape functions: $H_3(\mathbf{x^p})$ as described in LMF3 Landscape functions: $g_i(\mathbf{x}^{\mathbf{d},i}) = \begin{cases} \sum_{j=1}^{T_i} w_j b_1(\mathbf{x}^{\mathbf{d},i,j}) & \text{if } i \bmod 2 = 0 \\ \sum_{j=1}^{T_i} w_j b_2(\mathbf{x}^{\mathbf{d},i,j}) & \text{otherwise} \end{cases}$ $b_1(\mathbf{x}^{\mathbf{d},i,j}) = \begin{cases} bf_1(\mathbf{x}^{\mathbf{d},i,j}) & \text{if } j \bmod 3 = 0 \\ bf_4(\mathbf{x}^{\mathbf{d},i,j}) & \text{if } j \bmod 3 = 1 \\ bf_6(\mathbf{x}^{\mathbf{d},i,j}) & \text{otherwise} \end{cases}$ $b_2(\mathbf{x}^{\mathbf{d},i,j}) = \begin{cases} bf_3(\mathbf{x}^{\mathbf{d},i,j}) & \text{if } j \bmod 3 = 0 \\ bf_5(\mathbf{x}^{\mathbf{d},i,j}) & \text{if } j \bmod 3 = 1 \\ bf_6(\mathbf{x}^{\mathbf{d},i,j}) & \text{otherwise} \end{cases}$
LMF9	Inverted concave	Mixed	Partially separable	Shape functions: $H_4(\mathbf{x^p})$ $\begin{cases} h_1(\mathbf{x^p}) & = 1 - \cos(0.5\pi y_1)\cdots\cos(0.5\pi y_{m-2})\cos(0.5\pi y_{m-1}) \\ h_2(\mathbf{x^p}) & = 1 - \cos(0.5\pi y_1)\cdots\cos(0.5\pi y_{m-2})\sin(0.5\pi y_{m-1}) \\ & \cdots \\ h_{m-1}(\mathbf{x^p}) & = 1 - \cos(0.5\pi y_1)\sin(0.5\pi y_2) \\ h_m(\mathbf{x^p}) & = 1 - \sin(0.5\pi y_1) \end{cases}$

(Continued)

Table 2.6 (Continued)

Test problem	Features of Pareto front	Features of landscape	Separability of objective functions	Problem formula
				Landscape functions: $g_i(\mathbf{x}^{\mathbf{d},i}) = \begin{cases} \sum_{j=1}^{T_i} w_j b_1(\mathbf{x}^{\mathbf{d},i,j}) & \text{if } i \bmod 3 = 0 \\ \sum_{j=1}^{T_i} w_j b_2(\mathbf{x}^{\mathbf{d},i,j}) & \text{if } i \bmod 3 = 1 \\ \sum_{j=1}^{T_i} w_j b_3(\mathbf{x}^{\mathbf{d},i,j}) & \text{otherwise} \end{cases}$ $b_1(\mathbf{x}^{\mathbf{d},i,j}) = \begin{cases} bf_1(\mathbf{x}^{\mathbf{d},i,j}) & \text{if } j \bmod 2 = 0 \\ bf_2(\mathbf{x}^{\mathbf{d},i,j}) & \text{otherwise} \end{cases}$ $b_2(\mathbf{x}^{\mathbf{d},i,j}) = \begin{cases} bf_2(\mathbf{x}^{\mathbf{d},i,j}) & \text{if } j \bmod 2 = 0 \\ bf_3(\mathbf{x}^{\mathbf{d},i,j}) & \text{otherwise} \end{cases}$ $b_3(\mathbf{x}^{\mathbf{d},i,j}) = \begin{cases} bf_3(\mathbf{x}^{\mathbf{d},i,j}) & \text{if } j \bmod 2 = 0 \\ bf_4(\mathbf{x}^{\mathbf{d},i,j}) & \text{otherwise} \end{cases}$
LMF10	Inverted linear	Mixed	Partially separable	Shape functions: $\mathrm{H}_5(\mathbf{x}^{\mathbf{p}})$ $\begin{cases} h_1(\mathbf{x}^{\mathbf{p}}) & = 1 - y_1 y_2 \cdots y_{m-1} \\ h_2(\mathbf{x}^{\mathbf{p}}) & = 1 - y_1 y_2 \cdots (1 - y_{m-1}) \\ & \cdots \\ h_{m-1}(\mathbf{x}^{\mathbf{p}}) & = 1 - y_1(1 - y_2) \\ h_m(\mathbf{x}^{\mathbf{p}}) & = y_1 \end{cases}$ Landscape functions: $g_i(\mathbf{x}^{\mathbf{d},i}) = \begin{cases} \sum_{j=1}^{T_i} w_j b_1(\mathbf{x}^{\mathbf{d},i,j}) & \text{if } i \bmod 3 = 0 \\ \sum_{j=1}^{T_i} w_j b_2(\mathbf{x}^{\mathbf{d},i,j}) & \text{if } i \bmod 3 = 1 \\ \sum_{j=1}^{T_i} w_j b_3(\mathbf{x}^{\mathbf{d},i,j}) & \text{otherwise} \end{cases}$ $b_1(\mathbf{x}^{\mathbf{d},i,j}) = \begin{cases} bf_2(\mathbf{x}^{\mathbf{d},i,j}) & \text{if } j \bmod 2 = 0 \\ bf_4(\mathbf{x}^{\mathbf{d},i,j}) & \text{otherwise} \end{cases}$ $b_2(\mathbf{x}^{\mathbf{d},i,j}) = \begin{cases} bf_2(\mathbf{x}^{\mathbf{d},i,j}) & \text{if } j \bmod 2 = 0 \\ bf_5(\mathbf{x}^{\mathbf{d},i,j}) & \text{otherwise} \end{cases}$ $b_3(\mathbf{x}^{\mathbf{d},i,j}) = \begin{cases} bf_2(\mathbf{x}^{\mathbf{d},i,j}) & \text{if } j \bmod 2 = 0 \\ bf_3(\mathbf{x}^{\mathbf{d},i,j}) & \text{otherwise} \end{cases}$

LMF11	Inverted concave	Multimodal	Separable	Shape functions: $H_4(\mathbf{x^p})$ as described in LMF9 Landscape functions: $g_i(\mathbf{x^{d,\mathit{i}}}) = \begin{cases} \sum_{j=1}^{T_i} w_j b_1(\mathbf{x^{d,\mathit{i,j}}}) & \text{if } i \bmod 3 = 0 \\ \sum_{j=1}^{T_i} w_j b_2(\mathbf{x^{d,\mathit{i,j}}}) & \text{if } i \bmod 3 = 1 \\ \sum_{j=1}^{T_i} w_j b_3(\mathbf{x^{d,\mathit{i,j}}}) & \text{otherwise} \end{cases}$ $b_1(\mathbf{x^{d,\mathit{i,j}}}) = \begin{cases} bf_4(\mathbf{x^{d,\mathit{i,j}}}) & \text{if } j \bmod 2 = 0 \\ bf_5(\mathbf{x^{d,\mathit{i,j}}}) & \text{otherwise} \end{cases}$ $b_2(\mathbf{x^{d,\mathit{i,j}}}) = \begin{cases} bf_1(\mathbf{x^{d,\mathit{i,j}}}) & \text{if } j \bmod 2 = 0 \\ bf_4(\mathbf{x^{d,\mathit{i,j}}}) & \text{otherwise} \end{cases}$ $b_3(\mathbf{x^{d,\mathit{i,j}}}) = \begin{cases} bf_1(\mathbf{x^{d,\mathit{i,j}}}) & \text{if } j \bmod 2 = 0 \\ bf_5(\mathbf{x^{d,\mathit{i,j}}}) & \text{otherwise} \end{cases}$
LMF12	Inverted linear	Multimodal	Partially separable	Shape functions: $H_5(\mathbf{x^p})$ as described in LMF10 Landscape functions: $g_i(\mathbf{x^{d,\mathit{i}}}) = \begin{cases} \sum_{j=1}^{T_i} w_j b_1(\mathbf{x^{d,\mathit{i,j}}}) & \text{if } i \bmod 3 = 0 \\ \sum_{j=1}^{T_i} w_j b_2(\mathbf{x^{d,\mathit{i,j}}}) & \text{if } i \bmod 3 = 1 \\ \sum_{j=1}^{T_i} w_j b_3(\mathbf{x^{d,\mathit{i,j}}}) & \text{otherwise} \end{cases}$ $b_1(\mathbf{x^{d,\mathit{i,j}}}) = \begin{cases} bf_1(\mathbf{x^{d,\mathit{i,j}}}) & \text{if } j \bmod 2 = 0 \\ bf_6(\mathbf{x^{d,\mathit{i,j}}}) & \text{otherwise} \end{cases}$ $b_2(\mathbf{x^{d,\mathit{i,j}}}) = \begin{cases} bf_3(\mathbf{x^{d,\mathit{i,j}}}) & \text{if } j \bmod 2 = 0 \\ bf_5(\mathbf{x^{d,\mathit{i,j}}}) & \text{otherwise} \end{cases}$ $b_3(\mathbf{x^{d,\mathit{i,j}}}) = \begin{cases} bf_2(\mathbf{x^{d,\mathit{i,j}}}) & \text{if } j \bmod 2 = 0 \\ bf_4(\mathbf{x^{d,\mathit{i,j}}}) & \text{otherwise} \end{cases}$

Table 2.7 Detailed Information of the Six Basic Single-Objective Optimization Functions Which Are Used in the LMF Benchmark Test Suite.

Problem	Modality	Separability	Problem formula
Sphere function	Unimodal	Separable	$bf_1(\mathbf{x}) = \sum_{i=1}^{\vert\mathbf{x}\vert} (x_i)^2$
Schwefel function 2.12	Unimodal	Nonseparable	$bf_2(\mathbf{x}) = \sum_{i=1}^{\vert\mathbf{x}\vert} (\sum_{j=1}^{i} x_i)^2$
Schwefel function 2.21	Unimodal	Nonseparable	$bf_3(\mathbf{x}) = \max_i \{\vert x_i\vert, 1 \leq i \leq \vert\mathbf{x}\vert\}$
Ackley's function	Multimodal	Separable	$bf_4(\mathbf{x}) = -20\exp\left(-0.2\sqrt{\frac{1}{\vert\mathbf{x}\vert}\sum_{i=1}^{\vert\mathbf{x}\vert} x_i^2}\right) - \exp\left(\frac{1}{\vert\mathbf{x}\vert}\sum_{i=1}^{\vert\mathbf{x}\vert}\cos(2\pi x_i)\right) + 20 + e$
Rastrigin's function	Multimodal	Separable	$bf_5(\mathbf{x}) = \sum_{i=1}^{\vert\mathbf{x}\vert}(x_i^2 - 10\cos(2\pi x_i) + 10)$
Rosenbrock's function	Multimodal	Nonseparable	$bf_6(\mathbf{x}) = \sum_{i=1}^{\vert\mathbf{x}\vert-1}[100(x_i^2 - x_{i+1})^2 + (x_i - 1)^2]$

where $x_j^{\mathbf{p},i} \in [-1, 1], i = 1, \dots, m-1$. The shape function $\mathbf{H}(\mathbf{x}^{\mathbf{p}})$ is defined based on $(y_1, \dots, y_{m-1})$. The $bf_1, bf_2, \dots, bf_6$ are six basic single-objective optimization functions, and their detailed information is given in Table 2.7. The six single-objective optimization problems are nearly the same as those used in the LSMOP benchmark test suite. The only difference lies in that Griewank's function is replaced with the Schwefel function 2.12, since Griewank's function becomes easier with the increasing of the number of decision variables.

2.3 Performance Indicators

With the wide application of MOEAs, a variety of new MOEAs continue to be proposed. In order to investigate their performance, the quality of the solution sets obtained by them needs to be quantified by some performance indicators [26]. In this section, they are divided into three categories, namely, the indicators assessing only convergence, the indicators assessing only diversity, and the indicators assessing both convergence and diversity. These indicators are widely adopted to evaluate the performance of MOEAs tailored for both large-scale multi-objective optimization and non-large-scale multi-objective optimization.

2.3.1 Performance Indicators Assessing Only Convergence

The ratio of non-dominated individuals (RNI) [27], Purity [28], and generational distance (GD) [29] are three representative performance indicators that assess only the convergence of a solution set. Specifically, given a solution set A, RNI is mathematically formulated as

$$\text{RNI}(A) = \frac{|A_1|}{|A|}, \tag{2.6}$$

where $|A_1|$ is the number of non-dominated solutions in solution set A and $|A|$ represents the size of solution set A. Clearly, the larger the RNI, the better the convergence of solution set A.

Purity is a performance indicator that can compare multiple solution sets simultaneously. Given N solution sets $A^1, A^2, \ldots, A^N$ $(N > 2)$, let $r_i = |A_1^i| (i = 1, 2, \ldots, N)$ denotes the number of non-dominated solutions in the ith solution set. The union of all the non-dominated solutions in the N solution sets is represented as $A^* = \cup_{i=1}^{N} \{A_1^i\}$. Thereafter, a new solution set A_1^* can be obtained by performing non-dominated sorting procedure [30] on A^*, where A_1^* saves the non-dominated solutions. Let r_i^* denotes the number of non-dominated solutions of A^i which are presented in A_1^*. That is

$$r_i^* = | \{\gamma | \gamma \in A_1^i \text{ and } \gamma \in A_1^*\} |. \tag{2.7}$$

Then, the purity value of the ith solution set can be obtained by

$$P_i = \frac{r_i^*}{r_i}, i = 1, 2, \ldots, N. \tag{2.8}$$

Obviously, the larger the P_i, the better the ith solution set.

GD evaluates the quality of a solution set by using a set of reference points on the Pareto front. Specifically, given a reference point set $Z = \{z_1, z_2, \ldots, z_{|Z|}\}$ and a non-dominated solution set $A = \{a_1, a_2, \ldots, a_{|A|}\}$, where $|Z|$ and $|A|$ denote the number of reference points and solutions in Z and A, respectively. The GD value of solution set A is calculated as

$$\text{GD}(A) = \frac{1}{|A|}\left(\sum_{i=1}^{|A|}(d_i)^p\right)^{1/p}, \tag{2.9}$$

where d_i is the Euclidean distance from solution a_i to its nearest reference point in Z, and p is an integer parameter which is usually set to 1. Clearly, the smaller the GD value, the better the convergence of solution set A.

2.3.2 Performance Indicators Assessing Only Diversity

The spacing (SP) [31], maximum spread (MS) [19], CL_μ [32], and CPF [33] are the most representative performance indicators that assess only the diversity of a solution set. Specifically, SP calculates the standard deviation of the minimum distance of each solution to the others in a solution set A as

$$\mathrm{SP}(A) = \sqrt{\frac{1}{|A|}\sum_{i=1}^{|A|}(d_i - \overline{d})^2}, \tag{2.10}$$

where $d_i = \min_{k\in A \text{ and } k\neq i} \sum_{i=1}^{M} |f_m^i - f_m^k|$, f_m^i and f_m^k are the mth objective value of the ith solution and the kth solution, respectively. The $\overline{d}$ is the mean value of all the d_i, $1 \leq i \leq |A|$. It is worth noting that a value of SP near to zero is desirable, which means that the solutions are uniformly distributed over the Pareto front.

MS measures the range of a solution set A by considering the maximum extension on each objective. It is defined as

$$\mathrm{MS}(A) = \sqrt{\sum_{i=1}^{M} \max_{a,a'\in A}(a_i - a'_i)^2}, \tag{2.11}$$

where M denotes the number of objectives and a, a' are two different solutions in solution set A. Typically, the larger the MS value, the better the extensity of the solution set A.

CL_μ divides the solutions in solution set A into a number of hypercubes and assesses the diversity of solution set A by

$$CL_\mu(A) = \frac{|A|}{NDC_\mu(A)}, \tag{2.12}$$

where $NDC_\mu(A)$ is the number of hypercubes that have at least one solution, and μ is a parameter which denotes the number of divisions on each objective. Clearly, the smaller the $CL_\mu(A)$, the better the diversity of solution set A.

2.3.3 Performance Indicators Assessing Both Convergence and Diversity

The goal of solving an MOP is to obtain a set of uniformly distributed solutions that converge to the Pareto front as closely as possible, thus the indicators that assess both convergence and diversity are usually more desirable. As two representative performance indicators that belong to this category, the inverted generation distance (IGD) [34] and hypervolume (HV) [35] are widely used for performance assessment of MOEAs.

The IGD was first suggested in 2004 [36], and now it has been one of the most commonly used performance indicators. In fact, similar ideas were presented in earlier years [37, 38]. As its name suggested, IGD calculates the average distance from each reference point on the Pareto front to its nearest solution in the solution set. The left of Figure 2.4 provides a brief example of calculating IGD in

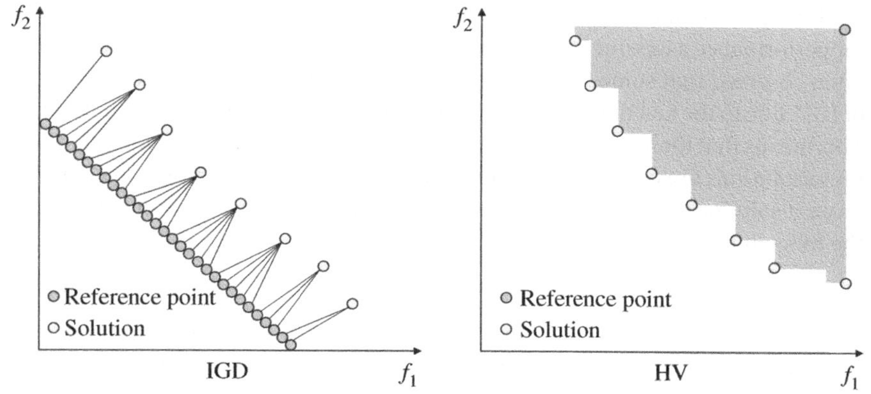

Figure 2.4 Brief examples of calculating IGD and HV in two-dimensional objective space.

the two-dimensional objective space. Given a set of reference points denoted by $Z = \{z_1, z_2, \dots, z_{|Z|}\}$ and a solution set A represented by $A = \{a_1, a_2, \dots, a_{|A|}\}$, where $|Z|$ and $|A|$ denote the number of reference points and solutions in Z and A, respectively, the IGD of solution set A is calculated as

$$\mathrm{IGD}(A) = \frac{1}{|Z|}\sum_{j=1}^{|Z|} \min\{d(z_j, a_k)|a_k \in A\}, \tag{2.13}$$

where z_j and a_k are the jth reference point and kth solution, respectively, $d(z_j, a_k)$ is the Euclidean distance between reference point z_j and solution a_k.

The idea of HV was initially proposed by Zitzler and Thiele [39]. In fact, there are also some other terms such as S-metric [40], Lebesgue measure [41], and hyperarea metric [42], which have a similar meaning as hypervolume. Given a solution set A and a reference point r, HV calculates the area covered by solution set A with respect to reference point r.

$$\mathrm{HV}(A, r) = \lambda\left(\bigcup_{a\in A}\{b|a \prec b \prec r\}\right), \tag{2.14}$$

where $\lambda(\cdot)$ is the Lebesgue measure of a solution set: $\lambda(Z) = \int_{z\in Z} 1Z(z)dz$ with $1Z$ being the characteristic function of Z that equals **1** if $z \in Z$ and 0 otherwise, $a \succeq b$ means that a dominates b. It is worth noting that, compared with IGD, HV has the property of Pareto compliance. That is to say, if a solution set A is better than another solution set B in terms of Pareto dominance, it can be ensured that A has a better HV value than B, but it is not the case in terms of IGD. The right of Figure 2.4 gives an example of calculating HV in the two-dimensional objective space, where the value of HV equals the size of the gray region.

Although IGD and HV have been the most popular performance indicators used for performance assessment, they have also some shortcomings. For example, it is often observed that some non-dominated solutions are ignored in the calculation of IGD, due to the fact that they are not the nearest neighbor of any reference point. This means that these solutions do not have any contribution to the IGD value, and these solutions are called as noncontributing solutions (NSs) in [43]. Specifically, given a solution y' in the non-dominated solution set A, it is regarded as a NS if the following condition is satisfied:

$$\nexists x \in Z : dis(x, y') = \min_{y \in A} dis(x, y), \tag{2.15}$$

where Z is a set of reference points that are uniformly sampled from the Pareto front, and $dis(x, y')$ represents the Euclidean distance between the reference point x and solution y'. By taking NSs into consideration, Tian *et al.* [43] proposed an improved indicator, namely, IGD with noncontributing solution detection (IGD-NS).

$$\text{IGD-NS}(A, Z) = \sum_{x \in Z} \min_{y \in A} dis(x, y) + \sum_{y' \in A'} \min_{x \in Z} dis(x, y'), \tag{2.16}$$

where A' is the set of NSs in population A. As can be seen from the definition, the first term in the above formula is similar to IGD. The second term calculates the sum of the minimum distance from each NS to the reference points in Z. In this way, a population can obtain a good IGD-NS value in the case that it not only shows good convergence and diversity but also contains as few NSs as possible.

There are also other improved indicators that are proposed to address the shortcomings of IGD and HV, such as $\text{IGD}_\mathbf{p}$ [44], IGD^+ [45], I_{SDE} [46], and volume measure [47]. Although many indicators are available for performance assessment, it is still worth noting that the assessment results of these indicators on a solution set may be inconsistent. While a well-convergent population with slightly worse diversity has a bad HV value, it may have the best IGD value due to the biased distribution of reference points [48]. By the contrary, for a population with good diversity but slightly worse convergence, while it has a bad IGD value, it may have a good value in terms of HV. To address this issue, multiple performance indicators are usually used simultaneously for performance measurement.

2.4 Test Problems for Sparse Large-Scale Multi-Objective Optimization

2.4.1 Synthetic Test Problems

In the academic research and industrial applications, a large number of problems are modeled as LSMOPs, where some of them have a special characteristic, i.e.,

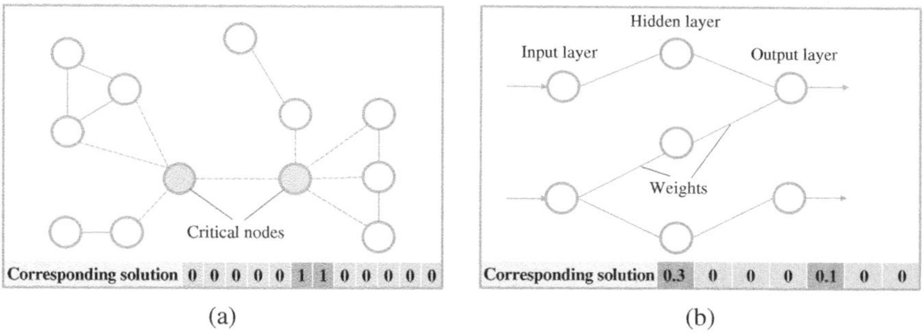

Figure 2.5 Brief examples of two sparse LSMOPs in practical applications. (a) Critical node detection problem. (b) Neural network training problem.

most of the decision variables in Pareto optimal solutions are zero-valued. This class of LSMOPs are herein named as sparse large-scale multi-objective optimization problems (sparse LSMOPs). Figure 2.5 gives examples of two sparse LSMOPs in the practical applications, namely, the critical node detection problem [49] and neural network training problem [50]. The critical node detection problem aims to find the critical nodes in a complex network, the deletion of which will maximize the destruction of the network. If we represent each critical node in the network by 1 and each noncritical node by 0, most decision variables in the optimal solution are zero-valued. In the neural network training problem, it aims to optimize the weights for improving the prediction accuracy and reducing the network complexity. If we represent each weight in the neural network by a real number, most decision variables in the optimal solution are zero-valued for sparse network training. The sparse LSMOPs exist extensively in many fields, such as feature selection [51], portfolio optimization [52], sparse signal reconstruction [53], and adversarial attack [54].

Despite that sparse LSMOPs are very common in many applications, most of the existing test problems do not consider the sparse nature of Pareto optimal solutions. There exist a few test problems having sparse Pareto optimal solutions, but the majority of them do not have a suitable hardness to assess the ability of MOEAs in solving sparse LSMOPs. For example, although the ZDT1–ZDT4 and ZDT6 test problems have a Pareto optimal set of $[0,1] \times \{0\}^{D-1}$ with D denoting the number of decision variables, they are easy to be solved since the lower boundary of their decision variables is 0. The DTLZ6 and DTLZ7 test problems have a Pareto optimal set of $[0,1]^{M-1} \times \{0\}^{D-M+1}$, where M and D represent the number of objectives and decision variables, respectively, but they are too difficult to be solved due to their complicated Pareto fronts instead of the challenge caused by the large number of decision variables. To this end, a sparse multi-objective optimization benchmark test suite, namely, SMOP, is proposed in [55], which can comprehensively assess the ability of MOEAs in solving sparse LSMOPs.

Each test problem in the SMOP benchmark suite is obtained by the following uniform design formulation:

$$\begin{cases} f_1(\mathbf{x}) & = h_1(\mathbf{x})(1 + g_{ns}(\mathbf{x}) + g_s(\mathbf{x})), \\ f_2(\mathbf{x}) & = h_2(\mathbf{x})(1 + g_{ns}(\mathbf{x}) + g_s(\mathbf{x})), \\ & \cdots \\ f_M(\mathbf{x}) & = h_M(\mathbf{x})(1 + g_{ns}(\mathbf{x}) + g_s(\mathbf{x})), \\ & \text{s.t. } \mathbf{x} = (x_1, x_2, \dots, x_D), \\ & x_1, x_2, \dots, x_{M-1} \in [0, 1], \\ & x_M, x_{M+1}, \dots, x_{M+K-1} \in [-1, 2], \\ & x_{M+K}, x_{M+K+1}, \dots, x_D \in [-1, 2], \end{cases} \tag{2.17}$$

where $f_1, f_2, \dots, f_M$ are the objective functions and $\mathbf{x} = (x_1, x_2, \dots, x_D)$ is the decision vector containing D decision variables. The decision vector $\mathbf{x}$ is divided into three parts, where $x_1, x_2, \dots, x_{M-1}$ are related to the shape functions $h_1, h_2, \dots, h_M$, $x_M, x_{M+1}, \dots, x_{M+K-1}$ are related to the landscape function g_{ns}, and $x_{M+K}, x_{M+K+1}, \dots, x_D$ are related to the landscape function g_s. Besides, the lower boundary of the last $D - M + 1$ decision variables is set to -1, which is different from the ZDT test problems that set the lower boundary to 0.

For the SMOP test problems, the solutions are Pareto optimal if and only if $g_{ns} = 0$ and $g_s = 0$, where g_{ns} can be 0 when $x_M, x_{M+1}, \dots, x_{M+K-1}$ are set to specific values, and g_s can be 0 only when $x_{M+K}, x_{M+K+1}, \dots, x_D$ are set to 0. Thus, the Pareto optimal solutions of the eight test problems are sparse when the decision variables $x_{M+K}, x_{M+K+1}, \dots, x_D$ are zero. Besides, the parameter K is determined by

$$K = \lceil \theta(D - M + 1) \rceil, \tag{2.18}$$

where $\theta \in [0, 1]$ is a parameter that controls the sparsity of Pareto optimal solutions.

Table 2.8 presents detailed information of the eight SMOP benchmark test problems. Three kinds of Pareto fronts, namely, the linear Pareto front, convex Pareto front, and concave Pareto front are involved. In terms of the features of landscape, unimodal, multimodal, deceptive, and mixed functions are employed, which pose a challenge to most of the existing MOEAs in obtaining sparse optimal solutions.

SMMOP [56] is another benchmark test suite tailored for sparse large-scale multi-objective optimization. Each SMMOP test problem has multiple equivalent sparse Pareto optimal solutions that are similar in the objective space but completely different in the decision space. This means that SMMOP benchmark test suit is designed for sparse large-scale multi-model multi-objective

Table 2.8 Detailed Information of the Eight SMOP Test Problems.

Test problem	Features of Pareto front	Features of landscape	Separability of objective functions	Problem formula
SMOP1	Linear	Mixed	Separable	Shape functions: $\begin{cases} h_1(\mathbf{x}) &= x_1x_2\cdots x_{M-1} \\ h_2(\mathbf{x}) &= x_1x_2\cdots(1-x_{M-1}) \\ \cdots & \\ h_M(\mathbf{x}) &= 1-x_1 \end{cases}$ Landscape functions: $\begin{cases} g_{ns}(\mathbf{x}) &= \frac{1}{D-M+1}\sum_{i=M}^{M+K-1} g_1(x_i,\pi/3) \\ g_s(\mathbf{x}) &= \frac{1}{D-M+1}\sum_{i=M+K}^{D} g_2(x_i,0) \end{cases}$ where: $\begin{cases} g_1(x,t) &= (x-t)^2 \\ g_2(x,t) &= 2*(x-t)^2+\sin(2\pi(x-t))^2 \end{cases}$
SMOP2	Linear	Multimodal, deceptive	Separable	Shape functions: they are the same as those of SMOP1 Landscape functions: $\begin{cases} g_{ns}(\mathbf{x}) &= \frac{1}{D-M+1}\sum_{i=M}^{M+K-1} g_2(x_i,\pi/3) \\ g_s(\mathbf{x}) &= \frac{1}{D-M+1}\sum_{i=M+K}^{D} g_3(x_i,0) \end{cases}$ where: $\begin{cases} g_2(x,t) &= 2*(x-t)^2+\sin(2\pi(x-t))2 \\ g_3(x,t) &= 4-(x-t)-\frac{4}{e^{100(x-t)^2}} \end{cases}$
SMOP3	Linear	Unimodal, deceptive	Separable	Shape functions: they are the same as those of SMOP1 Landscape functions: $\begin{cases} g_{ns}(\mathbf{x}) &= \frac{1}{D-M+1}\sum_{i=M}^{M+K-1} g_1(x_i,\pi/3) \\ g_s(\mathbf{x}) &= \frac{1}{D-M+1}\sum_{i=1}^{\lfloor\frac{D-M-K+1}{10}\rfloor} g_{dec}\left(\sum_{j=M+K+10(i-1)}^{M+K+10i-1} g_1(x_j,0)\right) \end{cases}$ where: $\begin{cases} g_{dec}(t) = 50-t, & \text{if } t>0 \\ g_{dec}(t) = 0, & \text{if } t=0 \end{cases}$ $g_1(x,t) = (x-t)^2$

(Continued)

Table 2.8 (Continued)

Test problem	Features of Pareto front	Features of landscape	Separability of objective functions	Problem formula
SMOP4	Convex	Deceptive	Separable	Shape functions: $\begin{cases} h_1(\mathbf{x}) &= (1-\cos(x_1\pi/2))(1-\cos(x_2\pi/2))\cdots(1-\cos(x_{M-1}\pi/2)) \\ h_2(\mathbf{x}) &= (1-\cos(x_1\pi/2))(1-\cos(x_2\pi/2))\cdots(1-\sin(x_{M-1}\pi/2)) \\ & \cdots \\ h_M(\mathbf{x}) &= 1-\sin(x_1\pi/2) \end{cases}$ Landscape functions: $\begin{cases} g_{ns}(\mathbf{x}) &= \frac{1}{D-M+1}\sum_{i=M}^{M+K-1} g_{int}(g_3(x_i,\pi/3)) \\ g_s(\mathbf{x}) &= \frac{1}{D-M+1}\sum_{i=M+K}^{D} g_{int}(g_3(x_i,\pi/3)) \end{cases}$ where $g_{int}(t)$ is set to t if t is smaller than the Kth largest value of $g_3(x_M,\pi/3)$, $\ldots, g_3(x_D,\pi/3)$, and 0 otherwise. $g_3(x,t) = 4-(x-t)-\frac{4}{e^{100(x-t)^2}}$
SMOP5	Convex	Unimodal	Separable	Shape functions: they are the same as those of SMOP4 Landscape functions: $\begin{cases} g_{ns}(\mathbf{x}) &= \frac{1}{D-M+1}\sum_{i=M}^{D} g_1(x_i,\pi/3)\times g_2(x_i,0) \\ g_s(\mathbf{x}) &= \frac{1}{D-M+1}\lvert K-\sum_{i=M}^{D}\lvert\mathrm{sign}(x_i)\rvert\rvert \end{cases}$ where $\lvert\mathrm{sign}(x_i)\rvert$ is 0 if x_i equals to 0 and 1 otherwise. $\begin{cases} g_1(x,t) &= (x-t)^2 \\ g_2(x,t) &= 2*(x-t)^2+\sin(2\pi(x-t))^2 \end{cases}$
SMOP6	Convex	Multimodal	Separable	Shape functions: they are the same as those of SMOP4 Landscape functions: $\begin{cases} g_{ns}(\mathbf{x}) &= \frac{1}{D-M+1}\sum_{i=M}^{M+K-1} g_{mul}(g_4(x_i,\frac{i-M}{D-M})) \\ g_s(\mathbf{x}) &= \frac{1}{D-M+1}\sum_{i=M+K}^{D} g_{mul}(g_4(x_i,\frac{i-M}{D-M})) \end{cases}$ where $g_{mul}(g_4(x_i,t))$ is set to 0 if $x_i=0$ and $g_4(x_i,t)$ is larger than the Kth smallest value of $g_4(x_M,\frac{M-M}{D-M}),\ldots,g_4(x_M,\frac{D-M}{D-M})$, and $g_4(x_i,t)$ otherwise. $g_4(x,t)=(x-\pi/3)^2+t\sin(6\pi(x-\pi/3))^2$

SMOP7	Concave	Multimodal	Mixed	Shape functions: $\begin{cases} h_1(\mathbf{x}) & = \cos(x_1\pi/2)\cos(x_2\pi/2)\cdots\cos(x_{M-1}\pi/2) \\ h_2(\mathbf{x}) & = \cos(x_1\pi/2)\cos(x_2\pi/2)\cdots\sin(x_{M-1}\pi/2) \\ & \cdots \\ h_M(\mathbf{x}) & = \sin(x_1\pi/2) \end{cases}$ Landscape functions: $\begin{cases} g_{ns}(\mathbf{x}) & = \frac{1}{D-M+1}\sum_{i=M}^{M+K-1} g_2(x_i, \pi/3) \\ g_s(\mathbf{x}) & = \frac{1}{D-M+1}(\sum_{i=M+K}^{D-1} g_2(x_i, 0.9x_{i+1}) + g_2(x_D, 0.9x_{M+K})) \end{cases}$ where $g_2(x,t) = 2 * (x-t)^2 + \sin(2\pi(x-t))^2$
SMOP8	Concave	Deceptive	Nonseparable	Shape functions: they are the same as those of SMOP7 Landscape functions: $\begin{cases} g_{ns}(\mathbf{x}) & = \frac{1}{D-M+1}\sum_{i=M}^{M+K-1} g_3(x_i, \mathrm{mod}(x_{i+1}+\pi, 2)) \\ g_s(\mathbf{x}) & = \frac{1}{D-M+1}(\sum_{i=M+K}^{D-1} g_3(x_i, 0.9x_{i+1}) + g_3(x_D, 0.9x_{M+K})) \end{cases}$ where $g_3(x,t) = 4 - (x-t) - \frac{4}{e^{100(x-t)^2}}$

optimization. Equation (2.19) gives the uniform design formulation of the SMMOP test problems.

$$\begin{cases} f_1(\mathbf{x}) &= h_1(\mathbf{x})(1+\min(g_1(\mathbf{x}), g_2(\mathbf{x}), \dots, g_{n_p}(\mathbf{x}))), \\ f_2(\mathbf{x}) &= h_2(\mathbf{x})(1+\min(g_1(\mathbf{x}), g_2(\mathbf{x}), \dots, g_{n_p}(\mathbf{x}))), \\ \dots & \\ f_M(\mathbf{x}) &= h_M(\mathbf{x})(1+\min(g_1(\mathbf{x}), g_2(\mathbf{x}), \dots, g_{n_p}(\mathbf{x}))), \\ & \text{s.t. } \mathbf{x} = (x_1, x_2, \dots, x_D), \\ & x_1, x_2, \dots, x_{M-1} \in [0,1], \\ & x_M, x_{M+1}, \dots, x_D \in [-1,2], \end{cases} \tag{2.19}$$

where f_1,f_2, … ,f_M are the objective functions, $h_1, h_2, \dots, h_M$ are the shape functions that are related to $x_1, x_2, \dots, x_{M-1}$, and $g_1, g_2, \dots, g_{n_p}$ are the landscape functions that are related to $x_M, x_{M+1}, \dots, x_D$. The $\mathbf{x}$ is the decision vector consisting of D decision variables, and M is the number of objectives. The component $\min(g_1(\mathbf{x}), g_2(\mathbf{x}), \dots, g_{n_p}(\mathbf{x}))$ introduces n_p equivalent sparse Pareto optimal solution sets, due to the fact that a solution can be Pareto optimal once one of $g_1, g_2, \dots, g_{n_p}$ is minimized.

Besides, to create equivalent Pareto optimal solutions with different sparse distribution, the landscape functions are defined as

$$\begin{aligned} g_i(\mathbf{x}) = {} & g_{one}(x_{M+(i-1)S}, x_{M+(i-1)S+1}, \dots, x_{M+iS-1}) \\ & + g_{zero}(x_M, \dots, x_{M+(i-1)S-1}, x_{M+iS}, \dots, x_D), \end{aligned} \tag{2.20}$$

where $g_{one} \geq 0$ can be minimized when the related decision variables are set to specified nonzero values, $g_{zero} \geq 0$ can be minimized when all the related decision variables are set to zero, and S controls the sparsity of the Pareto optimal solutions. In each landscape function, g_{one} and g_{zero} cannot be minimized simultaneously, which ensures that the equivalent Pareto optimal solutions have different sparse distribution and the number of equivalent Pareto optimal sets is always equal to the number of landscape functions n_p.

In the SMMOP benchmark test suite, three types of shape functions and five types of basic landscape functions are defined. Based on these functions, eight benchmark test problems are proposed, namely SMMOP1–SMMOP8. For each SMMOP test problem, the number of decision variables, the number of objectives, and the sparsity of Pareto optimal solutions can be specified flexibly, which are very similar to the SMOP test problems. The main difference lies in that the SMMOP test problems have n_p equivalent Pareto optimal sets, where n_p can also be specified flexibly.

Table 2.9 provides detailed information of the eight SMMOP test problems. $g_i(x, t)$ is the basic landscape function, where $x \in [-1, 2]$ is a decision variable

Table 2.9 Detailed Information of the Eight SMMOP Test Problems.

Test problem	Features of Pareto front	Features of landscape	Separability of objective functions	Problem formula
SMMOP1	Linear	Mixed	Separable	Shape functions: $\begin{cases} h_1(\mathbf{x}) &= x_1 x_2 \cdots x_{M-1} \\ h_2(\mathbf{x}) &= x_1 x_2 \cdots (1 - x_{M-1}) \\ & \cdots \\ h_M(\mathbf{x}) &= 1 - x_1 \end{cases}$ Landscape functions: $\begin{cases} g_{one} &= \frac{1}{D-M+1} \sum_x g_1(x, \pi/3) \\ g_{zero} &= \frac{1}{D-M+1} \sum_x g_2(x, 0) \end{cases}$ where: $\begin{cases} g_1(x,t) &= (x-t)^2 \\ g_2(x,t) &= 2 * (x-t)^2 + \sin(2\pi(x-t))^2 \end{cases}$
SMMOP2	Linear	Unimodal, deceptive	Separable	Shape functions: they are the same as those of SMMOP1 Landscape functions: $\begin{cases} g_{one} &= \frac{1}{D-M+1} \sum_x g_1(x, \pi/3) \\ g_{zero} &= \frac{1}{D-M+1} \sum_x g_3(x, 0) \end{cases}$ where: $\begin{cases} g_1(x,t) &= (x-t)^2 \\ g_3(x,t) &= 4 - (x-t) - \frac{4}{e^{100(x-t)^2}} \end{cases}$
SMMOP3	Linear	Multimodal, deceptive	Separable	Shape functions: they are the same as those of SMMOP1 Landscape functions: $\begin{cases} g_{one} &= \frac{1}{D-M+1} \sum_x g_2(x, \pi/3) \\ g_{zero} &= \frac{1}{D-M+1} \sum_x g_3(x, 0) \end{cases}$ where: $\begin{cases} g_2(x,t) &= 2 * (x-t)^2 + \sin(2\pi(x-t))^2 \\ g_3(x,t) &= 4 - (x-t) - \frac{4}{e^{100(x-t)^2}} \end{cases}$

(Continued)

Table 2.9 (Continued)

Test problem	Features of Pareto front	Features of landscape	Separability of objective functions	Problem formula
SMMOP4	Convex	Mixed	Separable	Shape functions: $\begin{cases} h_1(\mathbf{x}) &= (1-\cos(x_1\pi/2))(1-\cos(x_2\pi/2))\cdots(1-\cos(x_{M-1}\pi/2)) \\ h_2(\mathbf{x}) &= (1-\cos(x_1\pi/2))(1-\cos(x_2\pi/2))\cdots(1-\sin(x_{M-1}\pi/2)) \\ & \cdots \\ h_M(\mathbf{x}) &= 1-\sin(x_1\pi/2) \end{cases}$ Landscape functions: $\begin{cases} g_{one} &= \frac{1}{D-M+1}\sum_x g_1(x,\pi/3) \\ g_{zero} &= \frac{1}{D-M+1}\sum_x g_4(x,0) \end{cases}$ where: $\begin{cases} g_1(x,t) &= (x-t)^2 \\ g_4(x,t) &= \sqrt{(x-t)^2} + \sin(2\pi(x-t))^2 \end{cases}$
SMMOP5	Convex	Mixed	Separable	Shape functions: they are the same as those of SMMOP4 Landscape functions: $\begin{cases} g_{one} &= \frac{1}{D-M+1}\sum_x g_1(x,\pi/3) \\ g_{zero} &= \frac{1}{D-M+1}\sum_x g_5(x,0) \end{cases}$ where: $\begin{cases} g_1(x,t) &= (x-t)^2 \\ g_5(x,t) &= e^{\log(2)\cdot(x-t)^2}\cdot\sin(6\pi(x-t))^2 + (x-t)^2 \end{cases}$
SMMOP6	Convex	Multimodal	Separable	Shape functions: they are the same as those of SMMOP4 Landscape functions: $\begin{cases} g_{one} &= \frac{1}{D-M+1}\sum_x g_2(x,\pi/3) \\ g_{zero} &= \frac{1}{D-M+1}\sum_x g_4(x,0) \end{cases}$ where: $\begin{cases} g_2(x,t) &= 2*(x-t)^2 + \sin(2\pi(x-t))^2 \\ g_4(x,t) &= \sqrt{(x-t)^2} + \sin(2\pi(x-t))^2 \end{cases}$

SMMOP7	Concave	Multimodal	Mixed	Shape functions: $\begin{cases} h_1(\mathbf{x}) &= \cos(x_1\pi/2)\cos(x_2\pi/2)\cdots\cos(x_{M-1}\pi/2) \\ h_2(\mathbf{x}) &= \cos(x_1\pi/2)\cos(x_2\pi/2)\cdots\sin(x_{M-1}\pi/2) \\ & \cdots \\ h_M(\mathbf{x}) &= \sin(x_1\pi/2) \end{cases}$ Landscape functions: $\begin{cases} g_{one} &= \frac{1}{D-M+1}\sum_x g_2(x,\pi/3) \\ g_{zero} &= \frac{1}{D-M+1}\sum_x g_5(x,0) \end{cases}$ $\begin{cases} g_2(x,t) &= 2*(x-t)^2+\sin(2\pi(x-t))^2 \\ g_5(x,t) &= e^{\log(2)*(x-t)^2}\cdot\sin(6\pi(x-t))^2+(x-t)^2 \end{cases}$
SMMOP8	Concave	Multimodal	Nonseparable	Shape functions: they are the same as those of SMMOP7 Landscape functions: $\begin{cases} g_{one} &= \frac{1}{D-M+1}\sum_x g_4(x,\pi/3) \\ g_{zero} &= \frac{1}{D-M+1}\sum_x g_5(x,0) \end{cases}$ $\begin{cases} g_4(x,t) &= \sqrt{(x-t)^2}+\sin(2\pi(x-t))^2 \\ g_5(x,t) &= e^{\log(2)*(x-t)^2}\cdot\sin(6\pi(x-t))^2+(x-t)^2 \end{cases}$

and $t \in [-1, 2]$ is a parameter defining the optimal value of x, i.e., the landscape functions can be minimized when $x = t$. It is worth noting that, in the eight test problems, the parameter t for g_{one} is set to $\pi/3$, which is a value that cannot be easily found by MOEAs. The parameter t for g_{zero} is set to 0, which is neither the boundary nor the median of the range $[-1, 2]$ of the decision variable. Thus, it is also difficult to be found by MOEAs. By the contrary, if MOEAs take the sparse nature of Pareto optimal solutions into consideration, they may find the optimal decision variables in Pareto optimal solutions easily.

2.4.2 Practical Test Problems

Due to the fact that the sparse LSMOPs are generalized from the real problems in real world, a number of problems from different practical applications can also be used to test the performance of MOEAs in addressing sparse LSMOPs, e.g., the fields of machine learning, network science, software engineering, signal processing, data mining, and economics. In [57], 17 sparse LSMOPs in real-world applications are summarized, and 8 of them are used as practical test problems in the experimental studies, which include feature selection, instance selection, neural network training, community detection, critical node detection, sparse signal reconstruction, pattern mining, and portfolio optimization.

2.4.2.1 Feature Selection

This problem aims to select the fewest features for the lowest classification error [58]. Thus, it has two optimization objectives, i.e., minimizing the ratio of selected features and minimizing the classification error rate with respect to the selected features. The corresponding problem formula is given in the following:

$$\text{Minimize} \begin{cases} f_1(\mathbf{x}) &= \dfrac{|\mathbf{x}|}{D}, \\ f_2(\mathbf{x}) &= ErrorRate(\mathbf{x}), \end{cases} \tag{2.21}$$

where $\mathbf{x} = (x_1, x_2, \ldots, x_D)$ is the decision vector with each decision variable denoting whether the corresponding feature is selected and D denotes the total number of features. *ErrorRate*($\mathbf{x}$) denotes the error rate of a classifier.

2.4.2.2 Instance Selection

This problem aims to select the fewest samples for the lowest classification error [59]. Thus, its two optimization objectives are minimizing the ratio of selected samples and the classification error rate with respect to the selected samples. The corresponding problem formula is given in the following:

$$\text{Minimize} \begin{cases} f_1(\mathbf{x}) &= \dfrac{|\mathbf{x}|}{D}, \\ f_2(\mathbf{x}) &= ErrorRate(\mathbf{x}), \end{cases} \tag{2.22}$$

where $\mathbf{x} = (x_1, x_2, \dots, x_D)$ is the decision vector with each decision variable denoting whether the corresponding sample is selected and D denotes the total number of samples. *ErrorRate*($\mathbf{x}$) denotes the error rate of a classifier.

2.4.2.3 Neural Network Training

This problem aims to obtain the optimal weights of a neural network for the lowest classification error and network complexity [50]. Thus, it has two optimization objectives, i.e., minimizing the ratio of nonzero weights and minimizing the error rate of the neural network. The corresponding problem formula is given in the following:

$$\text{Minimize} \begin{cases} f_1(\mathbf{x}) &= \dfrac{\|\mathbf{x}\|_0}{D}, \\ f_2(\mathbf{x}) &= \mathit{ErrorRate}(\mathbf{x}), \end{cases} \tag{2.23}$$

where $\mathbf{x} = (x_1, x_2, \dots, x_D)$ is the decision vector with each decision variable denoting a weight, D denotes the total number of weights, $\|\mathbf{x}\|_0$ denotes the number of nonzero weights, and *ErrorRate*($\mathbf{x}$) denotes the error rate.

2.4.2.4 Community Detection

This problem aims to find the best community structure, where the nodes within the same community have dense connections and the nodes in different communities have sparse connections [60]. Its two optimization objectives are minimizing the negative value of the intra-link density and minimizing the inter-link density. The corresponding problem formula is given in the following:

$$\text{Minimize} \begin{cases} f_1(\mathbf{x}) &= 2(D - |\mathbf{x}|) - \displaystyle\sum_{i=1}^{|\mathbf{x}|} \frac{L(C_i, C_i)}{|C_i|}, \\ f_2(\mathbf{x}) &= \displaystyle\sum_{i=1}^{|\mathbf{x}|} \frac{L(C_i, \overline{C_i})}{|C_i|}, \end{cases} \tag{2.24}$$

where $\mathbf{x}$ is the decision vector denoting a set of central nodes in the network, D denotes the total number of nodes in the network, C_i denotes the community centered at node x_i, $\overline{C_i}$ denotes all the nodes that do not belong to C_i, and $L(C_i, C_j) = \sum_{p \in C_i, q \in C_j} A_{pq}$ with A being the adjacent matrix.

2.4.2.5 Critical Node Detection

This problem aims to delete the fewest nodes for the lowest pairwise connectivity of the remaining network [49]. Thus, it has two optimization objectives, i.e., minimizing the ratio of deleted nodes and minimizing the ratio of the pairwise connectivity of the remaining network to that of the original network.

The corresponding problem formula is given in the following:

$$\text{Minimize}\begin{cases} f_1(\mathbf{x}) &= \dfrac{|\mathbf{x}|}{D}, \\ f_2(\mathbf{x}) &= \dfrac{Connect(G[V\backslash\mathbf{x}])}{Connect(G)}, \end{cases} \tag{2.25}$$

where $\mathbf{x}$ is the decision vector denoting a set of deleted nodes, D denotes the total number of nodes in the network, and $Connect(G)$ denotes the pairwise connectivity of network G:

$$Connect(G) = \sum_{i \in V} \sum_{j \in V, j \neq i} Con(G, i, j), \tag{2.26}$$

where $Con(G, i, j)$ is set to 1 if node i is reachable from node j, otherwise, $Con(G, i, j)$ is set to 0.

2.4.2.6 Sparse Signal Reconstruction

This problem aims to find the most sparse signal for the lowest loss [53]. Its two optimization objectives are minimizing the sparsity of the reconstructed signal and minimizing the loss of the reconstructed signal. The corresponding problem formula is given in the following:

$$\text{Minimize}\begin{cases} f_1(\mathbf{x}) &= \dfrac{\|\mathbf{x}\|_0}{D}, \\ f_2(\mathbf{x}) &= \|\mathbf{y} - A\mathbf{x}\|_2^2, \end{cases} \tag{2.27}$$

where $\mathbf{x}$ is the decision vector denoting the reconstructed signal, D is the length of the reconstructed signal, $\mathbf{y}$ is the observed signal, and A is the sensing matrix.

2.4.2.7 Pattern Mining

This problem aims to find the most frequent and complete patterns from a transaction dataset [61], which is usually formulated as the following bi-objective optimization problem:

$$\text{Minimize}\begin{cases} f_1(\mathbf{x}) &= 1 - \dfrac{|\tau_{\mathbf{x}}|}{|\tau|}, \\ f_2(\mathbf{x}) &= 1 - \dfrac{1}{|\tau_{\mathbf{x}}|} \displaystyle\sum_{\mathbf{t} \in \tau_{\mathbf{x}}} \dfrac{|\mathbf{x}|}{|\mathbf{t}|}, \end{cases} \tag{2.28}$$

where $\mathbf{x}$ is the decision vector denoting the selected items (i.e., a pattern), τ denotes the whole transaction dataset, and $\tau_{\mathbf{x}}$ denotes the set of transactions $\mathbf{x}$ in τ:

$$\tau_{\mathbf{x}} = \{\mathbf{t} \in \tau | \mathbf{x} \subseteq \mathbf{t}\}. \tag{2.29}$$

The objective f_1 minimizes the negative value of the frequency $\mathbf{x}$ occurring in τ, the objective f_2 minimizes the negative value of the occupancy rate of $\mathbf{x}$ in $\tau_{\mathbf{x}}$.

2.4.2.8 Portfolio Optimization

This problem aims to find the portfolio of instruments that have the largest expected return and the lowest risk [52]. It is usually formulated as the following bi-objective optimization problem:

$$\text{Minimize}\begin{cases} f_1(\mathbf{x}) = \sum_{i=1}^{D}\sum_{j=1}^{D}\mathbf{x}_i\sigma_{i,j}\mathbf{x}_j, \\ f_2(\mathbf{x}) = 1 - \sum_{i=1}^{D}\mathbf{x}_i r_i, \end{cases} \tag{2.30}$$

where $\mathbf{x} = (x_1, x_2, \ldots, x_D)$ is the decision vector with each decision variable denoting the ratio of each instrument in the portfolio, $\sigma_{i,j}$ denotes the covariance between the ith instrument and the jth instrument, and r_i denotes the expected return of the ith instrument. The objective f_1 indicates the total risk of the portfolio, and f_2 indicates the negative value of the expected return.

Table 2.10 gives the datasets of the eight sparse LSMOPs in real-world applications. As can be seen, three datasets are used in each application, which result in three sparse LSMOPs with approximately 1000, 5000, and 10,000 decision variables.

2.5 Performance Indicator for Sparse Large-Scale Multi-Objective Optimization

In the past few years, when some MOEAs have been customized for sparse LSMOPs, their performance is usually assessed by the IGD indicator and HV indicator. But, these two indicators are somewhat unsuitable for assessing the performance of MOEAs on sparse LSMOPs, due to the following two facts. On the one hand, a set of reference points on the true Pareto front are needed for calculating IGD, but the true Pareto fronts of real-world applications are usually unknown. Although HV requires only a single reference point which is not on the true Pareto front, it is still difficult to set this reference point for fair performance comparison [65]. On the other hand, the two indicators assess only convergence and diversity of solutions in the objective space, while the sparsity of solutions in the decision space is not considered.

To address the issue as mentioned above, a comprehensive performance indicator, namely, CSD (convergence-sparsity–diversity) [66], is specifically designed for assessing the performance of MOEAs on sparse LSMOPs. In CSD, both the convergence and diversity of solutions in the objective space and the sparsity of solutions in the decision space are considered. Besides, the CSD indicator does not require

Table 2.10 Datasets of Eight Sparse LSMOPs in Real-World Applications.

Feature selection problem	Type of variables	No. of variables	Dataset	No. of samples	No. of features	No. of classes
FS1	Binary	800	Gse72526[a)]	61	800	4
FS2		5,966	Prostate GE[b)]	102	5,966	2
FS3		10,000	Arcene[b)]	200	10,000	2

Instance selection problem	Type of variables	No. of variables	Dataset	No. of samples	No. of features	No. of classes
IS1	Binary	862	Fouclass[c)]	862	3	2
IS2		4,177	Abalone[d)]	4,177	9	2
IS3		11,055	phishing[c)]	11,055	68	2

Neural network training problem	Type of variables	No. of variables	Dataset	No. of samples	No. of features	No. of classes
NN1	Real	1,181	Spambase[d)]	4,597	57	2
NN2		5,161	Semeion Handwritten Digit[d)]	1,593	256	10
NN3		10,041	Madelon[d)]	2,600	500	2

Community detection problem	Type of variables	No. of variables	Dataset	No. of nodes	No. of edges
CD1	Binary	1,133	Email[e)]	1,133	5451
CD2		4,039	Facebook [62]	4,039	88,234
CD3		9,885	Duke14 [62]	9,885	506,437

Critical node detection problem	Type of variables	No. of variables	Dataset	No. of nodes	No. of edges
CN1	Binary	1,176	Power Network [62]	1,176	8,688
CN2		4,039	Facebook [62]	4,039	88,234
CN3		9,885	Duke14 [62]	9,885	506,437

Table 2.10 (Continued)

Sparse signal reconstruction problem	Type of variables	No. of variables	Dateset	Length of signal	Length of received signal	Sparsity of signal
SR1	Real	1024	Synthetic [63]	1,024	480	260
SR2		5,120	Synthetic [63]	5,120	2,400	1,300
SR3		10,240	Synthetic [63]	10,240	4,800	2,600

Pattern mining problem	Type of variables	No. of variables	Dateset	No. of transactions	No. of items	Avg. length of transactions
PM1	Binary	1,000	Synthetic [64]	10,000	1,000	500
PM2		5,000	Synthetic [64]	50,000	5,000	2,500
PM3		10,000	Synthetic [64]	100,000	10,000	5,000

Portfolio optimization problem	Type of variables	No. of variables	Dataset	No. of instruments	Length of each instrument
PO1	Real	1,000	EURCHF[f)]	1,000	50
PO2		5,000	EURCHF[f)]	5,000	50
PO3		10,000	EURCHF[f)]	1,0000	50

a) https://www.ncbi.nlm.nih.gov/geo/query/acc.cgi.
b) http://featureselection.asu.edu/datasets.php.
c) https://www.csie.ntu.edu.tw/~ecjlin/libsvmtools/datasets/binary.html.
d) https://archive.ics.uci.edu/ml.
e) http://deim.urv.cat/~ealexandre.arenas/data/welcome.htm.
f) https://www.metatrader5.com/en.

any reference point or parameter, which enables it to have wide suitability not only on benchmark problems but also on practical applications.

Considering that the primary goal of solving sparse LSMOPs is to minimize the objectives, and the sparsity is an important factor in decision-making, CSD first assigns solution sets into different levels according to their convergence and sparsity. Then, the solution sets in the same level are sorted according to their diversity. Algorithm 2.1 presents the procedure of calculating CSD, which consists of four steps, i.e., assigning solution sets according to their convergence, assigning solution sets according to their sparsity, calculating the diversity of each solution set, and calculating the CSD value of each solution set.

Algorithm 2.1: General Procedure of Calculating CSD

Input: $A, B, \ldots$ (solution sets to be compared)
Output: CSD(A), CSD(B), … (CSD values of solution sets)

```
//assigning solution sets according to their
  convergence
```

Remove dominated solutions from each solution set temporarily and determine the superiority between each two solution sets in terms of convergence by Eq. (2.31);
$[\text{level}_1^c, \text{level}_2^c] \leftarrow$ Obtain the level of each solution set in terms of convergence;

```
//assigning solution sets according to their sparsity
```

Determine the superiority between each two solution sets in terms of sparsity by Eq. (2.32);
$[\text{level}_1^s, \text{level}_2^s] \leftarrow$ Obtain the level of each solution set in terms of sparsity;

```
//calculating the diversity of each solution set
```

Normalize the objective vectors of all solutions by Eq. (2.34);
$R \leftarrow$ Generate a set of uniformly distributed reference vectors;
$[\text{diversity}(A), \text{diversity}(B), \cdots] \leftarrow$ Calculate the diversity of each solution set by Eq. (2.36)

```
//calculating the CSD value of each solution set
```

$[\text{CSD}(A), \text{CSD}(B), \ldots] \leftarrow$ Calculate the CSD value of each solution set by Eq. (2.38);
return CSD(A), CSD(B), …;

In the first step, the superiority between each two solution sets in terms of convergence is first determined by the following formula:

$$\begin{cases} |\{\mathbf{y} \in B | \exists \mathbf{x} \in A : \mathbf{x} \prec \mathbf{y}\}| > 0.5|B| \\ |\{\mathbf{x} \in A | \exists \mathbf{y} \in A : \mathbf{x} \prec \mathbf{y}\}| > 0.5|A|' \end{cases} \tag{2.31}$$

where a solution set A is said to be superior over a solution set B if more than half the solutions in B are dominated by at least half the solutions in A. Then, the solution sets are assigned to different levels.

In the second step, the superiority between each two solution sets in terms of sparsity is determined by the following formula:

$$sparsity(A) \geq sparsity(B) + \frac{1}{D}, \tag{2.32}$$

where $sparsity(A)$ and $sparsity(B)$ indicate the sparsity of solution set A and solution set B, respectively. The solution set A is said to be superior over solution set B if and only if the sparsity of solution set A is larger than B, besides, the sparsity difference between the two solution sets is not smaller than $1/D$. The sparsity of a solution set A is calculated as:

$$sparsity(A) = 1 - \frac{1}{|A|}\frac{1}{D}\sum_{\mathbf{x}\in A} ||\mathbf{x}||_0, \quad (2.33)$$

where D denotes the total number of decision variables and $||\mathbf{x}||_0$ denotes the number of nonzero variables in solution $\mathbf{x}$.

As for the step of assessing the diversity of each solution set, a reference vector-based diversity criterion is used. The value of each objective function is first normalized by the following formula:

$$f_i(\mathbf{x}) = \frac{f_i(\mathbf{x}) - z_i^*}{z_i^{nad} - z_i^*}, \quad i = 1, 2, \dots, M, \quad (2.34)$$

where z^* is the ideal point containing the minimum objective values in all the solution sets and z^{nad} is the nadir point containing the maximum objective values in all the solution sets. Then, a set of uniformly distributed reference vectors are generated in the objective space, and each reference vector is associated with the solution closest to it. Next, the number $c_{\mathbf{x}}$ of reference vectors associated with each solution $\mathbf{x}$ is counted:

$$c_{\mathbf{x}} = | \{\mathbf{r} \in R | \mathbf{x} = argmin_{\mathbf{a}\in A} dis(\mathbf{f}(\mathbf{a}), r)\} |, \quad (2.35)$$

where $dis(\mathbf{f}(\mathbf{a}), r)$ denotes the distance between solution $\mathbf{a}$ and reference vector $\mathbf{r}$. Finally, the standard deviation of all $c_{\mathbf{x}}$ is regarded as the diversity of the solution set A:

$$diversity(A) = \sqrt{\frac{1}{|A|}\sum_{\mathbf{x}\in A}(c_{\mathbf{x}} - \overline{c})^2}, \quad (2.36)$$

where $\overline{c}$ is the mean value of all $c_{\mathbf{x}}$.

In the step of calculating the CSD value of each solution set, the harmonic average of the levels of each solution set in terms of convergence and sparsity is first calculated by

$$ha(A) \frac{2}{\frac{1}{\text{level}_A^c} + \frac{1}{\text{level}_A^s}}, \quad (2.37)$$

where $level_A^c$ and $level_A^s$ denote the level of solution set A in terms of convergence and sparsity, respectively. Then, the ranking $rank_A$ of each solution set A is obtained by sorting all the solution sets according to their harmonic averages of levels. Finally, the CSD value of each solution set is calculated by

$$\text{CSD}(A) = diversity(A) + \sum_{i=1}^{rank_A - 1} \max_{rank_B = i} diversity(B), \quad (2.38)$$

where $\max_{rank_B=i} diversity(B)$ denotes the maximum diversity value among the solution sets with a ranking of i.

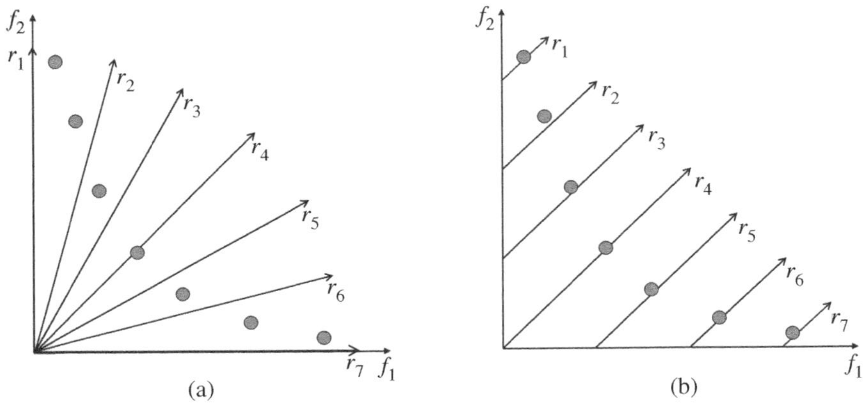

Figure 2.6 Difference between the diversity assessment based on two different sets of reference vectors. (a) $c_x = 1, 0, 2, 1, 2, 0, 1$, respectively. (b) $c_x = 1, 1, 1, 1, 1, 1, 1$, respectively.

It is worth noting that, in the step of assessing the diversity of each solution set, the used reference vectors are parallel instead of starting from the origin. As illustrated in Fig. 2.6, for the seven solutions uniformly distributed on a convex curve, they are associated with 1, 0, 2, 1, 2, 0, 1 reference vectors starting from the origin but 1, 1, 1, 1, 1, 1, 1 parallel reference vectors. Clearly, the parallel reference vectors can better assess the diversity of solution sets lying on convex or concave surfaces.

2.6 Summary

In this chapter, a brief introduction to evolutionary large-scale multi-objective optimization is first given. Then, some primary concepts are introduced, followed by the introduction of some commonly used large-scale multi-objective optimization test problems. Next, three kinds of performance indicators are elaborated, namely, the indicators assessing only convergence, the indicators assessing only diversity, and the indicators assessing both convergence and diversity. Afterward, this chapter introduces the test problems that are tailored for sparse large-scale multi-objective optimization. Finally, a dedicated performance indicator which assesses convergence, sparsity, and diversity simultaneously is introduced in detail.

In the following chapters, the MOEAs customized for LSMOPs and sparse LSMOPs are elaborated in Chapters 3 and 4, respectively, followed by the introduction of MOEAs tailored for some representative applications, mainly including the fields of community detection, logistic scheduling, power systems, deep learning, and radiotherapy planning.

References

1 X. Zhang, K. Zhou, H. Pan, L. Zhang, X. Zeng, and Y. Jin, "A network reduction-based multiobjective evolutionary algorithm for community detection in large-scale complex networks," *IEEE Transactions on Cybernetics*, vol. 50, no. 2, pp. 703–716, 2018.

2 Y. Su, K. Zhou, X. Zhang, R. Cheng, and C. Zheng, "A parallel multi-objective evolutionary algorithm for community detection in large-scale complex networks," *Information Sciences*, vol. 576, pp. 374–392, 2021.

3 S. Yang, Y. Tian, C. He, X. Zhang, K. C. Tan, and Y. Jin, "A gradient-guided evolutionary approach to training deep neural networks," *IEEE Transactions on Neural Networks and Learning Systems*, vol. 33, no. 9, pp. 4861–4875, 2021.

4 J. Xiao, T. Zhang, J. Du, and X. Zhang, "An evolutionary multiobjective route grouping-based heuristic algorithm for large-scale capacitated vehicle routing problems," *IEEE Transactions on Cybernetics*, vol. 51, no. 8, pp. 4173–4186, 2019.

5 X. Xiang, Y. Tian, J. Xiao, and X. Zhang, "A clustering-based surrogate-assisted multiobjective evolutionary algorithm for shelter location problem under uncertainty of road networks," *IEEE Transactions on Industrial Informatics*, vol. 16, no. 12, pp. 7544–7555, 2019.

6 C. He, R. Cheng, C. Zhang, Y. Tian, Q. Chen, and X. Yao, "Evolutionary large-scale multiobjective optimization for ratio error estimation of voltage transformers," *IEEE Transactions on Evolutionary Computation*, vol. 24, no. 5, pp. 868–881, 2020.

7 Y. Tian, Y. Feng, C. Wang, R. Cao, X. Zhang, X. Pei, K. C. Tan, and Y. Jin, "A large-scale combinatorial many-objective evolutionary algorithm for intensity-modulated radiotherapy planning," *IEEE Transactions on Evolutionary Computation*, vol. 26, no. 6, pp. 1511–1525, 2022.

8 Y. Tian, X. Zheng, X. Zhang, and Y. Jin, "Efficient large-scale multiobjective optimization based on a competitive swarm optimizer," *IEEE Transactions on Cybernetics*, vol. 50, no. 8, pp. 3696–3708, 2019.

9 W. Hong, P. Yang, and K. Tang, "Evolutionary computation for large-scale multi-objective optimization: a decade of progresses," *International Journal of Automation and Computing*, vol. 18, no. 2, pp. 155–169, 2021.

10 K. Tang, X. Yao, P. N. Suganthan, C. MacNish, Y. P. Chen, C. M. Chen, and Z. Yang, "Benchmark functions for the CEC'2008 special session and competition on large-scale global optimization," *Technical Report*, 2007.

11 K. Tang, X. Li, P. N. Suganthan, Z. Yang, and T. Weise, "Benchmark functions for the CEC'2010 special session and competition on large-scale global optimization," *Technical Report*, 2009.

12 X. Li, K. Tang, M. N. Omidvar, Z. Yang, and K. Qin, "Benchmark functions for the CEC'2013 special session and competition on large-scale global optimization," *Technical Report*, 2012.

13 M. N. Omidvar, X. Li, and K. Tang, "Designing benchmark problems for large-scale continuous optimization," *Information Sciences*, vol. 316, pp. 419–436, 2015.

14 P. Xu, W. Luo, X. Lin, J. Zhang, and X. Wang, "A large-scale continuous optimization benchmark suite with versatile coupled heterogeneous modules," *Swarm and Evolutionary Computation*, vol. 78, p. 101280, 2023.

15 M. N. Omidvar, X. Li, and X. Yao, "A review of population-based metaheuristics for large-scale black-box global optimizationPart I," *IEEE Transactions on Evolutionary Computation*, vol. 26, no. 5, pp. 802–822, 2021.

16 S. Mahdavi, M. E. Shiri, and S. Rahnamayan, "Metaheuristics in large-scale global continues optimization: a survey," *Information Sciences*, vol. 295, pp. 407–428, 2015.

17 M. N. Omidvar, X. Li, and X. Yao, "A review of population-based metaheuristics for large-scale black-box global optimizationPart II," *IEEE Transactions on Evolutionary Computation*, vol. 26, no. 5, pp. 823–843, 2021.

18 Y. Zhang, Y. Tian, and X. Zhang, "A comparison study of evolutionary algorithms on large-scale sparse multi-objective optimization problems," in *Proceedings of IEEE International Conference on Evolutionary Multi-Criterion Optimization*, 2021, pp. 424–437.

19 E. Zitzler, K. Deb, and L. Thiele, "Comparison of multiobjective evolutionary algorithms: empirical results," *Evolutionary Computation*, vol. 8, no. 2, pp. 173–195, 2000.

20 K. Deb, L. Thiele, M. Laumanns, and E. Zitzler, "Scalable test problems for evolutionary multiobjective optimization," in *Proceedings of IEEE International Conference on Evolutionary Multiobjective Optimization*, 2005, pp. 105–145.

21 S. Huband, P. Hingston, L. Barone, and L. While, "A review of multiobjective test problems and a scalable test problem toolkit," *IEEE Transactions on Evolutionary Computation*, vol. 10, no. 5, pp. 477–506, 2006.

22 H. Li, Q. Zhang, and J. Deng, "Biased multiobjective optimization and decomposition algorithm," *IEEE Transactions on Cybernetics*, vol. 47, no. 1, pp. 52–66, 2016.

23 R. Cheng, M. Li, Y. Tian, X. Zhang, S. Yang, Y. Jin, and X. Yao, "A benchmark test suite for evolutionary many-objective optimization," *Complex & Intelligent Systems*, vol. 3, pp. 67–81, 2017.

24 R. Cheng, Y. Jin, M. Olhofer, and B. sendhoff, "Test problems for large-scale multiobjective and many-objective optimization," *IEEE Transactions on Cybernetics*, vol. 47, no. 12, pp. 4108–4121, 2016.

25 S. Liu, Q. Lin, K.-C. Wong, Q. Li, and K. C. Tan, "Evolutionary large-scale multiobjective optimization: benchmarks and algorithms," *IEEE Transactions on Evolutionary Computation*, vol. 27, no. 3, pp. 401–415, 2021.

26 M. Li and X. Yao, "Quality evaluation of solution sets in multiobjective optimization: a survey," *ACM Computing Surveys*, vol. 52, pp. 1–38, 2019.

27 K. C. Tan, T. H. Lee, and E. F. Khor, "Evolutionary algorithms for multi-objective optimization: performance assessments and comparisons," *Artificial Intelligence Review*, vol. 17, no. 4, pp. 251–290, 2002.

28 S. Bandyopadhyay, S. Pal, and B. Aruna, "Multiobjective GAs, quantitative indices, and pattern classification," *IEEE Transactions on Systems, Man, and Cybernetics, Part B (Cybernetics)*, vol. 34, no. 5, pp. 2088–2099, 2004.

29 D. A. Van Veldhuizen and G. B. Lamont, "Evolutionary computation and convergence to a Pareto front," in *Proceedings of Late Breaking Papers at the 1998 International Conference on Genetic Programming*, 1998, pp. 221–228.

30 X. Zhang, Y. Tian, R. Cheng, and Y. Jin, "An efficient approach to nondominated sorting for evolutionary multiobjective optimization," *IEEE Transactions on Evolutionary Computation*, vol. 19, no. 2, pp. 201–213, 2014.

31 J. R. Schott, *Fault tolerant design using single and multi-criteria genetic algorithms*. Massachusetts Institute of Technology, 1995.

32 J. Wu and S. Azarm, "Metrics for quality assessment of a multiobjective design optimization solution set," *Journal of Mechanical Design*, vol. 123, no. 1, pp. 18–25, 2001.

33 Y. Tian, R. Cheng, X. Zhang, M. Li, and Y. Jin, "Diversity assessment of multi-objective evolutionary algorithms: performance metric and benchmark problems [research frontier]," *IEEE Computational Intelligence Magazine*, vol. 14, no. 3, pp. 61–74, 2019.

34 P. Bosman and D. Thierens, "The balance between proximity and diversity in multiobjective evolutionary algorithms," *IEEE Transactions on Evolutionary Computation*, vol. 7, no. 2, pp. 174–188, 2003.

35 K. Shang, H. Ishibuchi, L. He, and L. M. Pang, "A survey on the hypervolume indicator in evolutionary multiobjective optimization," *IEEE Transactions on Evolutionary Computation*, vol. 25, no. 1, pp. 1–20, 2020.

36 C. A. C. Coello Coello and M. Reyes Sierra, "A study of the parallelization of a coevolutionary multi-objective evolutionary algorithm," in *Proceedings of the 2004 Mexican International Conference on Artificial Intelligence*, 2004, pp. 688–697.

37 H. Ishibuchi, T. Yoshida, and T. Murata, "Balance between genetic search and local search in memetic algorithms for multiobjective permutation flowshop scheduling," *IEEE Transactions on Evolutionary Computation*, vol. 7, no. 2, pp. 204–223, 2003.

38 P. Czyzżak and A. Jaszkiewicz, "Pareto simulated annealinga metaheuristic technique for multiple-objective combinatorial optimization," *Journal of Multi-Criteria Decision Analysis*, vol. 7, no. 1, pp. 34–47, 1998.

39 E. Zitzler and L. Thiele, "Multiobjective evolutionary algorithms: a comparative case study and the strength Pareto approach," *IEEE Transactions on Evolutionary Computation*, vol. 3, no. 4, pp. 257–271, 1999.

40 E. Zitzler, *Evolutionary algorithms for multiobjective optimization: methods and applications*. Swiss Federal Institute of Technology Zurich, 1999.

41 M. Fleischer, "The measure of Pareto optima applications to multi-objective metaheuristics," in *Proceedings of IEEE International Conference on Evolutionary Multi-Criterion Optimization*, 2003, pp. 519–533.

42 D. A. Van Veldhuizen, *Multiobjective evolutionary algorithms: classifications, analyses, and new innovations*. Air Force Institute of Technology, 1999.

43 Y. Tian, X. Zhang, R. Cheng, and Y. Jin, "A multi-objective evolutionary algorithm based on an enhanced inverted generational distance metric," in *Proceedings of the 2016 IEEE Congress on Evolutionary Computation*, 2016, pp. 5222–5229.

44 O. Schutze, X. Esquivel, A. Lara, and C. A. Coello Coello, "Using the averaged Hausdorff distance as a performance measure in evolutionary multiobjective optimization," *IEEE Transactions on Evolutionary Computation*, vol. 16, no. 4, pp. 504–522, 2012.

45 H. Ishibuchi, H. Masuda, Y. Tanigaki, and Y. Nojima, "Modified distance calculation in generational distance and inverted generational distance," in *Proceedings of IEEE International Conference on Evolutionary Multi-Criterion Optimization*, 2015, pp. 110–125.

46 M. Li, S. Yang, and X. Liu, "Shift-based density estimation for Pareto-based algorithms in many-objective optimization," *IEEE Transactions on Evolutionary Computation*, vol. 18, no. 3, pp. 348–365, 2014.

47 J. E. Fieldsend, R. M. Everson, and S. Singh, "Using unconstrained elite archives for multiobjective optimization," *IEEE Transactions on Evolutionary Computation*, vol. 7, no. 3, pp. 305–323, 2003.

48 H. Ishibuchi, R. Imada, Y. Setoguchi, and Y. Nojima, "Reference point specification in inverted generational distance for triangular linear Pareto front," *IEEE Transactions on Evolutionary Computation*, vol. 22, no. 6, pp. 961–975, 2018.

49 M. Lalou, M. A. Tahraoui, and H. Kheddouci, "The critical node detection problem in networks: a survey," *Computer Science Review*, vol. 28, pp. 92–117, 2018.

50 Y. Jin and B. Sendhoff, "Pareto-based multiobjective machine learning: an overview and case studies," *IEEE Transactions on Systems, Man, and Cybernetics, Part C (Applications and Reviews)*, vol. 38, no. 3, pp. 397–415, 2008.

51 G. Roffo, S. Melzi, U. Castellani, A. Vinciarelli, and M. Cristani, "Infinite feature selection: a graph-based feature filtering approach," *IEEE Transactions on Pattern Analysis and Machine Intelligence*, vol. 43, no. 12, pp. 4396–4410, 2020.

52 A. Ponsich, A. L. Jaimes, and C. A. Coello Coello, "A survey on multiobjective evolutionary algorithms for the solution of the portfolio optimization problem and other finance and economics applications," *IEEE Transactions on Evolutionary Computation*, vol. 17, no. 3, pp. 321–344, 2012.

53 H. Li, Q. Zhang, J. Deng, and Z.-B. Xu, "A preference-based multiobjective evolutionary approach for sparse optimization," *IEEE Transactions on Neural Networks and Learning Systems*, vol. 29, no. 5, pp. 1716–1731, 2017.

54 Y. Tian, J. Pan, S. Yang, X. Zhang, S. He, and Y. Jin, "Imperceptible and sparse adversarial attacks via a dual-population-based constrained evolutionary algorithm," *IEEE Transactions on Artificial Intelligence*, vol. 4, no. 2, pp. 268–281, 2022.

55 Y. Tian, X. Zhang, C. Wang, and Y. Jin, "An evolutionary algorithm for large-scale sparse multiobjective optimization problems," *IEEE Transactions on Evolutionary Computation*, vol. 24, no. 2, pp. 380–393, 2019.

56 Y. Tian, R. Liu, X. Zhang, H. Ma, K. C. Tan, and Y. Jin, "A multipopulation evolutionary algorithm for solving large-scale multimodal multiobjective optimization problems," *IEEE Transactions on Evolutionary Computation*, vol. 25, no. 3, pp. 405–418, 2020.

57 Y. Tian, C. Lu, X. Zhang, K. C. Tan, and Y. Jin, "Solving large-scale multiobjective optimization problems with sparse optimal solutions via unsupervised neural networks," *IEEE Transactions on Cybernetics*, vol. 51, no. 6, pp. 3115–3128, 2020.

58 B. Xue, M. Zhang, and W. N. Browne, "Particle swarm optimization for feature selection in classification: a multi-objective approach," *IEEE Transactions on Cybernetics*, vol. 43, no. 6, pp. 1656–1671, 2012.

59 N. Verbiest, J. Derrac, C. Cornelis, S. Garcia, and F. Herrera, "Evolutionary wrapper approaches for training set selection as preprocessing mechanism for support vector machines: experimental evaluation and support vector analysis," *Applied Soft Computing*, vol. 38, pp. 10–22, 2016.

60 L. Zhang, H. Pan, Y. Su, X. Zhang, and Y. Niu, "A mixed representation-based multiobjective evolutionary algorithm for overlapping community detection," *IEEE Transactions on Cybernetics*, vol. 47, no. 9, pp. 2703–2716, 2017.

61 X. Zhang, F. Duan, L. Zhang, F. Cheng, Y. Jin, and K. Tang, "Pattern recommendation in task-oriented applications: a multi-objective perspective [application notes]," *IEEE Computational Intelligence Magazine*, vol. 12, no. 3, pp. 43–53, 2017.

62 R. Rossi and N. Ahmed, "The network data repository with interactive graph analytics and visualization," in *Proceedings of the AAAI Conference on Artificial Intelligence*, vol. 29, no. 1, 2015.

63 J. Liang, M. Gong, H. Li, C. Yue, and B. Qu, "Problem definitions and evaluation criteria for the CEC special session on evolutionary algorithms for sparse optimization," *Technical Report*, 2018.

64 R. Agrawal and R. Srikant, "Fast algorithms for mining association rules," in *Proceedings of the 20th International Conference on Very Large Data Bases*, 1994, pp. 487–499.

65 H. Ishibuchi, R. Imada, Y. Setoguchi, and Y. Nojima, "How to specify a reference point in hypervolume calculation for fair performance comparison," *Evolutionary Computation*, vol. 26, no. 3, pp. 411–440, 2018.

66 Y. Su, Z. Jin, Y. Tian, X. Zhang, and K. C. Tan, "Comparing the performance of evolutionary algorithms for sparse multi-objective optimization via a comprehensive indicator [research frontier]," *IEEE Computational Intelligence Magazine*, vol. 17, no. 3, pp. 34–53, 2022.

3

Evolutionary Algorithms for Large-Scale Multi-Objective Optimization

3.1 Introduction

The research of evolutionary computation in many areas focuses on the innovation of selection strategies, such as multi-objective optimization [1], many-objective optimization [2], and constrained optimization [3]. This is because evolutionary algorithms can generate promising solutions for these problems but are unable to distinguish them to retain the better ones. Thus, novel selection strategies are developed to preserve the population diversity for multi-objective optimization [4], alleviate the dominance resistance phenomenon for many-objective optimization [5], and balance between objective optimization and constraint satisfaction for constrained optimization [6]. These selection strategies are combined with the variation operators in genetic algorithms, differential evolution, particle swarm optimization, and other metaheuristics, which are effective enough for obtaining a set of well-converged and diversified solutions.

However, evolutionary algorithms can hardly generate promising solutions in high-dimensional decision spaces due to the curse of dimensionality [7]. Even if many metaheuristics and mathematical programming methods have shown fast convergence speed on single-objective optimization problems, they are inefficient to generate a set of diversified solutions for multi-objective optimization [8]. Therefore, the research of large-scale multi-objective evolutionary algorithms (MOEAs) focuses on the innovation of variation operators rather than selection strategies, which should be able to generate high-quality solutions approximating the Pareto fronts of large-scale multi-objective optimization problems (LSMOPs) with good diversity.

3.1.1 Ideas of Searching in High-Dimensional Spaces

For large-scale single-objective optimization, many metaheuristics have shown outstanding performance on benchmark problems and competitions, such as

Evolutionary Large-Scale Multi-Objective Optimization and Applications: Integrating Evolutionary Computation, Machine Learning, and Data Science, First Edition. Xingyi Zhang, Ran Cheng, Ye Tian, and Yaochu Jin.

the advanced variants of differential evolution and particle swarm optimization [9]. Intuitively, the variation operators suggested in these metaheuristics can be equipped with multi-objective selection strategies for handling LSMOPs. However, the difficulties of single-objective optimization and multi-objective optimization are fundamentally different. First, the goal of single-objective evolutionary algorithms is to approximate the only optimal solution of a problem, which cannot make all solutions in the population well-converged when solving LSMOPs [10]. Second, it is difficult to distinguish non-dominated solutions quantitatively, which hinders the parameter adaptation in differential evolution [11, 12] and the learning strategies in particle swarm optimization [13, 14]. As a matter of fact, the popular variation operators for single-objective optimization and multi-objective optimization are different. For instance, genetic operators are effective for the diversity preservation in multi-objective optimization, but they converge slower than many other operators on a single objective [15]. Most variation operators in MOEAs (including those based on differential evolution and particle swarm optimization) are tailed by polynomial mutation for local exploitation [16, 17], while it is rarely used in single-objective optimization.

First-order gradient descent methods are particularly effective for the optimization problems with massive variables, such as the training of deep neural networks [18, 19]. Besides, the gradients of black-box objectives can be well approximated by finite difference [20] and solved by gradient descent methods. However, an inescapable issue is the high computational complexity of gradient calculation, where D function evaluations are consumed to calculate the gradients of a solution with D decision variables. When solving LSMOPs, gradient descent methods turn out to be less efficient than evolutionary algorithms, since the update of one solution in gradient descent methods takes more function evaluations than the update of a whole population in evolutionary algorithms. This fact seems counterintuitive since all the partial derivatives of some objective functions can be calculated together (e.g., the backpropagation of deep neural networks), which reduces the number of consumed function evaluations from D to 1 [21]. Nevertheless, it is assumed that all the partial derivatives cannot be calculated together for generic problems, and the number of consumed function evaluations remains D when using finite difference. In addition, gradient descent methods have to aggregate multiple objectives when solving LSMOPs, which can only obtain boundary solutions (i.e., only one objective is minimized) on non-convex Pareto fronts regardless of the weights for objective aggregation [8].

In short, single-objective optimizers cannot be directly used for solving LSMOPs. Besides, the convergence speed of evolutionary algorithms cannot be unlimitedly improved due to the curse of dimensionality, which is inevitable in high-dimensional decision spaces. Considering that LSMOPs are generally NP-hard problems whose global optimal solutions cannot be found in polynomial time, novel strategies have to be designed to pursue other goals alternatively, such as solving LSMOPs with particular characteristics and finding local optimal solutions.

3.1.2 Taxonomy of Large-Scale MOEAs

Large-scale MOEAs can be roughly divided into three categories, including decision variable grouping-based MOEAs, decision space reduction-based MOEAs, and novel search strategy-based MOEAs [22]. As plotted in Fig. 3.1(a), decision variable grouping-based MOEAs aim to solve LSMOPs in a divide-and-conquer manner, where the large number of decision variables is divided into multiple groups and optimized alternately. Assuming that an LSMOP has D decision variables with each having k candidate values, the total number of candidate solutions can be reduced from k^D to $Gk^{\frac{D}{G}}$ if the decision variables are divided into G groups with equal size. That is, the number of candidate solutions is exponentially reduced with the increment of the number of groups G. Extremely, each variable is assigned to an independent group making $G = D$, and thus the number of candidate solutions is minimized to Dk. However, the arbitrary division of decision variables is questionable due to the presence of variable interactions, where the optimal value of a decision variable may depend on the values of other decision variables, and the alternate optimization of these decision variables gets stuck in local optimums.

Taking the following single-objective optimization problem as an example:

$$\begin{aligned} \text{Minimize } & f(x_1, x_2) = -x_1^2 - x_2^2, \\ \text{s.t. } & x_1, x_2 \in [0, 1], \end{aligned} \tag{3.1}$$

where the two decision variables x_1, x_2 of the problem are optimized in sequence. First, since the partial derivative $\frac{\partial f}{\partial x_1} = -2x_1 < 0$, the optimal value of x_1 is 1. Second, since the minimum value of $f(1, x_2) = -1 - x_2^2$ is -2 when $x_2 = 1$, the global optimum of the problem is -2 when $(x_1, x_2) = (1, 1)$. On the other hand, considering the following single-objective optimization problem:

$$\begin{aligned} \text{Minimize } & f(x_1, x_2) = 1.5\sqrt{x_1 + x_2} - x_1^2 - x_2^2, \\ \text{s.t. } & x_1, x_2 \in [0, 1], \end{aligned} \tag{3.2}$$

where the two decision variables x_1, x_2 of the problem are also optimized in sequence. First, since the partial derivative $\frac{\partial f}{\partial x_1} = \frac{0.75}{\sqrt{x_1+x_2}} - 2x_1$ decreases with the increment of x_1, the optimal value of x_1 can be either 0 or 1. Second, since the minimum value of $f(0, x_2) = 1.5\sqrt{x_2} - x_2^2$ is 0 when $x_2 = 0$ and the minimum value of $f(1, x_2) = 1.5\sqrt{1 + x_2} - 1 - x_2^2$ is 0.1213 when $x_2 = 1$, the global optimum of the problem is 0 when $(x_1, x_2) = (0, 0)$ and the local optimum of the problem is

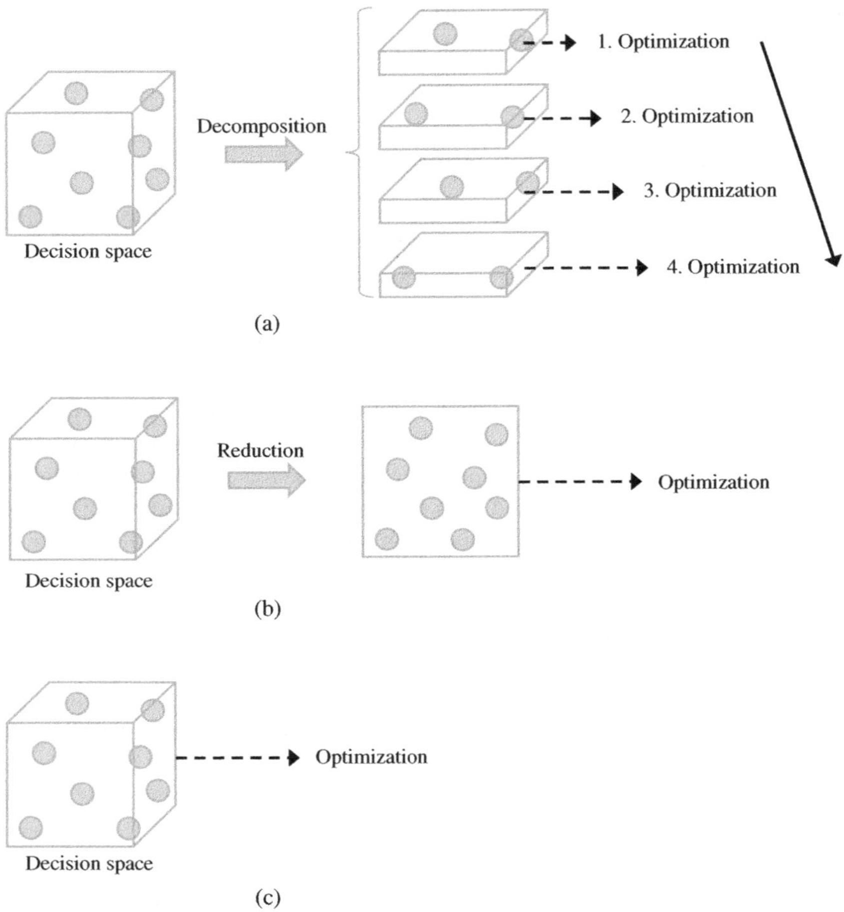

Figure 3.1 Core ideas of three categories of large-scale MOEAs. (a) Decision variable grouping-based MOEAs. (b) Decision space reduction-based MOEAs. (c) Novel search strategy-based MOEAs.

0.1213 when $(x_1, x_2) = (1, 1)$. However, since $f(0, x_2) > f(1, x_2)$ when $x_2 > \frac{25}{144}$, the first step optimizes x_1 to 1 with a probability of $1 - \frac{25}{144} \approx 83\%$, which means that the local optimal solution is much more likely to be found than the global optimal solution.

In short, the decision variables in Eq. (3.1) can be separately optimized while the decision variables in Eq. (3.2) cannot. The difference mainly lies in the partial derivatives, where the $\frac{\partial f}{\partial x_1}$ of Eq. (3.1) is independent of x_2 but the $\frac{\partial f}{\partial x_1}$ of Eq. (3.2) is not. In other words,

$$\arg\min_{x_1,x_2} f(x_1,x_2) = \left(\arg\min_{x_1} f(x_1,\ldots), \arg\min_{x_2} f(\ldots,x_2)\right), \tag{3.3}$$

holds for Eq. (3.1) and thus the global optimal solution can be found by alternately optimizing x_1 and x_2, whereas it does not hold for Eq. (3.2) and thus the global optimal solution cannot be found by alternately optimizing x_1 and x_2. The x_1 and x_2 in Eq. (3.2) are known as interacted decision variables, which can only be divided into the same group and optimized simultaneously. To address this dilemma, a variety of variable grouping methods have been developed for large-scale optimization, such as random grouping [23], ordered grouping [24], and differential grouping [25]. While these grouping methods were originally tailored for converging to the single optimum of single-objective optimization problems, they can hardly find multiple solutions spreading over the whole Pareto fronts due to the lack of the consideration of diversity. Therefore, some large-scale MOEAs divide the decision variables by considering their contributions to convergence and diversity [26–28], where the convergence-related variables are further divided according to their interactions for converging to the Pareto front, and the diversity-related variables are simultaneously optimized to enhance the population diversity. Obviously, decision variable grouping-based MOEAs are only suitable for LSMOPs with separable decision variables. If all the decision variables are interacted with each other like Eq. (3.2), decision variable grouping-based MOEAs will degenerate into general MOEAs since all the decision variables have to be optimized together.

As plotted in Fig. 3.1(b), decision space reduction-based MOEAs aim to reduce the number of decision variables of LSMOPs, and thus the total number of candidate solutions can be directly reduced from k^D to $k^{D'}$, where D' is the number of reduced decision variables. Considering the following single-objective optimization problem:

$$\begin{aligned} \text{Minimize} \;\; & f(x_1,x_2) = -x_1^2 - (x_1 - x_2)^2, \\ \text{s.t.} \;\; & x_1, x_2 \in [0,1], \end{aligned} \tag{3.4}$$

where the global optimum of the problem is -1 when $(x_1,x_2) = (1,1)$. During the optimization procedure, the desirable setting $x_2 = x_1$ may be concluded from the non-dominated solutions in the current population, and thus only x_1 needs to be optimized. At each generation, dimensionality reduction methods are used to reduce the number of decision variables of parent solutions according to the current population. Then, offspring solutions are generated by the shortened parent solutions, and they are recovered by dimensionality reduction methods for function evaluation. To reduce the dimensionality of decision spaces, some large-scale MOEAs develop heuristic methods (e.g., problem transformation [29] and problem reformulation [30]) or adopt machine learning methods (e.g., random embedding [31], principal component analysis [32], neural networks [33], and clustering [34]) to mine the interrelations between decision variables. The dimensionality reduction methods work in an unsupervised manner, which focus on the decision variables of solutions without considering their objective values. Thus, they do not need to consider the interactions between decision variables and are much more efficient than decision variable grouping methods. Nevertheless, the decision variables of many problems like Eqs. (3.1) and (3.2) are unrelated to each other, the elimination of any decision variable is likely to lose the global optimums. Apart from specific LSMOPs such as those with low intrinsic dimensions [31] or sparse optimal solutions [33], decision space reduction-based MOEAs can quickly converge to the local optimums of LSMOPs, but can hardly approximate the global optimums of LSMOPs even if massive function evaluations are available.

As plotted in Fig. 3.1(c), there exist some large-scale MOEAs suggesting novel search strategies to directly search in the high-dimensional decision space. While decision variable grouping-based MOEAs and decisions space reduction-based MOEAs employ conventional variation operators (e.g., simulated binary crossover [35], polynomial mutation [36], and differential evolution [37]) to generate offspring solutions after dividing and reducing the dimensionality of decision spaces, respectively, novel search strategy-based MOEAs aim to enhance the convergence speed by suggesting new methods for offspring generation, including new variation operators and probability models. The new variation operators include particle swarm optimization [38], competitive swarm optimizer [10], memetic evolution [39], and so on [40, 41], and the probability models include the Gaussian process [42], direct sampling [43], covariance matrix adaptation evolution strategy [44], generative adversarial networks [45], and so on [46, 47]. In addition, since gradient descent methods converge quickly in high-dimensional decision spaces, the ideas of gradient-assisted [8] and pseudo-gradient-assisted [48] evolutionary computation are also adopted by MOEAs, where the gradient search is responsible for enhancing the population convergence and the evolutionary search is responsible for retaining the population diversity and escaping from local optimums. As a consequence, novel search strategy-based MOEAs are very versatile since they make no assumption about the characteristics of LSMOPs and strike a balance between exploitation and exploration in high-dimensional decision spaces as well as low-dimensional decision spaces. Novel search strategy-based MOEAs may not be as efficient as decision variable grouping-based and decisions space reduction-based MOEAs, but they are suitable for handling the LSMOPs whose decision variables cannot be divided or decision spaces cannot be reduced.

3.2 Random Grouping-Based Evolutionary Algorithm

The cooperative coevolution generalized differential evolution 3 (CCGDE3) is the first MOEA tailored for solving LSMOPs, which is a decision variable grouping-based MOEA with a random grouping method [23]. Although the random grouping method does not consider the interactions between decision variables, it is very simple and does not introduce additional function evaluations. Moreover, the randomness of the division of decision variables can maintain the interactions to some extent [49], which has inspired many subsequent MOEAs with more advanced grouping methods [50–52].

3.2.1 Random Grouping Method of CCGDE3

Before the optimization procedure, CCGDE3 randomly divides D decision variables into S groups with equal size and assigns one subpopulation to each variable group to be evolved independently. As described in Algorithm 3.1, the initialization operation begins with the random grouping of all the decision variables (Lines 1–5), where each variable group $Group_i$ contains approximately D/S decision variables. Then, a population Pop with N solutions is randomly initialized (Line 6), and the decision variables belonging to each group $Group_i$ constitute a subpopulation P_i (Line 8). CCGDE3 will evolve the S subpopulations

Algorithm 3.1: *Initialization*(N, S)

Input: N (population size), S (number of groups)
Output: P (S subpopulations), *Group* (S variable groups)

`//Random grouping of decision variables`
1 $V \leftarrow \{1, \dots, D\}$; `//D is the number of decision variables`
2 **for** $i = 1, \dots, S$ **do**
3     $v \leftarrow$ Randomly select at most $\lceil D/S \rceil$ variables from V;
4     $V \leftarrow V \setminus v$;
5     $Group_i \leftarrow v$;

`//Generation of subpopulations`
6 $Pop \leftarrow$ Randomly generate N solutions;
7 **for** $i = 1, \dots, S$ **do**
8     $P_i \leftarrow Combine(Pop, Pop, Group_i)$; `//Algorithm 3.2`
9 **return** P and *Group*;

separately, where only the decision variables in $Group_i$ will be optimized for subpopulation P_i.

It should be noted that a solution with partial decision variables cannot be evaluated on objective functions. Thus, the solutions in each subpopulation should be combined with the solutions in other subpopulations to form complete decision vectors. For this aim, CCGDE3 randomly pairs the solutions as described in Algorithm 3.2. For each solution $\mathbf{x}$ in subpopulation P_i, a solution $\mathbf{y}$ is randomly selected from the non-dominated solutions in all the subpopulations (Line 3), and the decision variables of $\mathbf{x}$ that do not belong to $Group_i$ are set to those of $\mathbf{y}$ (Line 6). That is, the solutions in subpopulation P_i actually have complete decision vectors, but only the decision variables in $Group_i$ mate with each other to generate offspring solutions.

Algorithm 3.2: *Combine(Q, P, Index)*

Input: Q (new subpopulation to be complemented), P (solutions in all subpopulations), *Index* (variables belonging to Q)

Output: Q (complete subpopulation)

1 $P \leftarrow$ Delete the dominated solutions in P;
2 **foreach** solution $\mathbf{x} \in Q$ **do**
3     $\mathbf{y} \leftarrow$ Randomly select a solution from P;
4   **for** $i = 1, \dots, D$ **do**
5   **if** $i \notin Index$ **then**
6             $x_i \leftarrow y_i$;
7 **return** Q;

3.2.2 Optimization via GDE3

Except for the coevolution of multiple subpopulations focusing on different decision variables, CCGDE3 does not suggest new variation operators or selection strategies. To evolve each subpopulation, GDE3 [53] is adopted to generate offspring solutions and truncate populations. The environmental selection strategy of GDE3 is an enhancement of that of NSGA-II, where the crowding distances are iteratively calculated to alleviate the limitation of one-off selection [4]. As described in Algorithm 3.3, for each offspring solution $\mathbf{y}$ and its main parent $\mathbf{x}$, $\mathbf{y}$ is used to replace $\mathbf{x}$ directly if $\mathbf{y}$ dominates $\mathbf{x}$ (Line 5). Otherwise, $\mathbf{y}$ is put into the subpopulation if it is not dominated by $\mathbf{x}$ (Line 7). After all the offspring solutions are checked, the enlarged population is sorted by non-dominated sorting (Line 8),

Algorithm 3.3: *EnvironmentalSelection(P, Q)*

Input: P (a subpopulation), Q (offspring solutions generated by P)
Output: P (subpopulation for next generation)

1 $N \leftarrow |P|$;
2 **foreach** $\mathbf{y} \in Q$ **do**
3     $\mathbf{x} \leftarrow$ The solution in P which has the same index as $\mathbf{y}$ in Q;
4   **if** $\mathbf{y}$ dominates $\mathbf{x}$ **then**
5         $\mathbf{x} \leftarrow \mathbf{y}$;
6   **else if** $\mathbf{x}$ does not dominate $\mathbf{y}$ **then**
7         $P \leftarrow P \cup \{\mathbf{y}\}$;
8 *Front* $\leftarrow$ *NondominatedSorting(P)*; `//Algorithm 1.2`
9 $k \leftarrow 1$;
10 **while** $\{P_j | Front_j = k\} \neq \emptyset$ **do**
11     $F_k \leftarrow \{P_j | Front_j = k\}$; `//Set of solutions in the kth front`
12     $k \leftarrow k + 1$;
13 $k \leftarrow \text{argmin}_k |F_1 \bigcup \cdots \bigcup F_k| \geq N$;
14 **while** $|F_k| > N - |F_1 \bigcup \cdots \bigcup F_{k-1}|$ **do**
15   *Crowd* $\leftarrow$ *CrowdingDistance*(F_k, $1 \times |F_k|$ vector of ones); `//Algorithm 1.3`
16     $F_k \leftarrow$ Delete a solution from F_k with the smallest *Crowd*;
17 $P \leftarrow F_1 \bigcup \cdots \bigcup F_k$;
18 **return** P;

and all the solutions in the first k non-dominated fronts are retained, where k is the minimum value ensuring that the number of retained solutions is not smaller than the population size N (Line 13). To eliminate excessive solutions, the crowding distances of the solutions are calculated (Line 15) and the solution with the smallest crowding distance is eliminated from the kth non-dominated front (Line 16). This operation repeats until the number of retained solutions is equal to N (Line 14), and the retained solutions constitute the subpopulation for next generation (Line 17).

3.2.3 Procedure of CCGDE3

With the random grouping method and the optimization procedure of GDE3, the procedure of CCGDE3 is described in Algorithm 3.4. First, the decision variables

Algorithm 3.4: Procedure of CCGDE3

Input: N (population size), S (number of groups)
Output: P (final population)

1 $[P, Group] \leftarrow Initialization(N, S)$; `//Algorithm 3.1`
2 **while** termination criterion not fulfilled **do**
3   **for** $i = 1, \ldots, S$ **do**
4         $Q \leftarrow \emptyset$;
5   **foreach** $\mathbf{x} \in P_i$ **do**
6             $[\mathbf{y}, \mathbf{z}] \leftarrow$ Randomly select two solutions from P_i;
7             $\mathbf{q} \leftarrow$ Perform differential evolution and polynomial mutation based on $\mathbf{x}, \mathbf{y}, \mathbf{z}$; `//Only the variables in` $Group_i$ `are generated`
8             $Q \leftarrow Q \cup \{\mathbf{q}\}$;
9         $Q \leftarrow Combine(Q, P_1 \cup \cdots \cup P_S, Group_i)$; `//Algorithm 3.2`
10         $P_i \leftarrow EnvironmentalSelection(P_i, Q)$; `//Algorithm 3.3`
11 $P \leftarrow EnvironmentalSelection(P_1 \cup \cdots \cup P_S, N)$; `//Algorithm 1.6`
12 **return** P;

are divided into S groups and S subpopulations with N solutions are initialized (Line 1). At each generation, each subpopulation is evolved one by one (Line 3), and each solution in the subpopulation is selected as the main parent (Line 5) to generate an offspring solution by using differential evolution and polynomial mutation (Line 7). Note that here only the decision variables corresponding to the subpopulation are changed in the offspring solution. To evaluate the newly generated offspring solutions, they are combined with the solutions in other subpopulations to form complete decision vectors (Line 9). To select N solutions from the combination of the subpopulation and its offspring solutions, the environmental selection strategy of GDE3 is adopted (Line 10). Finally, a population is obtained as the output of the algorithm, which consists of N solutions selected from all the subpopulations by using the environmental selection strategy of NSGA-II (Line 11).

At each generation of CCGDE3, the time complexity is $O(SND)$ for offspring generation and $O(SMN^2 \log N)$ for environmental selection, where S denotes the number of groups, D denotes the number of decision variables, and M denotes the number of objectives. Therefore, the time complexity of one generation of CCGDE3 is $O(SND + SMN^2 \log N)$.

3.3 Decision Variable Clustering-Based Evolutionary Algorithm

The alternate optimization of randomly divided variable groups is likely to get stuck in local optimums, and the population diversity cannot be explicitly maintained among multiple subpopulations. Thus, the large-scale many-objective evolutionary algorithm (LMEA) uses two decision variable grouping strategies, namely, differential grouping and decision variable clustering [27], to address the above two issues. At the expense of additional function evaluations, LMEA first divides the decision variables into convergence-related ones and diversity-related ones via decision variable clustering, and then divides the convergence-related variables via differential grouping, thus maintaining population diversity and global optimums. Besides, LMEA updates partial decision variables of a single population at a time, which does not require the random combination of solutions that may generate unpromising decision vectors. The idea of decision variable clustering in LMEA is robust in handling complex relations between decision variables and population convergence/diversity, which has been leveraged by many advanced MOEAs [54–56].

3.3.1 Decision Variable Clustering Method of LMEA

The core contribution of LMEA is the decision variable clustering strategy, which aims to divide all the decision variables into convergence-related ones and diversity-related ones to be optimized via different strategies. Convergence-related variables indicate those mainly affecting the population convergence, i.e., the change of such decision variables changes the distance of a solution to the Pareto front. By contrast, diversity-related variables indicate those mainly affecting the population diversity, i.e., the change of such decision variables changes the position of a solution on the Pareto front.

Considering the above natures of convergence-related variables and diversity-related variables, the decision variable clustering strategy identifies them based on the perturbations of decision variables. As detailed in Algorithm 3.5, given *nSel* randomly selected solutions for perturbation (Line 2), each decision variable of each solution is randomly perturbed (i.e., assigning a random value within the decision space) for *nPer* times (Line 4). Thus, there will be $nSel \times D$ solution sets with each having *nPer* solutions as depicted in Fig. 3.2(a), where D denotes the total number of decision variables. Then, a line is fitted for each solution set in the M-dimensional objective space (Line 5) as depicted in Fig. 3.2(b), where the mean square error of the line is calculated (Line 6) and the angle between the line

Algorithm 3.5: *VariableClustering(P, nSel, nPer)*

Input: P (population), $nSel$ (number of selected solutions for perturbation), $nPer$ (number of perturbations on each solution)

Output: CV (convergence-related variables), DV (diversity-related variables)

1 **for** $i = 1, \ldots, D$ **do**
 //D is the number of decision variables
2 $P' \leftarrow$ Randomly select $nSel$ solutions from P;
3 **for** $j = 1, \ldots, nSel$ **do**
4   $P'_j \leftarrow$ Perturb the ith decision variable of the jth solution in P' for $nPer$ times to generate a population;
5   $L \leftarrow$ Fit a line according to the solutions in P'_j in the objective space;
6   $MSE_{ij} \leftarrow$ Mean square error of the line;
7   $Angle_{ij} \leftarrow$ The angle between L and $f_1 + \ldots + f_M = 1$; //M is the number of objectives
8 $[S_1, S_2] \leftarrow$ Group the decision variables into two sets based on $Angle$ by using k-means;
9 $CV \leftarrow \{i = 1, \ldots, D | \frac{1}{nSel} \sum_{j=1}^{nSel} MSE_{ij} < 0.01\}$;
10 $CV \leftarrow CV \cap S_i$; //$S_i$ is one of S_1 and S_2 having a smaller mean of $Angle$
11 $DV \leftarrow \{1, \ldots, D\} \setminus CV$;
12 **return** CV and DV;

and $f_1 + \cdots + f_M = 1$ is calculated (Line 7), as depicted in Fig. 3.2(c). Obviously, a smaller angle indicates that the solutions generated by perturbing a decision variable have greater difference in terms of distances to the Pareto front, and thus the decision variable is more likely to be a convergence-related variable. By contrast, a larger angle indicates that the solutions generated by perturbing a decision variable have greater difference in terms of positions on the Pareto front, and thus the decision variable is more likely to be a diversity-related variable.

Therefore, to distinguish between convergence-related variables and diversity-related variables, the angles of the lines generated by the perturbations on each decision variable can be used as features for clustering. As depicted in Fig. 3.2(d), k-means is used to divide the decision variables into two groups according to the angles of lines (Line 8), where the group with decision variables corresponding to smaller angles are identified as convergence-related variables, and the group with decision variables corresponding to larger angles are identified as diversity-related variables (Lines 9–11).

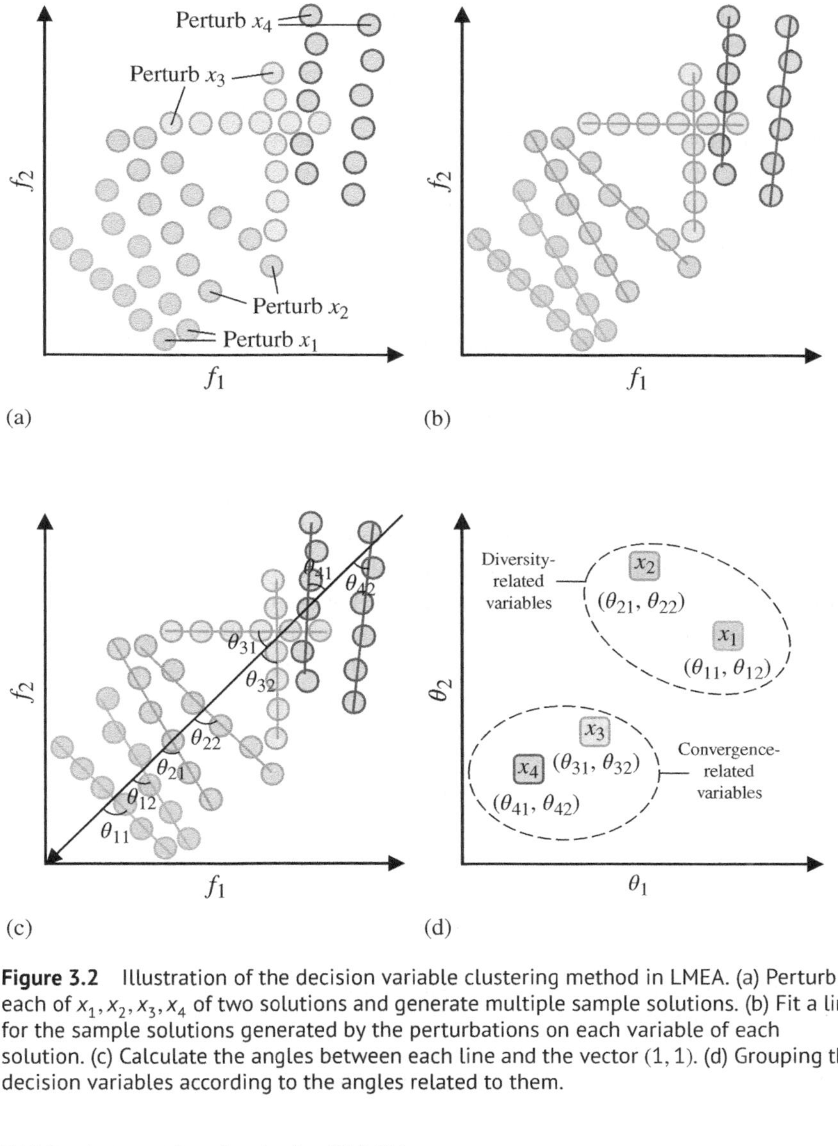

Figure 3.2 Illustration of the decision variable clustering method in LMEA. (a) Perturb each of x_1, x_2, x_3, x_4 of two solutions and generate multiple sample solutions. (b) Fit a line for the sample solutions generated by the perturbations on each variable of each solution. (c) Calculate the angles between each line and the vector $(1, 1)$. (d) Grouping the decision variables according to the angles related to them.

3.3.2 Interaction Analysis of LMEA

While the population diversity can be enhanced by fine-tuning diversity-related variables, the population convergence has to be ensured by fully optimizing convergence-related variables. That is, the optimal values of convergence-related variables are likely to be unique but the optimal values of diversity-related variables are not. Thus, LMEA further divides convergence-related variables into multiple groups to optimize them in a divide-and-conquer manner, where the convergence-related variables in different groups are not interacted with any others to ensure that the global optimums are not lost. For this aim, a differential grouping method [25] is adopted to detect variable interactions based on perturbations. More specifically, decision variable x_i is interacted with decision variable x_j if there exist a solution $\mathbf{x}$ and four variable values a_1, a_2, b_1, b_2 such that

$$\begin{cases} f(\mathbf{x})|_{x_i=a_1,x_j=b_1} > f(\mathbf{x})|_{x_i=a_2,x_j=b_1}, \\ f(\mathbf{x})|_{x_i=a_1,x_j=b_2} < f(\mathbf{x})|_{x_i=a_2,x_j=b_2}. \end{cases} \tag{3.5}$$

Obviously, $\frac{\partial f}{\partial x_i}$ must depend on x_j if the above conditions hold, and thus x_i is said to be interacted with x_j. In practice, the solution $\mathbf{x}$ is randomly selected from the population, the variable values a_1, b_1 are the original ones of $\mathbf{x}$, and the variable values a_2, b_2 are randomly sampled. Therefore, three additional function evaluations are needed to calculate the objective values of $f(\mathbf{x})|_{x_i=a_2,x_j=b_1}$, $f(\mathbf{x})|_{x_i=a_1,x_j=b_2}$, and $f(\mathbf{x})|_{x_i=a_2,x_j=b_2}$. Moreover, in terms of multi-objective optimization, x_i is interacted with x_j if the above conditions hold on any objective $f_m(\mathbf{x})$.

In order to maintain the global optimums, all the interacted convergence-related variables should be assigned to the same group. For example, if x_1 is interacted with x_2, x_2 is interacted with x_3, but x_3 is not interacted with x_1, then the three variables should be assigned to the same group. Thus, the detection between x_1 and x_3 is unnecessary if x_1 has been verified to have interacted with x_2 and x_2 has been verified to have interacted with x_3, and many unnecessary function evaluations can be saved. As detailed in Algorithm 3.6, LMEA checks the interactions between each convergence-related variable v and those in existing groups (Line 9), on the basis of *nCor* solutions randomly selected from the population (Line 8). Then, if v is not interacted with any variable in existing groups, it is assigned to a new group (Line 16). While if v is interacted with some variables in multiple groups, all these groups together with v constitute a new group (Lines 18–20). In this case, LMEA need not to detect the interactions between all pairs of convergence-related variables, thus saving many function evaluations especially when many variables are interacted with each other.

3.3.3 Convergence Optimization and Diversity Optimization

After performing decision variable clustering and interaction analysis, some groups of convergence-related variables and one group of diversity-related variables are obtained. To fully optimize convergence-related variables for good convergence and fine-tune diversity-related variables for good diversity, LMEA suggests two different optimization procedures carried out on convergence-related variables and diversity-related variables, respectively. The optimization of

Algorithm 3.6: *InteractionAnalysis(P, CV, nCor)*

Input: *P* (population), *nCor* (number of selected solutions)
Output: *CVs* (convergence-related variable groups)

```
CVs ← ∅;
foreach v ∈ CV do
    CorSet ← ∅;
    foreach Group ∈ CVs do
        foreach u ∈ Group do
            flag ← false;
            foreach i = 1, ..., nCor do
                x ← Randomly select a solution from P;
                if variable v is interacted with variable u in x then
                    flag ← true;
                    CorSet ← CorSet ∪ {Group};
                    break;
            if flag then
                break;
    if CorSet = ∅ then
        CVs ← CVs ∪ {{v}};
    else
        CVs ← CVs \ CorSet;
        Group ← All variables in CorSet and v;
        CVs ← CVs ∪ {Group};
return CVs;
```

convergence-related variables is elaborated in Algorithm 3.7, which begins with the non-dominated sorting on the current population (Line 1) and the calculation of the distances between solutions in the current population to the origin in the objective space (Line 2). Thus, both the non-dominated front number and the distance to the origin can measure the convergence degree of a solution, and the convergence-related variables of a solution with better convergence degree are more likely to be optimized. For each group of convergence-related variables, LMEA generates N new offspring solutions by changing only the variables in the group (Line 5), where N is the population size. Based on the non-dominated front numbers, some solutions are selected from the current population and put into the mating pool S (Line 9), and $|S|$ offspring solutions

Algorithm 3.7: *ConvergenceOptimization*(P, *CVs*)

Input: P (population), *CVs* (convergence-related variable groups)
Output: P (population for next generation)

1 *Front* ← *NondominatedSorting*(P); `//Algorithm 1.2`
2 *Con* ← Calculate the distance between each solution in P and the origin in the objective space;
3 **foreach** *Group* ∈ *CVs* **do**
4   *nEvaluated* ← 0;
5   **while** *nEvaluated* < $|P|$ **do**
6         $S \leftarrow \emptyset$;
7   **foreach** $\mathbf{x} \in P$ **do**
8   **if** *rand* < $Front_{\mathbf{x}} / \max(Front)$ **then**
9                 $S \leftarrow S \cup \{\mathbf{x}\}$;
10         $O \leftarrow \emptyset$;
11   **foreach** $\mathbf{s} \in S$ **do**
12             $[\mathbf{x}, \mathbf{y}]$ ← *TournamentSelection*(P, 2, *Front*, *Con*); `//Algorithm 1.4`
13             $\mathbf{o} \leftarrow \mathbf{s}$;
14   o_{Group} ← Perform simulated binary crossover and polynomial mutation based on x_{Group} and y_{Group};
15             $O \leftarrow O \cup \{\mathbf{o}\}$; `//`$o_{Group}$ `denotes the decision variables in` *Group* `of solution` $\mathbf{o}$
16   *nEvaluated* ← *nEvaluated* + $|O|$;
17   *Front* ← *NondominatedSorting*($P \cup O$); `//Algorithm 1.2`
18   *Con* ← Calculate the distance between each solution in O and the origin in the objective space;
19   **foreach** $\mathbf{o} \in O$ **do**
20   P ← Use $\mathbf{o}$ to replace its main parent in P according to *Front* and *Con*;
21 **return** P;

are generated by simulated binary crossover and polynomial mutation (Line 14) based on two solutions selected via binary tournament selection (Line 12). Then, each offspring solution **o** is associated with a parent **s** in S, where all the decision variables outside the current group of **o** are set to the same as those of **s** (Line 13). This way, only the decision variables in the current group are changed. After the generation of offspring solutions, each of them is used to replace its parent in the population if it has a smaller non-dominated front number or distance to the origin (Line 20). Since LMEA focuses on improving the population convergence without considering the population diversity, the selection strategies consider the convergence degrees rather than the diversity degrees of solutions. In fact, since diversity-related variables keep unchanged during the optimization procedure, the population diversity is also unchanged in most cases.

The optimization of diversity-related variables is elaborated in Algorithm 3.8, which is similar to general MOEAs as both the population convergence and population diversity are taken into consideration. By generating N offspring solutions by simulated binary crossover and polynomial mutation (Line 5) based

---

Algorithm 3.8: *DiversityOptimization*(P, DV)

Input: P (population), DV (diversity-related variables)
Output: P (population for next generation)

1 $O \leftarrow \emptyset$;
2 **foreach** $\mathbf{s} \in P$ **do**
3     $[\mathbf{x}, \mathbf{y}] \leftarrow$ Randomly select two solutions from P;
4     $\mathbf{o} \leftarrow \mathbf{s}$;
5     $o_{DV} \leftarrow$ Perform simulated binary crossover and polynomial mutation based on x_{DV} and y_{DV};
6     $O \leftarrow O \cup \{\mathbf{o}\}$; `//`$o_{DV}$ `denotes the decision variables in` DV `of solution` $\mathbf{o}$
7 $Front \leftarrow NondominatedSorting(P \cup O)$; `//Algorithm 1.2`
8 $i \leftarrow 1$;
9 **while** $\{P_j | Front_j = i\} \neq \emptyset$ **do**
10     $F_i \leftarrow \{P_j | Front_j = i\}$; `//Set of solutions in the` i`-th front`
11     $i \leftarrow i + 1$;
12 $k \leftarrow \text{argmin}_i |F_1 \bigcup \cdots \bigcup F_i| \geq N$;
13 $Q \leftarrow F_1 \bigcup \cdots \bigcup F_{k-1}$;
14 $Angle \leftarrow$ Calculate the angles between each two solutions in $Q \cup F_k$ in the objective space;
15 **while** $|Q| < |P|$ **do**
16     $\mathbf{s} \leftarrow \text{argmax}_{\mathbf{x} \in F_k} \min_{\mathbf{y} \in Q} Angle_{\mathbf{xy}}$;
17     $Q \leftarrow Q \cup \{\mathbf{s}\}$;
18 $P \leftarrow Q$;
19 **return** P;

---

on two solutions randomly selected from the current population, each offspring solution **o** is associated with a unique parent **s** in the population. That is, all the convergence-related variables of **o** are set to the same as those of **s** (Line 4); in other words, only the diversity-related variables are changed. After the generation of offspring solutions, they are combined with the current population and N solutions are selected and survive to the next generation. Here, LMEA selects solutions with better non-dominated front numbers (Line 13) and selects solutions from the last non-dominated front one by one according to their angles to the solutions having been selected (Line 16). This selection strategy adopts the non-dominated sorting of NSGA-II [57] and the truncation strategy of SPEA2 [58], which inherits the merits of both the two MOEAs for retaining N solutions with good convergence and diversity.

3.3.4 Procedure of LMEA

With the decision variable clustering, interaction analysis, convergence optimization, and diversity optimization, the procedure of LMEA is described in Algorithm 3.9. First, a population with N solutions is randomly initialized (Line 1), which is then used to divide all the decision variables into convergence-related variables and diversity-related variables (Line 2), and divide the convergence-related variables into some groups (Line 3). At each generation, the convergence-related variables and diversity-related variables are alternately optimized (Lines 5 and 6), where only a single population is evolved during the

Algorithm 3.9: Procedure of LMEA

Input: N (population size), *nSel* (number of selected solutions for perturbation), *nPer* (number of perturbations on each solution), *nCor* (number of selected solutions)

Output: P (final population)

1 $P \leftarrow$ Randomly generate N solutions;
2 $[CV, DV] \leftarrow$ *VariableClustering*$(P, nSel, nPer)$; `//Algorithm 3.5`
3 $CVs \leftarrow$ *InteractionAnalysis*$(P, CV, nCor)$; `//Algorithm 3.6`
4 **while** termination criterion not fulfilled **do**
5 $P \leftarrow$ *ConvergenceOptimization*(P, CVs); `//Algorithm 3.7`
6 $P \leftarrow$ *DiversityOptimization*(P, DV); `//Algorithm 3.8`
7 **return** P;

whole optimization procedure. Finally, the population is returned as the output of the algorithm (Line 7).

At each generation of LMEA, the time complexity is $O(MN^2)$ for both convergence optimization and diversity optimization, where M denotes the number of objectives. Therefore, the time complexity of one generation of LMEA is $O(MN^2)$.

3.4 Problem Reformulation-Based Evolutionary Algorithm

Different from decision variable grouping, the large-scale multi-objective framework (LSMOF) aims to directly reduce the dimensionality of decision spaces, so as to solve a small-scale optimization problem rather than a large-scale one. Instead of optimizing all the D decision variables of an LSMOP, LSMOF optimizes two weight variables for each of r solutions, where the combination of a weight variable and a solution leads to a new solution for approximating the Pareto set. Thus, the dimensionality of the decision space is reduced from D to $2r$. By maximizing the HV [59] value of the $2r$ new solutions, LSMOF reformulates the LSMOP into a small-scale single-objective optimization problem. LSMOF can quickly converge to local optimums due to the dimensionality reduction of decision spaces, but can hardly approximate the global optimums due to the inexistence of the global optimums in the reduced decision spaces. Owing to the rapid convergence speed of LSMOF, the idea of problem reformulation has been carried forward in many other MOEAs for solving LSMOPs [60–62].

3.4.1 Problem Reformulation of LSMOF

The problem reformulation method of LSMOF converts an LSMOP into a small-scale single-objective optimization problem to be solved by differential evolution. For the reformulated problem, the decision variables are a number of weight variables generating new solutions based on the existing ones, and the objective is the HV value of all the new solutions. As described in Algorithm 3.10, the nadir point (i.e., a point containing the maximum values of all the objectives) of the current population P is first obtained as the reference point for HV calculation (Line 1). Then, r reference solutions are selected from P via an environmental selection strategy (Line 2). It is worth noting that LSMOF is a framework compatible with many existing MOEAs, where NSGA-II with non-dominated sorting and crowding distance-based selection strategy is adopted here. For each reference solution $\mathbf{x}_i$, two weight variables λ_{i1} and λ_{i2} are introduced to generate two new

Algorithm 3.10: *Reformulation(P, r)*

Input: P (population), r (number of reference solutions)
Output: f' (reformulated problem), $\mathbf{x}_1, \dots, \mathbf{x}_r$ (reference solutions)

1 $Nad \leftarrow$ Calculate the nadir point of P;
2 $P' \leftarrow EnvironmentalSelection(P, r)$; `//Algorithm 1.6`
3 **for** $i = 1, \dots, r$ **do**
4     $\mathbf{x}_i^l(\lambda_{i1}) \leftarrow$ Build a function based on $\mathbf{x}_i$ with variable λ_{i1} by Eq. (3.6);
5     $\mathbf{x}_i^u(\lambda_{i2}) \leftarrow$ Build a function based on $\mathbf{x}_i$ with variable λ_{i2} by Eq. (3.7); `//`$\mathbf{x}_i$ `is the` i`th solution in` P'
6 $\lambda \leftarrow (\lambda_{11}, \lambda_{12}, \dots, \lambda_{r1}, \lambda_{r2})$; `//Decision vector of the reformulated problem`
7 $f' \leftarrow HV(\{\mathbf{x}_1^l(\lambda_{11}), \mathbf{x}_1^u(\lambda_{12}), \dots, \mathbf{x}_r^l(\lambda_{r1}), \mathbf{x}_r^u(\lambda_{r2})\}, Nad)$; `//Objective function of the reformulated problem`
8 **return** f' and $\mathbf{x}_1, \dots, \mathbf{x}_r$;

solutions $\mathbf{x}_i^l(\lambda_{i1})$ and $\mathbf{x}_i^u(\lambda_{i2})$, that is,

$$\mathbf{x}_i^l(\lambda_{i1}) = \mathbf{o} + \lambda_{i1}\frac{\mathbf{x}_i - \mathbf{o}}{\|\mathbf{x}_i - \mathbf{o}\|}\|\mathbf{t} - \mathbf{o}\| \tag{3.6}$$

and

$$\mathbf{x}_i^u(\lambda_{i2}) = \mathbf{t} - \lambda_{i2}\frac{\mathbf{t} - \mathbf{x}_i}{\|\mathbf{t} - \mathbf{x}_i\|}\|\mathbf{t} - \mathbf{o}\|, \tag{3.7}$$

where $\mathbf{o}$ is the lower boundary point and $\mathbf{t}$ is the upper boundary point of the decision space. As illustrated in Fig. 3.3, $\mathbf{x}_i^l$ locates on the line through $\mathbf{o}$ and $\mathbf{x}$, where the distance between $\mathbf{x}_i^l$ and $\mathbf{x}$ is determined by the weight variable λ_{i1}. Similarly, $\mathbf{x}_i^u$ locates on the line through $\mathbf{t}$ and $\mathbf{x}$, where the distance between $\mathbf{x}_i^u$ and $\mathbf{x}$ is determined by the weight variable λ_{i2}. Thus, the intersections of the two lines and the Pareto set are Pareto optimal solutions, which can be found by finding the optimal values of λ_{i1} and λ_{i2}. That is, the optimization of the D decision variables of an LSMOP is converted into the optimization of two weight variables.

Since the optimization of the weight variables of a reference solution cannot explicitly specify the location of the new solution on the Pareto front, LSMOF optimizes the weight variables of r reference solutions together to enhance both the population convergence and population diversity (Line 6). To evaluate the new solutions generated by weight variable optimization, LSMOF calculates the HV value of all the new solutions (Line 7), where a higher HV value indicates a better convergence and diversity of the new solutions. As a consequence, LSMOF

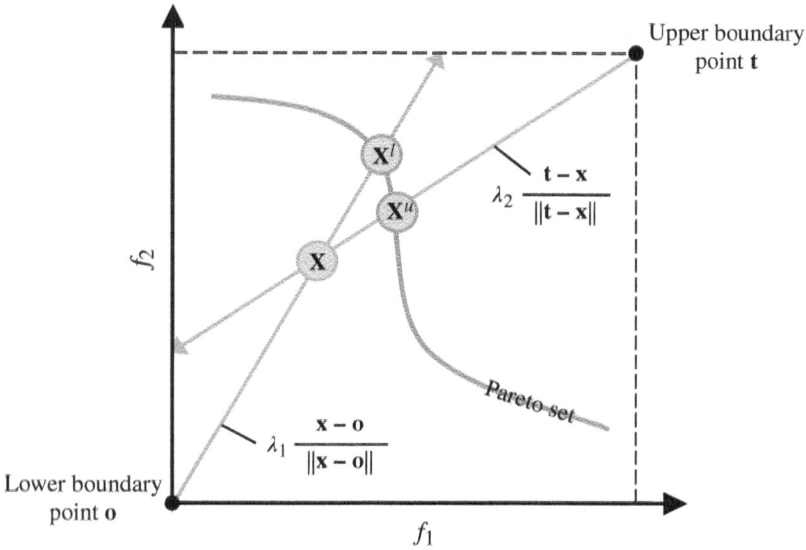

Figure 3.3 Illustration of the problem reformulation method in LSMOF.

reformulates an LSMOP into the following single-objective optimization problem:

$$\begin{aligned} &\text{Maximize } f'(\lambda) = \text{HV}(\{\mathbf{x}_1^l(\lambda_{11}), \mathbf{x}_1^u(\lambda_{12}), \dots, \mathbf{x}_r^l(\lambda_{r1}), \mathbf{x}_r^u(\lambda_{r2})\}, Nad), \\ &\text{s.t. } \lambda = (\lambda_{11}, \lambda_{12}, \dots, \lambda_{r1}, \lambda_{r2}) \in [0, 0.5]^{2r}, \end{aligned} \tag{3.8}$$

where Nad is the nadir point of the current population used as the reference point for HV calculation. This way, the number of variables to be optimized is decreased from D to $2r$, and the dimensionality of the decision space is considerably reduced. The problem reformulation significantly increases the convergence speed on LSMOPs, but it can hardly approximate the global optimums since the one-dimensional lines may not intersect the $(M-1)$-dimensional Pareto set in the D-dimensional decision space, where the number of objectives M is much smaller than the number of decision variables D.

3.4.2 Single-Objective Optimization of LSMOF

To fully optimize the weight variables of r best solutions in the current population, LSMOF adopts differential evolution to maximize the reformulated problem f', where the detailed procedure is presented in Algorithm 3.11. More specifically, N' solutions are first generated as the initial population (Line 1), where each solution is a vector of $\lambda = (\lambda_{11}, \lambda_{12}, \dots, \lambda_{r1}, \lambda_{r2}) \in [0, 0.5]^{2r}$. Then, the initial population is evaluated on f' (Line 5) and optimized for g generations (Line 6).

Algorithm 3.11: *Optimization*($f', \mathbf{x}_1, \dots, \mathbf{x}_r, N', g$)

Input: f' (reformulated problem), $\mathbf{x}_1, \dots, \mathbf{x}_r$ (reference solutions), N' (population size for single-objective optimization), g (number of generations for single-objective optimization)

Output: A (all the generated solutions)

1 $\Lambda \leftarrow$ Randomly generate N' solutions;
2 $A \leftarrow \emptyset$;
3 **foreach** $\lambda \in \Lambda$ **do**
4     $A \leftarrow A \cup \{\mathbf{x}_1^l(\lambda_{11}), \mathbf{x}_1^u(\lambda_{12}), \dots, \mathbf{x}_r^l(\lambda_{r1}), \mathbf{x}_r^u(\lambda_{r2})\}$;
5 Evaluate Λ on f';
6 **for** *generation* $= 1, \dots, g$ **do**
7   **foreach** $\lambda \in \Lambda$ **do**
8         $[\boldsymbol{\lambda}', \boldsymbol{\lambda}'', \boldsymbol{\lambda}'''] \leftarrow$ Randomly select three solutions from Λ;
9         $\gamma \leftarrow$ Perform differential evolution on $\boldsymbol{\lambda}'$, $\boldsymbol{\lambda}''$, and $\boldsymbol{\lambda}'''$;
10         $A \leftarrow A \cup \{\mathbf{x}_1^l(\gamma_{11}), \mathbf{x}_1^u(\gamma_{12}), \dots, \mathbf{x}_r^l(\gamma_{r1}), \mathbf{x}_r^u(\gamma_{r2})\}$;
11   Evaluate γ on f';
12   **if** $f'(\gamma) > f'(\lambda)$ **then**
13   Replace λ in Λ with γ;
14 **return** A;

Note that here the evaluation of a solution consumes $2r$ function evaluations on the original LSMOP. To update each solution λ at each generation, a new solution γ is generated by differential evolution based on three solutions randomly selected from the current population (Lines 8 and 9), and γ is used to replace λ in the population if the former has a better objective value on f' (Lines 12 and 13). During the optimization procedure, LSMOF saves all the newly generated solutions for the original LSMOP in an archive A (Lines 4 and 10), which is the output of the single-objective optimization procedure (Line 14). The solutions in the archive will be combined with the current population, and then truncated by an environmental selection strategy for the next round of optimization.

3.4.3 Procedure of LSMOF

With the problem reformulation and single-objective optimization, the procedure of LSMOF is described in Algorithm 3.12, which begins with the random

Algorithm 3.12: Procedure of LSMOF

Input: N (population size), r (number of reference solutions), N' (population size for single-objective optimization), g (number of generations for single-objective optimization), tr (threshold)

Output: P (final population)

```
1  P ← Randomly generate N solutions;
2  while termination criterion not fulfilled do
3      if the ratio of consumed function evaluations is less than tr then
4          [f′, x_1, ..., x_r] ← Reformulation(P, r);
5          A ← Optimization(f′, x_1, ..., x_r, N′, g);
6          [P, Front, Crowd] ← EnvironmentalSelection(P ∪ A, N);
             //Algorithm 1.6
7      else
8          P′ ← TournamentSelection(P, N, Front, −Crowd); //Algorithm 1.4
9          O ← Variation(P′); //Algorithm 1.5
10         [P, Front, Crowd] ← EnvironmentalSelection(P ∪ O, N);
             //Algorithm 1.6
11 return P;
```

initialization of a population with N solutions (Line 1). At each generation, if the ratio of consumed function evaluations is less than a predefined threshold tr, LSMOF reformulates the LSMOP (Line 4) and optimizes the weight variables via single-objective differential evolution algorithm (Line 5), where the solutions generated during the single-objective optimization procedure are combined with the population and truncated by non-dominated sorting and crowding distance-based selection (Line 6). Otherwise, LSMOF optimizes the population by using the mating selection strategy, genetic operators, and environmental selection strategy of NSGA-II until termination (Lines 8–10). Finally, the population is returned as the output of the algorithm (Line 11).

At each generation of LSMOF, the time complexity is $O(MN^2)$ for problem reformulation and $O(N'gr^M)$ for single-objective optimization, where M denotes the number of objectives, r denotes the number of reference solutions, N' denotes the population size for single-objective optimization, and g denotes the number of generations for single-objective optimization. Therefore, the time complexity of one generation of LSMOF is $O(N'gr^M)$.

3.5 Competitive Swarm Optimizer-Based Evolutionary Algorithm

In addition to dividing the decision variables or reducing the number of decision variables, the large-scale multi-objective competitive swarm optimizer (LMOCSO) suggests a new variation operator for optimizing all the decision variables of LSMOPs together. With no assumption of the relations between decision variables, LMOCSO searches in the original decision spaces and follows a similar optimization procedure as general MOEAs. Although LMOCSO is less effective than decision variable group-based and decision space reduction-based MOEAs on particular LSMOPs, it is flexible to handle complex LSMOPs with a robust performance and has also been used in subsequent algorithms [63, 64].

3.5.1 Enhanced Competitive Swarm Optimizer of LMOCSO

In comparison to genetic algorithms that are widely adopted in MOEAs, particle swarm optimization converges faster in high-dimensional decision spaces due to the learning strategies between particles. However, particle swarm optimization is not popular for multi-objective optimization due to the difficulty in selecting best particles and maintaining the population diversity. By contrast, competitive swarm optimizer (CSO) is more flexible, which generates solutions in a similar way as genetic operators and is compatible with the selection strategies of many existing MOEAs. At the tth generation of CSO, two particles are selected each time and the worse one (i.e., loser particle) is updated by

$$\begin{aligned} \Delta(t) &= \mathbf{x}_w(t) - \mathbf{x}_l(t), \\ \mathbf{v}_l(t+1) &= r_0\mathbf{v}_l(t) + r_1\Delta(t), \\ \mathbf{x}_l(t+1) &= \mathbf{x}_l(t) + \mathbf{v}_l(t+1), \end{aligned} \tag{3.9}$$

where $\mathbf{x}_w$ is the winner particle (i.e., leader), $\mathbf{x}_l$ is the loser particle to be updated and $\mathbf{v}_l$ is its velocity, and r_0, r_1 are two random values within $[0, 1]$. Figure 3.4(a) depicts the convergence trajectories of CSO, where the particle is first updated by the difference between the positions of it and the leader $r_1\Delta(t)$, and then updated by its previous velocity $r_0\mathbf{v}_l(t)$. Therefore, the particle converges spirally toward the leader at a slow speed.

On the contrary, as illustrated in Fig. 3.4(b), if the particle is first updated by its previous velocity $r_0\mathbf{v}_l(t)$ and then updated by the difference between the positions of it and the leader $r_1\Delta(t)$, it can directly move toward the leader in a faster speed. For this aim, each particle should be first updated by its previous velocity:

$$\begin{aligned} \mathbf{x}'_l(t) &= \mathbf{x}_l(t) + r_0\mathbf{v}_l(t), \\ \mathbf{x}'_w(t) &= \mathbf{x}_w(t) + r_0\mathbf{v}_w(t), \end{aligned} \tag{3.10}$$

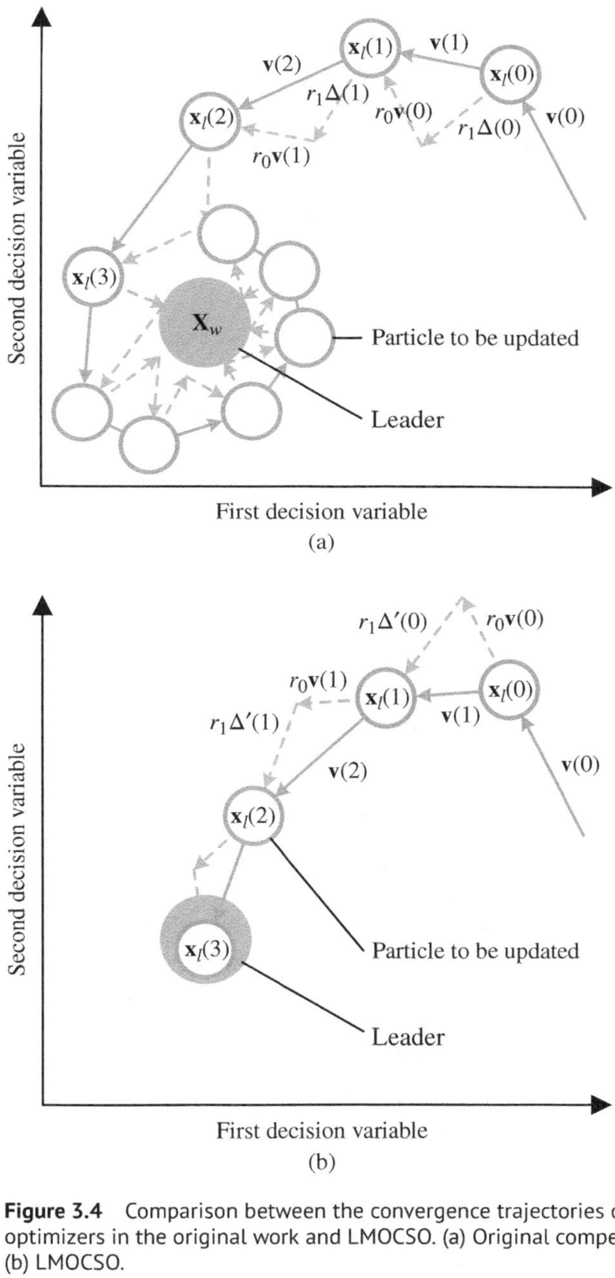

Figure 3.4 Comparison between the convergence trajectories of the competitive swarm optimizers in the original work and LMOCSO. (a) Original competitive swarm optimizer. (b) LMOCSO.

and then the loser particle is updated by

$$\begin{aligned}\Delta'(t) &= \mathbf{x}'_w(t) - \mathbf{x}'_l(t),\\ \mathbf{v}_l(t+1) &= r_0\mathbf{v}_l(t) + r_1\Delta'(t),\\ \mathbf{x}_l(t+1) &= \mathbf{x}_l(t) + \mathbf{v}_l(t+1).\end{aligned} \tag{3.11}$$

However, the decision variables of the two particles are changed after performing Eq. (3.10), which means that $\mathbf{x}'_l$ may become better than $\mathbf{x}'_w$ and they should be reevaluated. To avoid additional function evaluations, the above formulas can be written as

$$\begin{aligned}\mathbf{x}'_l(t+1) &= \mathbf{x}_l(t+1) + r_0\mathbf{v}_l(t+1)\\ &= \mathbf{x}_l(t) + \mathbf{v}_l(t+1) + r_0\mathbf{v}_l(t+1)\\ &= \mathbf{x}_l(t) + \mathbf{v}_l(t+1) + r_0\mathbf{v}_l(t+1) + r_0\mathbf{v}_l(t) - r_0\mathbf{v}_l(t)\\ &= (\mathbf{x}_l(t) + r_0\mathbf{v}_l(t)) + \mathbf{v}_l(t+1) + (r_0\mathbf{v}_l(t+1) - r_0\mathbf{v}_l(t))\\ &= \mathbf{x}'_l(t) + \mathbf{v}_l(t+1) + r_0(\mathbf{v}_l(t+1) - \mathbf{v}_l(t)).\end{aligned} \tag{3.12}$$

That is,

$$\begin{aligned}\mathbf{v}_l(t+1) &= r_0\mathbf{v}_l(t) + r_1(\mathbf{x}'_w(t) - \mathbf{x}'_l(t)),\\ \mathbf{x}'_l(t+1) &= \mathbf{x}'_l(t) + \mathbf{v}_l(t+1) + r_0(\mathbf{v}_l(t+1) - \mathbf{v}_l(t)).\end{aligned} \tag{3.13}$$

Since $\mathbf{x}_l$ is eliminated from the above formula, $\mathbf{x}'_l$ can be updated by itself without additional function evaluation. By replacing the symbols $\mathbf{x}'_l, \mathbf{x}'_w$ with $\mathbf{x}_l, \mathbf{x}_w$, the particle update strategy of LMOCSO is defined as

$$\begin{aligned}\mathbf{v}_l(t+1) &= r_0\mathbf{v}_l(t) + r_1(\mathbf{x}_w(t) - \mathbf{x}_l(t)),\\ \mathbf{x}_l(t+1) &= \mathbf{x}_l(t) + \mathbf{v}_l(t+1) + r_0(\mathbf{v}_l(t+1) - \mathbf{v}_l(t)),\end{aligned} \tag{3.14}$$

where $\mathbf{x}_w$ is the winner particle, $\mathbf{x}_l$ is the loser particle to be updated and $\mathbf{v}_l$ is its velocity, and r_0, r_1 are two random values within [0, 1].

Algorithm 3.13 details the procedure of generating offspring solutions in LMOCSO. First, the fitness value of each particle is calculated to determine loser and winner particles (Line 1). According to the minimum shift-based density estimation-based distance [65] between each particle $\mathbf{p}$ and others, LMOCSO quantitatively measures both the convergence and diversity of $\mathbf{p}$ by

$$\text{Fitness}(\mathbf{p}) = \min_{\mathbf{q}\in P\setminus\{\mathbf{p}\}} \sqrt{\sum_{i=1}^{M}(\max\{0, f_i(\mathbf{q}) - f_i(\mathbf{p})\})^2}, \tag{3.15}$$

where $f_i(\mathbf{p})$ denotes the ith objective value of $\mathbf{p}$ and M denotes the number of objectives. At each time, two solutions are randomly picked up from the current population (Line 4), where the one with a smaller fitness value is the loser particle and the one with a larger fitness value is the winner particle (Lines 6–11).

Algorithm 3.13: *UpdatingParticle(P)*

Input: P (current population)
Output: P' (new population)

```
1  Fitness ← Calculate the fitness of particles in P by Eq. (3.15);
2  P' ← ∅;
3  while P ≠ ∅ do
4      [p, q] ← Randomly select two particles from P;
5      P ← P \ {p, q};
6      if Fitness(p) < Fitness(q) then
7          x_l ← p; //Loser particle
8          x_w ← q; //Winner particle
9      else
10         x_l ← q; //Loser particle
11         x_w ← p; //Winner particle
12     x_l ← Update x_l by Eq. (3.14);
13     [x_l, x_w] ← Mutate x_l, x_w by polynomial mutation;
14     P' ← P' ∪ {x_l, x_w};
15 return P';
```

Then, the loser particle is updated by Eq. (3.14), and both the winner particle and the updated loser particle are tuned by polynomial mutation (Lines 12 and 13). The two mutated particles are regarded as offspring solutions (Line 14), and the algorithm repeatedly generates offspring solutions until all the particles in the current population have been picked up (Line 3).

3.5.2 Procedure of LMOCSO

With the enhanced CSO operator, the procedure of LMOCSO is described in Algorithm 3.14, which begins with the random initialization of a population with N particles (Line 1) and the generation of N uniformly distributed reference vectors (Line 2). At each generation, LMOCSO generates offspring solutions by using the proposed CSO operator (Line 4), which are further combined with the current population and truncated by the environmental selection strategy of RVEA [66] (Line 5). Finally, the population is returned as the output of the algorithm (Line 6).

At each generation of LMOCSO, the time complexity is $O(ND)$ for offspring generation and $O(MN^2)$ for environmental selection, where D denotes the number of

Algorithm 3.14: Procedure of LMOCSO

Input: N (population size)
Output: P (final population)

$P \leftarrow$ Randomly generate N solutions;
$V \leftarrow$ *UniformSampling*(N, M); `//Algorithm 1.8`
while termination criterion not fulfilled **do**
  $P' \leftarrow$ *UpdatingParticle*(P);
  $P \leftarrow$ *EnvironmentalSelection*$(P \cup P', V, N)$; `//Algorithm 1.9`
return P;

decision variables and M denotes the number of objectives. Therefore, the time complexity of one generation of LMOCSO is $O(ND + MN^2)$.

3.6 Experimental Comparisons

To investigate the performance of CCGDE3, LMEA, LSMOF, and LMOCSO, they are compared with two representative MOEAs on benchmark LSMOPs. The compared MOEAs include NSGA-II [57] and GDE3 [53], which are based on genetic algorithms and differential evolution, respectively. The tested benchmark LSMOPs include DTLZ1–DTLZ4 [67] and LSMOP1–LSMOP4 [68], where the number of objectives is set to 2 and the number of decision variables D is set to 100, 200, and 500. The population size is set to 100 and the number of function evaluations is set to $10,000 \times D$; besides, the number of function evaluations for LSMOF is set to 100,000. The populations obtained by the compared algorithms are assessed by IGD [69].

Table 3.1 presents the experimental results averaged over 30 runs, where the best results are shown in bold and the statistically similar ones are highlighted. It can be found that the three algorithms LMEA, LSMOF, and LMOCSO exhibit the best overall performance and obtain the best results on all the test instances, which verify the effectiveness of their search strategies (i.e., decision variable grouping, decision space reduction, and novel variation operator) for solving LSMOPs. By contrast, CCGDE3 is not effective due to the simple random grouping strategy. For the problems DTLZ1–DTLZ4 without variable interaction, LMEA and LSMOF obtain the best performance by using the decision variable clustering-based divide-and-conquer search and the problem reformulation-based decision space reduction method, respectively. For the problems LSMOP1–LSMOP4 with complex variable interactions, LMOCSO obtains the best performance by using an enhanced CSO operator.

Table 3.1 IGD Values Obtained by Six MOEAs on Eight Benchmark LSMOPs.

Problem (*D*)	NSGA-II	GDE3	CCGDE3	LMEA	LSMOF	LMOCSO
DTLZ1 (100)	6.4641e−3	2.5365e−2	1.3813e+3	**2.6312e−3**	2.8905e−3	1.0347e+2
DTLZ2 (100)	5.0945e−3	5.6725e−3	1.8383e+0	4.4037e−3	5.8698e−3	**3.9712e−3**
DTLZ3 (100)	1.4480e−2	3.9884e−2	3.0835e+3	7.3405e−3	**5.4832e−3**	2.1541e+2
DTLZ4 (100)	1.8954e−1	5.5866e−3	2.2559e+0	8.0055e−3	**5.0232e−3**	1.8850e−1
LSMOP1 (100)	2.7571e−1	6.2147e−3	2.1560e+0	4.7560e−2	4.1547e−1	**3.5745e−3**
LSMOP2 (100)	4.3101e−2	2.8685e−2	2.0254e−1	6.1241e−2	5.5942e−2	**6.3805e−3**
LSMOP3 (100)	1.1489e+0	1.4267e+0	1.1126e+1	2.7178e+0	1.0867e+0	**7.0710e−1**
LSMOP4 (100)	5.8359e−2	2.9452e−2	2.8232e−1	8.6353e−2	1.5108e−1	**4.2708e−3**
DTLZ1 (200)	1.5720e−2	9.5149e−1	2.8179e+3	3.1029e−3	**2.8849e−3**	3.3643e+2
DTLZ2 (200)	5.0094e−3	6.1236e−3	5.0360e+0	4.4560e−3	6.1075e−3	**3.9790e−3**
DTLZ3 (200)	3.2281e−2	9.5901e−2	7.7788e+3	1.1735e−2	**5.6668e−3**	5.0970e+2
DTLZ4 (200)	1.8936e−1	6.2517e−3	5.3000e+0	5.9216e−3	**5.2421e−3**	1.8850e−1
LSMOP1 (200)	3.2032e−1	7.7526e−3	2.6546e+0	1.4445e−2	5.1534e−1	**3.5750e−3**
LSMOP2 (200)	5.6428e−2	2.3440e−2	1.4495e−1	8.3842e−2	3.8909e−2	**6.0185e−3**
LSMOP3 (200)	8.7617e−1	1.4618e+0	1.5685e+1	**5.8280e−1**	1.5325e+0	7.0711e−1
LSMOP4 (200)	5.9652e−2	2.0127e−2	1.9623e−1	5.6249e−2	9.9746e−2	**3.7896e−3**
DTLZ1 (500)	5.4060e−2	6.3960e+1	9.6375e+3	1.3312e−2	**2.7453e−3**	1.3887e+3
DTLZ2 (500)	5.1267e−3	8.0086e−3	1.4485e+1	4.5287e−3	5.3777e−3	**4.0098e−3**
DTLZ3 (500)	9.8166e−2	1.2375e+3	2.6620e+4	4.1436e−2	**5.5054e−3**	1.9382e+3
DTLZ4 (500)	**5.0468e−3**	8.0196e−3	1.2614e+1	8.9537e−3	5.0949e−3	3.7304e−1
LSMOP1 (500)	3.1811e−1	1.2168e−2	3.6470e+0	1.1068e−1	5.9538e−1	**3.5760e−3**
LSMOP2 (500)	2.7927e−2	4.0673e−2	7.1917e−2	**1.8366e−2**	2.4553e−2	2.2377e−2
LSMOP3 (500)	7.7529e−1	1.9431e+0	1.9325e+1	**6.8525e−1**	1.5644e+0	7.0711e−1
LSMOP4 (500)	5.1863e−2	1.7382e−2	1.1073e−1	2.7795e−2	5.0988e−2	**3.6211e−3**

Moreover, Fig. 3.5 depicts the convergence profiles of IGD values obtained by the compared MOEAs on DTLZ1 and LSMOP1 with 100 decision variables. For CCGDE3, it converges more slowly than the other MOEAs, which indicates the ineffectiveness of the random grouping method for large-scale multi-objective optimization. For LMEA, it converges slowly at the beginning since many function evaluations are consumed for the sake of decision variable clustering, but it can obtain promising populations at last by using the divide-and-conquer

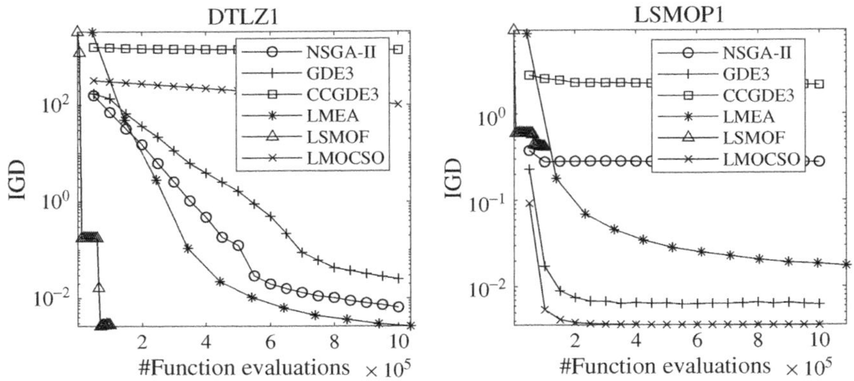

Figure 3.5 Solutions obtained by six MOEAs on DTLZ1 and LSMOP1 with 100 decision variables.

search. For LSMOF, it converges much more quickly than the other MOEAs, but it is likely to get stuck in local optimums, especially on LSMOP1 with complex variable interactions. As for LMOCSO, it converges fast on LSMOP1 with complex variable interactions, but it gets stuck in local optimums on DTLZ1 due to the ineffectiveness of particle swarm optimization in handling multimodal landscapes.

3.7 Summary

This chapter introduces four representative MOEAs for solving LSMOPs. The most significant feature of these MOEAs is that they suggest different ideas to handle the high-dimensional decision spaces, such as the decision variable grouping methods in CCGDE3 and LMEA, the decision space reduction method in LSMOF, and the novel variation operator in LMOCSO. According to the discussions and experimental results, different types of large-scale MOEAs are effective for different types of LSMOPs, while no MOEA can obtain the best performance on all types of LSMOPs. For example, decision variable grouping-based MOEAs are good at solving LSMOPs with few variable interactions, decision space reduction-based MOEAs can quickly converge to local optimums of LSMOPs with simple Pareto sets, and novel search strategy-based MOEAs are more robust in solving complex LSMOPs. Besides, most large-scale MOEAs are ineffective for solving small-scale problems. After all, if large-scale MOEAs dominate general MOEAs on small-scale problems, the latter will become meaningless and the research of large-scale multi-objective optimization will not be separated from the research of multi-objective optimization.

References

1 A. Zhou, B.-Y. Qu, H. Li, S.-Z. Zhao, P. N. Suganthan, and Q. Zhang, "Multi-objective evolutionary algorithms: a survey of the state of the art," *Swarm and Evolutionary Computation*, vol. 1, no. 1, pp. 32–49, 2011.

2 B. Li, J. Li, K. Tang, and X. Yao, "Many-objective evolutionary algorithms: a survey," *ACM Computing Surveys*, vol. 48, no. 1, pp. 1–35, 2015.

3 J. Liang, X. Ban, K. Yu, B. Qu, K. Qiao, C. Yue, K. Chen, and K. C. Tan, "A survey on evolutionary constrained multi-objective optimization," *IEEE Transactions on Evolutionary Computation*, vol. 27, no. 2, pp. 201–221, 2022.

4 Y. Tian, C. He, R. Cheng, and X. Zhang, "A multistage evolutionary algorithm for better diversity preservation in multiobjective optimization," *IEEE Transactions on Systems, Man, and Cybernetics: Systems*, vol. 51, no. 9, pp. 5880–5894, 2019.

5 Y. Tian, H. Wang, X. Zhang, and Y. Jin, "Effectiveness and efficiency of non-dominated sorting for evolutionary multi- and many-objective optimization," *Complex & Intelligent Systems*, vol. 3, pp. 247–263, 2017.

6 Y. Tian, Y. Zhang, Y. Su, X. Zhang, K. C. Tan, and Y. Jin, "Balancing objective optimization and constraint satisfaction in constrained evolutionary multiobjective optimization," *IEEE Transactions on Cybernetics*, vol. 52, no. 9, pp. 9559–9572, 2021.

7 T. Chugh, Y. Jin, K. Miettinen, J. Hakanen, and K. Sindhya, "A surrogate-assisted reference vector guided evolutionary algorithm for computationally expensive many-objective optimization," *IEEE Transactions on Evolutionary Computation*, vol. 22, no. 1, pp. 129–142, 2016.

8 Y. Tian, H. Chen, H. Ma, X. Zhang, K. C. Tan, and Y. Jin, "Integrating conjugate gradients into evolutionary algorithms for large-scale continuous multi-objective optimization," *IEEE/CAA Journal of Automatica Sinica*, vol. 9, no. 10, pp. 1801–1817, 2022.

9 J.-R. Jian, Z.-H. Zhan, and J. Zhang, "Large-scale evolutionary optimization: a survey and experimental comparative study," *International Journal of Machine Learning and Cybernetics*, vol. 11, no. 3, pp. 729–745, 2020.

10 Y. Tian, X. Zheng, X. Zhang, and Y. Jin, "Efficient large-scale multiobjective optimization based on a competitive swarm optimizer," *IEEE Transactions on Cybernetics*, vol. 50, no. 8, pp. 3696–3708, 2019.

11 R. Tanabe and A. Fukunaga, "Success-history based parameter adaptation for differential evolution," in *Proceedings of the 2013 IEEE Congress on Evolutionary Computation*, 2013, pp. 71–78.

12 K. M. Sallam, S. M. Elsayed, R. K. Chakrabortty, and M. J. Ryan, "Improved multi-operator differential evolution algorithm for solving unconstrained problems," in *Proceedings of the 2020 IEEE Congress on Evolutionary Computation*, 2020, pp. 1–8.

13 J. J. Liang, A. K. Qin, P. N. Suganthan, and S. Baskar, "Comprehensive learning particle swarm optimizer for global optimization of multimodal functions," *IEEE Transactions on Evolutionary Computation*, vol. 10, no. 3, pp. 281–295, 2006.

14 R. Cheng and Y. Jin, "A competitive swarm optimizer for large scale optimization," *IEEE Transactions on Cybernetics*, vol. 45, no. 2, pp. 191–204, 2014.

15 Y. Tian, S. Peng, X. Zhang, T. Rodemann, K. C. Tan, and Y. Jin, "A recommender system for metaheuristic algorithms for continuous optimization based on deep recurrent neural networks," *IEEE Transactions on Artificial Intelligence*, vol. 1, no. 1, pp. 5–18, 2020.

16 H. Li and Q. Zhang, "Multiobjective optimization problems with complicated Pareto sets, MOEA/D and NSGA-II," *IEEE Transactions on Evolutionary Computation*, vol. 13, no. 2, pp. 284–302, 2008.

17 M. Yuen, S. Ng, and M. Leung, "A competitive mechanism multi-objective particle swarm optimization algorithm and its application to signalized traffic problem," *Cybernetics and Systems*, vol. 52, no. 1, pp. 73–104, 2020.

18 D. P. Kingma and J. Ba, "Adam: a method for stochastic optimization," *arXiv preprint arXiv:1412.6980*, 2014.

19 S. Yang, Y. Tian, C. He, X. Zhang, K. C. Tan, and Y. Jin, "A gradient guided evolutionary approach to training deep neural networks," *IEEE Transactions on Neural Networks and Learning Systems*, vol. 33, no. 9, pp. 4861–4875, 2021.

20 J. C. Strikwerda, *Finite difference schemes and partial differential equations.* Chapman and Hall, 2004.

21 Y. Tian, H. Chen, X. Xiang, H. Jiang, and X. Zhang, "A comparative study on evolutionary algorithms and mathematical programming methods for continuous optimization," in *Proceedings of the 2022 IEEE Congress on Evolutionary Computation*, 2022, pp. 1–8.

22 Y. Tian, L. Si, X. Zhang, R. Cheng, C. He, K. C. Tan, and Y. Jin, "Evolutionary large-scale multi-objective optimization: a survey," *ACM Computing Surveys*, vol. 54, no. 8, pp. 1–34, 2021.

23 L. M. Antonio and C. A. Coello Coello, "Use of cooperative coevolution for solving large scale multiobjective optimization problems," in *Proceedings of the 2013 IEEE Congress on Evolutionary Computation*, 2013, pp. 2758–2765.

24 W. Chen, T. Weise, Z. Yang, and K. Tang, "Large-scale global optimization using cooperative coevolution with variable interaction learning," in *Proceedings of the 11th International Conference on Parallel Problem Solving from Nature*, 2010, pp. 300–309.

25 M. N. Omidvar, X. Li, Y. Mei, and X. Yao, "Cooperative co-evolution with differential grouping for large scale optimization," *IEEE Transactions on Evolutionary Computation*, vol. 18, no. 3, pp. 378–393, 2013.

26 X. Ma, F. Liu, Y. Qi, X. Wang, L. Li, L. Jiao, M. Yin, and M. Gong, "A multiobjective evolutionary algorithm based on decision variable analyses for multiobjective optimization problems with large-scale variables," *IEEE Transactions on Evolutionary Computation*, vol. 20, no. 2, pp. 275–298, 2015.

27 X. Zhang, Y. Tian, R. Cheng, and Y. Jin, "A decision variable clustering-based evolutionary algorithm for large-scale many-objective optimization," *IEEE Transactions on Evolutionary Computation*, vol. 22, no. 1, pp. 97–112, 2016.

28 Y. Xu, C. Xu, H. Zhang, L. Huang, Y. Liu, Y. Nojima, and X. Zeng, "A multi-population multi-objective evolutionary algorithm based on the contribution of decision variables to objectives for large-scale multi/many-objective optimization," *IEEE Transactions on Cybernetics*, vol. 53, no. 11, pp. 6998–7007.

29 H. Zille, H. Ishibuchi, S. Mostaghim, and Y. Nojima, "A framework for large-scale multiobjective optimization based on problem transformation," *IEEE Transactions on Evolutionary Computation*, vol. 22, no. 2, pp. 260–275, 2017.

30 C. He, L. Li, Y. Tian, X. Zhang, R. Cheng, Y. Jin, and X. Yao, "Accelerating large-scale multiobjective optimization via problem reformulation," *IEEE Transactions on Evolutionary Computation*, vol. 23, no. 6, pp. 949–961, 2019.

31 H. Qian and Y. Yu, "Solving high-dimensional multi-objective optimization problems with low effective dimensions," in *Proceedings of the AAAI Conference on Artificial Intelligence*, vol. 31, no. 1, 2017.

32 R. Liu, R. Ren, J. Liu, and J. Liu, "A clustering and dimensionality reduction based evolutionary algorithm for large-scale multi-objective problems," *Applied Soft Computing*, vol. 89, p. 106120, 2020.

33 Y. Tian, C. Lu, X. Zhang, K. C. Tan, and Y. Jin, "Solving large-scale multi-objective optimization problems with sparse optimal solutions via unsupervised neural networks,"*IEEE Transactions on Cybernetics*, vol. 51, no. 6, pp. 3115–3128, 2020.

34 Y. Tian, Y. Feng, X. Zhang, and C. Sun, "A fast clustering based evolutionary algorithm for super-large-scale sparse multi-objective optimization," *IEEE/CAA Journal of Automatica Sinica*, vol. 10, no. 4, pp. 1048–1063, 2022.

35 K. Deb and R. B. Agrawal, "Simulated binary crossover for continuous search space," *Complex Systems*, vol. 9, no. 2, pp. 115–148, 1995.

36 K. Deb and M. Goyal, "A combined genetic adaptive search (GeneAS) for engineering design," *Computer Science and Informatics*, vol. 26, no. 4, pp. 30–45, 1996.

37 R. Storn and K. Price, "Differential evolution—a simple and efficient heuristic for global optimization over continuous spaces," *Journal of Global Optimization*, vol. 11, no. 4, pp. 341–359, 1997.

38 A. Prajapati, "A particle swarm optimization approach for large-scale many-objective software architecture recovery," *Journal of King Saud University-Computer and Information Sciences*, vol. 34, no. 10, pp. 8501–8513, 2022.

39 H. Ge, N. Zhang, L. Sun, X. Wang, and Y. Hou, "A memetic evolution system with statistical variable classification for large-scale many-objective optimization," *Applied Soft Computing*, vol. 114, p. 108158, 2022.

40 S. Qi, J. Zou, S. Yang, and J. Zheng, "A level-based multi-strategy learning swarm optimizer for large-scale multi-objective optimization,"*Swarm and Evolutionary Computation*, vol. 73, p. 101100, 2022.

41 C. He, R. Cheng, and D. Yazdani, "Adaptive offspring generation for evolutionary large-scale multiobjective optimization," *IEEE Transactions on Systems, Man, and Cybernetics: Systems*, vol. 52, no. 2, pp. 786–798, 2020.

42 R. Cheng, Y. Jin, K. Narukawa, and B. Sendhoff, "A multiobjective evolutionary algorithm using Gaussian process-based inverse modeling,"*IEEE Transactions on Evolutionary Computation*, vol. 19, no. 6, pp. 838–856, 2015.

43 W. Liu, L. Chen, X. Hao, W. Zhou, X. Cao, and F. Xie, "Offspring regeneration method based on bi-level sampling for large-scale evolutionary multi-objective optimization," *Swarm and Evolutionary Computation*, vol. 75, p. 101152, 2022.

44 H. Chen, R. Cheng, J. Wen, H. Li, and J. Weng, "Solving large-scale many-objective optimization problems by covariance matrix adaptation evolution strategy with scalable small subpopulations," *Information Sciences*, vol. 509, pp. 457–469, 2020.

45 Z. Wang, H. Hong, K. Ye, G. Zhang, M. Jiang, and K. C. Tan, "Manifold interpolation for large-scale multiobjective optimization via generative adversarial networks," *IEEE Transactions on Neural Networks and Learning Systems*, vol. 34, no. 8, pp. 4631–4645, 2021.

46 W. Hong, K. Tang, A. Zhou, H. Ishibuchi, and X. Yao, "A scalable indicator-based evolutionary algorithm for large-scale multiobjective optimization," *IEEE Transactions on Evolutionary Computation*, vol. 23, no. 3, pp. 525–537, 2018.

47 H. Hong, K. Ye, M. Jiang, D. Cao, and K. C. Tan, "Solving large-scale multiobjective optimization via the probabilistic prediction model," *Memetic Computing*, vol. 14, no. 2, pp. 165–177, 2022.

48 S. Liu, J. Li, Q. Lin, Y. Tian, and K. C. Tan, "Learning to accelerate evolutionary search for large-scale multiobjective optimization,"*IEEE Transactions on Evolutionary Computation*, vol. 27, no. 1, pp. 67–81, 2022.

49 M. N. Omidvar, X. Li, Z. Yang, and X. Yao, "Cooperative co-evolution for large scale optimization through more frequent random grouping," in *Proceedings of the 2010 IEEE Congress on Evolutionary Computation*, 2010, pp. 1–8.

50 A. Song, Q. Yang, W.-N. Chen, and J. Zhang, "A random-based dynamic grouping strategy for large scale multi-objective optimization," in *Proceedings of the 2016 IEEE Congress on Evolutionary Computation*, 2016, pp. 468–475.

51 M. Li and J. Wei, "A cooperative co-evolutionary algorithm for large-scale multi-objective optimization problems," in *Proceedings of the 2018 Annual Conference on Genetic and Evolutionary Computation Conference*, 2018, pp. 1716–1721.

52 J. Liu, R. Liu, and X. Zhang, "Recursive grouping and dynamic resource allocation method for large-scale multi-objective optimization problem," *Applied Soft Computing*, vol. 130, p. 109651, 2022.

53 S. Kukkonen and J. Lampinen, "GDE3: The third evolution step of generalized differential evolution," in *Proceedings of the 2005 IEEE Congress on Evolutionary Computation*, vol. 1, 2005, pp. 443–450.

54 W. Du, W. Zhong, Y. Tang, W. Du, and Y. Jin, "High-dimensional robust multi-objective optimization for order scheduling: a decision variable classification approach," *IEEE Transactions on Industrial Informatics*, vol. 15, no. 1, pp. 293–304, 2018.

55 C. He, R. Cheng, L. Li, K. C. Tan, and Y. Jin, "Large-scale multiobjective optimization via reformulated decision variable analysis," *IEEE Transactions on Evolutionary Computation*, vol. 28, no. 1, pp. 47–61, 2022.

56 L. Ma, M. Huang, S. Yang, R. Wang, and X. Wang, "An adaptive localized decision variable analysis approach to large-scale multiobjective and many-objective optimization," *IEEE Transactions on Cybernetics*, vol. 52, no. 7, pp. 6684–6696, 2021.

57 K. Deb, A. Pratap, S. Agarwal, and T. Meyarivan, "A fast and elitist multiobjective genetic algorithm: NSGA-II," *IEEE Transactions on Evolutionary Computation*, vol. 6, no. 2, pp. 182–197, 2002.

58 E. Zitzler, M. Laumanns, and L. Thiele, "SPEA2: Improving the strength pareto evolutionary algorithm," *TIK Report*, vol. 103, 2001.

59 L. While, P. Hingston, L. Barone, and S. Huband, "A faster algorithm for calculating hypervolume," *IEEE Transactions on Evolutionary Computation*, vol. 10, no. 1, pp. 29–38, 2006.

60 C. He, R. Cheng, Y. Tian, and X. Zhang, "Iterated problem reformulation for evolutionary large-scale multiobjective optimization," in *Proceedings of the 2020 IEEE Congress on Evolutionary Computation*, 2020, pp. 1–8.

61 S. Qin, C. Sun, Y. Jin, Y. Tan, and J. Fieldsend, "Large-scale evolutionary multi-objective optimization assisted by directed sampling," *IEEE Transactions on Evolutionary Computation*, vol. 25, no. 4, pp. 724–738, 2021.

62 L. Chen, H. Wang, and W. Ma, "Two-stage multi-tasking transform framework for large-scale many-objective optimization problems,"*Complex & Intelligent Systems*, vol. 7, pp. 1499–1513, 2021.

63 S. Qi, J. Zou, S. Yang, Y. Jin, J. Zheng, and X. Yang, "A self-exploratory competitive swarm optimization algorithm for large-scale multiobjective optimization," *Information Sciences*, vol. 609, pp. 1601–1620, 2022.

64 S. Liu, Q. Lin, Q. Li, and K. C. Tan, "A comprehensive competitive swarm optimizer for large-scale multiobjective optimization," *IEEE Transactions on Systems, Man, and Cybernetics: Systems*, vol. 52, no. 9, pp. 5829–5842, 2021.

65 M. Li, S. Yang, and X. Liu, "Shift-based density estimation for Pareto-based algorithms in many-objective optimization," *IEEE Transactions on Evolutionary Computation*, vol. 18, no. 3, pp. 348–365, 2013.

66 R. Cheng, Y. Jin, M. Olhofer, and B. Sendhoff, "A reference vector guided evolutionary algorithm for many-objective optimization," *IEEE Transactions on Evolutionary Computation*, vol. 20, no. 5, pp. 773–791, 2016.

67 K. Deb, L. Thiele, M. Laumanns, and E. Zitzler, "Scalable test problems for evolutionary multiobjective optimization," in *Evolutionary multiobjective optimization: theoretical advances and applications*. Springer, 2005, pp. 105–145.

68 R. Cheng, Y. Jin, M. Olhofer, and B. Sendhoff, "Test problems for large-scale multiobjective and many-objective optimization," *IEEE Transactions on Cybernetics*, vol. 47, no. 12, pp. 4108–4121, 2016.

69 C. A. Coello Coello and N. C. Cortés, "Solving multiobjective optimization problems using an artificial immune system," *Genetic Programming and Evolvable Machines*, vol. 6, pp. 163–190, 2005.

4

Evolutionary Algorithms for Sparse Large-Scale Multi-Objective Optimization

4.1 Introduction

The "curse of dimensionality" encountered by evolutionary algorithms in huge decision spaces is inevitable, as they iteratively search for better solutions in a black-box manner. In addition to developing multi-objective evolutionary algorithms (MOEAs) for solving general LSMOPs, it is reasonable to improve the convergence speed by involving problem-specific information in offspring generation, such as the gradients used in mathematical programming methods [1]. While it is costly to customize an evolutionary algorithm for every single LSMOP through human expertise, developing effective search strategies for specific types of LSMOPs turns out to be desirable.

4.1.1 Characteristics of Sparse Large-Scale Multi-Objective Optimization Problems

In real-world applications, there exist many optimization problems having sparse optimal solutions, in which the optimal values of most decision variables are zero. For the neural network training task in supervised learning [2], the goal is to find the optimal weights between layers, where many weights are expected to be zero for alleviating overfitting. Hence, the optimal solutions for neural network training are sparse when real encoding is used, where each decision variable denotes a weight. For the frequent pattern mining task in data mining [3], the goal is to find the most frequent and complete pattern (e.g., a set of items) from a transaction dataset (e.g., a set of historical shopping lists having many candidate items), where only a small proportion of all the candidate items is preferred. Hence, the optimal solutions for frequent pattern mining are sparse when binary encoding is used, where each decision variable denotes whether an item is selected or not.

The above tasks are collectively known as sparse optimization problems, which are likely to contain a large number of decision variables. For the aforementioned

Evolutionary Large-Scale Multi-Objective Optimization and Applications: Integrating Evolutionary Computation, Machine Learning, and Data Science, First Edition. Xingyi Zhang, Ran Cheng, Ye Tian, and Yaochu Jin.

examples, a neural network should have enough weights for learning complex mappings between features and labels [4], and a transaction dataset usually consists of abundant candidate items for recommending the most complete patterns to customers [5]. Furthermore, sparse optimization problems generally have multiple conflicting objectives to be optimized. For some sparse LSMOPs like neural network training, the sparsity of optimal solutions is explicitly defined as an objective to be optimized (e.g., the network complexity). While for some other sparse LSMOPs like the frequent pattern mining, the sparsity of optimal solutions is implicitly determined by all the objectives (e.g., in Eq. (2.28), the selected items $\mathbf{x}$ should occupy a small proportion of all the items in a transaction dataset τ, otherwise $\tau_{\mathbf{x}}$ will be empty).

4.1.2 Ideas of Finding Sparse Solutions

It is possible to solve sparse LSMOPs using general large-scale MOEAs, but their efficiency cannot meet the requirement in real-world scenarios due to the relatively expensive objective functions. For example, evaluating a solution for neural network training needs to calculate the output of all the training samples, and evaluating a solution for frequent pattern mining should traverse the whole transaction dataset. Fortunately, the sparse nature of the optimal solutions of sparse LSMOPs is known in advance, making it possible to optimize only the nonzero variables. Since the nonzero variables are much fewer than the zero variables of sparse LSMOPs, the problems can be considerably eased and many function evaluations can be saved. In comparison to optimizing all the decision variables of general LSMOPs, there are two tasks in solving sparse LSMOPs, including finding the nonzero variables and optimizing the nonzero variables. Furthermore, for the sparse LSMOPs with binary encoding (e.g., frequent pattern mining), the task of optimizing the nonzero variables can be omitted.

Since large-scale MOEAs focus on continuous optimization, they are suitable for optimizing the nonzero variables but unsuitable for finding the nonzero variables of sparse LSMOPs, where the latter is essentially a binary optimization problem. On the other hand, binary-coded MOEAs (e.g., NSGA-II with uniform crossover and bitwise mutation [6]) are also inefficient for finding nonzero variables since they are not tailored for finding sparse solutions. Taking the random initialization strategy as an example, the expectation of the number of zero variables E_{zero} equals the expectation of the number of nonzero variables $E_{nonzero}$ in the initial solutions. However, since $E_{nonzero} \ll E_{zero}$ in the optimal solutions of sparse LSMOPs, the initial solutions are far away from the optimal solutions and many function evaluations should be consumed for the sake of convergence. Similarly, for the bitwise mutation flipping each decision variable with the same probability, assuming that a solution having d nonzero variables and $(D - d)$ zero variables is mutated with a probability of p on each dimension, the expectation of the number of nonzero variables after one mutation will be

$$\begin{aligned} E^1_{nonzero} &= d + p(D-d) - pd \\ &= (1-2p)d + pD. \end{aligned} \tag{4.1}$$

Analogously, the expectation of the number of nonzero variables after k mutations will be

$$\begin{aligned} E^k_{nonzero} &= (1-2p)^k d + pD + pD(1-2p) + \cdots + pD(1-2p)^{k-1} \\ &= (1-2p)^k d + pD\frac{1-(1-2p)^k}{1-(1-2p)}. \end{aligned} \tag{4.2}$$

Therefore, after a sufficient number of mutations, the expectation of the number of nonzero variables will be

$$\lim_{k\to\infty} E^k_{nonzero} = pD\frac{1}{1-(1-2p)} = \frac{D}{2}, \tag{4.3}$$

which means that $E^k_{zero} = D - E^k_{nonzero} = E^k_{nonzero}$ for $k \to \infty$, regardless of the number of original nonzero variables d and the mutation probability p. That is, the bitwise mutation can hardly make $E_{nonzero} \ll E_{zero}$ for generating sparse solutions.

To address the difficulties encountered by large-scale MOEAs and binary-coded MOEAs in finding the nonzero variables of sparse LSMOPs, several sparse MOEAs have been proposed in recent years, each of which develops novel strategies to find nonzero variables and adopts existing operators to optimize nonzero variables. In the following sections, three representative sparse MOEAs are introduced and empirically compared.

4.2 Bi-Level Encoding-Based Evolutionary Algorithm

The sparse evolutionary algorithm (SparseEA) is the first generic evolutionary algorithm for solving sparse LSMOPs [7]. The main contribution of SparseEA is to represent each solution by using a bi-level encoding, which differentiates the search of nonzero variables and the optimization of nonzero variables. Moreover, a population initialization strategy, a crossover operator, and a mutation operator are customized to address the shortcomings of those in binary-coded MOEAs, thus facilitating the search of nonzero variables.

4.2.1 Bi-Level Encoding of SparseEA

Formally, SparseEA represents a solution $\mathbf{x} = (x_1, x_2, \ldots, x_D) \in \Omega$ with a binary vector $\mathbf{xb} = (xb_1, xb_2, \ldots, xb_D) \in \{0,1\}^D$ and a real vector $\mathbf{xr} = (xr_1, xr_2, \ldots, xr_D) \in \Omega$, where the decision variables are calculated as

$$x_i = xb_i \times xr_i, \quad i = 1, 2, \ldots, D. \tag{4.4}$$

That is, SparseEA stores **xb** and **xr** instead of **x** during the optimization. For instance, a binary vector $\mathbf{xb} = (0, 1, 0, 1, 0)$ and a real vector $\mathbf{xr} = (0.2, 0.4, 0.6, 0.8, 1.0)$ correspond to a solution $\mathbf{x} = (0, 0.4, 0, 0.8, 0)$. In this way, the optimization of the binary vector is responsible for the search of nonzero variables, and the optimization of the real vector is responsible for the optimization of nonzero variables. It is worth noting that although the introduction of binary vector **xb** doubles the number of variables to be optimized, all the **xb** form a binary search space that is much smaller than the continuous decision space, and all the real variables xr_i corresponding to $xb_i = 0$ need not be optimized anymore. In other words, the search space is highly reduced by optimizing many variables in **xb** to zero. Besides, the bi-level encoding parallelizes the search and optimization of nonzero variables, which achieves high efficiency in searching for sparse optimal solutions.

As a matter of fact, the effectiveness of such a bi-level encoding has been verified on specific applications such as neural network training [8] and portfolio optimization [9], while SparseEA adopts it as a generic strategy for solving various sparse LSMOPs. In SparseEA, the optimization of real vectors is implemented by general strategies used in real-coded MOEAs, while the optimization of binary vectors is implemented by customized strategies for ensuring the sparsity of generated solutions. These customized strategies include a population initialization strategy, a crossover operator, and a mutation operator.

4.2.2 Population Initialization Strategy of SparseEA

In comparison to random initialization where the sparsity (i.e., the ratio of nonzero variables) of the generated solutions is approximately 0.5, the population initialization strategy of SparseEA aims to generate a set of solutions with different sparsity, thereby better approximating the optimal solutions whose sparsity is unknown. Moreover, SparseEA estimates the probability that each dimension of the binary vector should be zero, which can benefit the search of nonzero variables.

The pseudocode of the population initialization strategy of SparseEA is presented in Algorithm 4.1. To begin with, a solution set S with D solutions is generated for calculating the scores (Line 1). For the binary vector of the ith solution in S, all of its binary variables xb are set to zero (Line 3), except for the ith binary variable being one (Line 4). For the real vector of the ith solution in S, it is randomly sampled in the decision space for continuous sparse LSMOPs (Line 6) and set to a vector of ones for binary sparse LSMOPs (Line 8). That is, the real variables of all the solutions are fixed to one when solving binary sparse LSMOPs, while only the binary variables need to be optimized. Then, as illustrated in Fig. 4.1, the quality of all the solutions in S are measured by non-dominated sorting, where the non-dominated front numbers of all the solutions constitute a

Algorithm 4.1: *Initialization(N)*

```
Input: N (population size)
Output: P (initial population), score (scores of decision variables)
   //Calculate the scores of variables
1  S ← ∅; //A population for calculating the scores
2  for i = 1, ..., D do
3      xb ← A 1 × D vector of zeros; //D is the number of variables
4      xb_i ← 1;
5      if it is a continuous optimization problem then
6          xr ← Randomly generate a 1 × D vector in the decision space;
7      else if it is a binary optimization problem then
8          xr ← A 1 × D vector of ones;
9      S ← S ∪ {{xb, xr}};
10 score ← NondominatedSorting(S); //Algorithm 1.2
   //Generate the initial population
11 P ← ∅; //The initial population
12 for i = 1, ..., N do
13     xb ← A 1 × D vector of zeros;
14     for j = 1, ..., ⌈rand × D⌉ do
15         [m, n] ← Randomly select two dimensions;
16         if score_m < score_n then
17             xb_m ← 1;
18         else
19             xb_n ← 1;
20     if it is a continuous optimization problem then
21         xr ← Randomly generate a 1 × D vector in the decision space;
22     else if it is a binary optimization problem then
23         xr ← A 1 × D vector of ones;
24     P ← P ∪ {{xb, xr}};
25 return P and score;
```

vector **score** (Line 10). Obviously, a smaller non-dominated front number $score_i$ indicates a better quality of the solution whose ith decision variable is nonzero, which implies a lower probability that the ith dimension of the binary vector should be zero. Therefore, it is reasonable to use the vector **score** to guide the generation of the binary vectors of solutions, where a binary variable with a

A	0.3	0	0	0	0	0	0	0	$score_1 = 1$
B	0	0.5	0	0	0	0	0	0	$score_2 = 1$
C	0	0	0.9	0	0	0	0	0	$score_3 = 1$
D	0	0	0	0.1	0	0	0	0	$score_4 = 2$
E	0	0	0	0	0.3	0	0	0	$score_5 = 2$
F	0	0	0	0	0	0.6	0	0	$score_6 = 2$
G	0	0	0	0	0	0	0.2	0	$score_7 = 2$
H	0	0	0	0	0	0	0	0.8	$score_8 = 3$

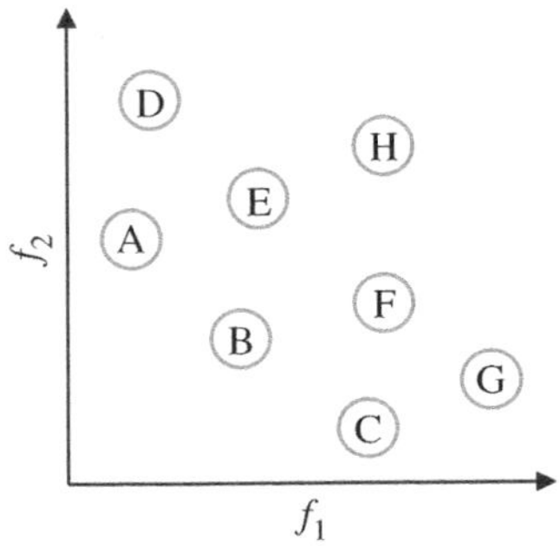

Figure 4.1 The strategy for calculating vector **score** in SparseEA.

lower *score* value is more likely to be set to one, and vice versa. In the population initialization, crossover operator, and mutation operator of SparseEA, the binary variables to be flipped are selected by binary tournament selection according to the vector **score**.

Afterward, a population P with N solutions is initialized by generating the solutions one by one (Line 12). The generation of a new solution begins with setting its binary vector to a vector of zeros (Line 13). Then, two dimensions are randomly selected (Line 15) and the one corresponding to a lower *score* value is flipped to one (Lines 16–19). This operation repeats for $\lceil rand \times D \rceil$ times to diversify the sparsity of the generated solutions (Line 14), where *rand* denotes a uniformly distributed random number within $[0, 1]$. As for the real vector of the new solution, it is randomly sampled in the decision space for continuous sparse LSMOPs (Line 21) and set to a vector of ones for binary sparse LSMOPs (Line 23). Lastly, both the initial population P and the vector **score** are returned (Line 25).

4.2.3 Crossover and Mutation Operators of SparseEA

In SparseEA, the real vectors of offspring solutions are generated through simulated binary crossover [10] and polynomial mutation [11], which are the same as most multi-objective genetic algorithms. On the contrary, the binary vectors of offspring solutions are generated through new crossover and mutation operators, which are designed on the basis of uniform crossover and bitwise mutation, respectively. Similar to the population initialization strategy of SparseEA, the vector **score** is additionally involved in the new crossover and mutation operators to ensure the sparsity of offspring. The pseudocode of the crossover and mutation operators of SparseEA is presented in Algorithm 4.2, where the binary vectors $\mathbf{xb}, \mathbf{yb}$ of two parents are used to generate the binary vector $\mathbf{ob}$ of one offspring solution each time, and the real vectors $\mathbf{xr}, \mathbf{yr}$ of the same parents are used to generate the real vector $\mathbf{or}$ of the same offspring solution (Line 3).

Algorithm 4.2: *Variation*(P', **score**)

Input: P' (a set of parents), **score** (scores of decision variables)
Output: O (offspring population)

```
O ← ∅;
while P' ≠ ∅ do
    [{xb, xr}, {yb, yr}] ← Randomly select two parents from P';
    P' ← P' \ {{xb, xr}, {yb, yr}};
    //Generate the binary vector ob of the offspring solution
    ob ← xb;
    //Crossover
    if rand < 0.5 then
        [m, n] ← Randomly select two dimensions from the nonzero elements in xb ∩ ¬yb;
        if score_m > score_n then
            ob_m ← 0;
        else
            ob_n ← 0;
    else
        [m, n] ← Randomly select two dimensions from the nonzero elements in ¬xb ∩ yb;
        if score_m < score_n then
            ob_m ← 1;
        else
            ob_n ← 1;
    //Mutation
    if rand < 0.5 then
        [m, n] ← Randomly select two dimensions from the nonzero elements in ob;
        if score_m > score_n then
            ob_m ← 0;
        else
            ob_n ← 0;
    else
        [m, n] ← Randomly select two dimensions from the nonzero elements in ¬ob;
        if score_m < score_n then
            ob_m ← 1;
        else
            ob_n ← 1;
    //Generate the real vector or of the offspring solution
    if it is a continuous optimization problem then
        or ← Perform simulated binary crossover and polynomial mutation based on xr and
        yr;
    else if it is a binary optimization problem then
        or ← A 1 × D vector of ones;
    O ← O ∪ {{ob, or}};
return O;
```

First, the binary vector **ob** of the offspring solution is set to the same as the first parent (Line 5), and one of the following two operations is performed with the same probability: randomly selecting two dimensions from the nonzero elements in $\mathbf{xb} \cap \overline{\mathbf{yb}}$ (Line 7), and flipping the binary variable corresponding to a higher *score* value to zero (Lines 8–11); or randomly selecting two dimensions from the nonzero elements in $\overline{\mathbf{xb}} \cap \mathbf{yb}$ (Line 13), and flipping the binary variable corresponding to a lower *score* value to one (Lines 14–17). In short, the new crossover operator flips a zero variable or a nonzero variable with the same probability, hence, the $E_{nonzero}$ and E_{zero} of each solution keep unchanged, and the sparsity of offspring solutions can be ensured.

Second, the binary vector **ob** of the offspring solution is then mutated by one of the following two operations with the same probability: randomly selecting two dimensions from the nonzero elements in **ob** (Line 19), and flipping the binary variable corresponding to a higher *score* value to zero (Lines 20–23); or randomly selecting two dimensions from the nonzero elements in $\overline{\mathbf{ob}}$ (Line 25), and flipping the binary variable corresponding to a lower *score* value to one (Lines 26–29). Obviously, the new mutation operator can also ensure the sparsity since it flips a zero variable or a nonzero variable of each offspring solution with the same probability.

Third, when solving continuous sparse LSMOPs, the real vector **or** of the offspring solution is generated by performing simulated binary crossover and polynomial mutation on **xr** and **yr** (Line 31). When solving binary sparse LSMOPs, the real vector **or** is always set to a vector of ones without performing crossover or mutation (Line 33). Lastly, the offspring population is returned (Line 35).

4.2.4 Procedure of SparseEA

With the new population initialization strategy, crossover operator, and mutation operator for generating sparse solutions, the procedure of SparseEA is detailed in Algorithm 4.3. First of all, a population P with N solutions is initialized and the vector **score** is calculated by the new population initialization strategy (Line 1). Then, the non-dominated front number and crowding distance of each solution in P are determined (Lines 2 and 3). At each generation, the mating pool P' is constructed by selecting $2N$ parents from the population P via binary tournament selection according to the non-dominated front numbers and crowding distances of solutions (Line 5). After that, N offspring solutions are generated by the new crossover and mutation operators (Line 6). To select N solutions from the combination of P and O, SparseEA adopts the environmental selection strategy of NSGA-II

Algorithm 4.3: Procedure of SparseEA

Input: N (population size)
Output: P (final population)

1 [P, **score**] ← *Initialization*(N); //Algorithm 4.1
2 *Front* ← *NondominatedSorting*(P); //Algorithm 1.2
3 *Crowd* ← *CrowdingDistance*(P, *Front*); //Algorithm 1.3
4 **while** termination criterion not fulfilled **do**
5 P' ← *TournamentSelection*(P, $2N$, *Front*, *Crowd*); //Algorithm 1.4
6 O ← *Variation*(P', **score**); //Algorithm 4.2
7 [P, *Front*, *Crowd*] ← *EnvironmentalSelection*($P \cup O$, N); //Algorithm 1.6
8 **return** P;

[6] for simplicity, where N solutions with better non-dominated front numbers and crowding distances survive to the next generation (Line 7). Finally, the population P is returned as the output of the algorithm (Line 8).

It can be concluded from Algorithm 4.1 that the calculation of **score** is relatively time-consuming for large-scale optimization due to the use of non-dominated sorting, which holds a time complexity of $O(MD^2)$ for solving a problem with M objectives and D decision variables [12]. Nevertheless, this operation is not in the main loop of the algorithm. At each generation of SparseEA, the time complexity is $O(N)$ for mating pool selection, $O(ND)$ for offspring generation, and $O(MN^2)$ for environmental selection. Therefore, the time complexity of one generation of SparseEA is $O(MN^2 + ND)$, which is the same as many other MOEAs.

4.3 Machine Learning-Assisted Evolutionary Algorithm

The multi-objective evolutionary algorithm based on Pareto optimal subspace learning (MOEA/PSL) inherits the bi-level encoding of SparseEA, where unsupervised neural networks are adopted to learn the Pareto optimal subspace according to the current population [13]. More specifically, a restricted Boltzmann machine (RBM) is used to learn a sparse distribution from the binary vectors, and a denoising auto-encoder (DAE) is used to learn a compact representation from the real vectors. By generating offspring solutions in the Pareto optimal subspace rather than in the original decision space, the search space is highly reduced and the optimization of the binary vectors and real vectors can be considerably eased.

4.3.1 Restricted Boltzmann Machine and Auto-Encoder

As the building block of deep belief network, RBM is a two-layer neural network for the dimensionality reduction of inputs with binary encoding [14]. Assuming that the weights between the input layer and hidden layer are stored in a matrix $\mathbf{w} \in \mathbb{R}^{D\times k}$, the probability that the hidden layer $\mathbf{h} \in \{0,1\}^{1\times k}$ being one can be calculated based on an input vector $\mathbf{x} \in \{0,1\}^{1\times D}$:

$$p(\mathbf{h} = \mathbf{1}|\mathbf{x}) = \sigma(\mathbf{a} + \mathbf{xw}), \tag{4.5}$$

where $\mathbf{a} \in \mathbb{R}^{1\times k}$ denotes the weights of the bias and $\sigma(x) = \frac{1}{1+e^{-x}}$ denotes the sigmoid function. After calculating the probability, the value of each element h_i can be sampled by comparing $p(h_i = 1|\mathbf{x})$ with a uniformly distributed random number within $[0,1]$:

$$h_i = p(h_i = 1|\mathbf{x}) > rand, \quad i = 1, \dots, k. \tag{4.6}$$

As for the reconstruction operation, the probability that the input layer $\mathbf{x}' \in \{0,1\}^{1\times D}$ being one can be calculated based on $\mathbf{h}$:

$$p(\mathbf{x}' = \mathbf{1}|\mathbf{h}) = \sigma(\mathbf{a}' + \mathbf{hw}^T), \tag{4.7}$$

where $\mathbf{a}' \in \mathbb{R}^{1\times D}$ denotes the weights of the bias. Similarly, the value of each element x_i' can be sampled by comparing $p(x_i = 1|\mathbf{h})$ with a uniformly distributed random number within $[0,1]$:

$$x_i' = p(x_i' = 1|\mathbf{h}) > rand, \quad i = 1, \dots, D. \tag{4.8}$$

Given a set of training samples as input vectors, an RBM can be trained by using the contrastive divergence algorithm [15] for minimizing the difference between each pair of $\mathbf{x}$ and $\mathbf{x}'$ in an unsupervised manner. Thus, the hidden layer $\mathbf{h}$ can be used as a reduced representation of $\mathbf{x}$ for $k < D$.

Auto-encoder serves as the building block of stacked auto-encoder, which is a three-layer neural network for the dimensionality of inputs with real encoding [16]. In fact, the output layer of auto-encoder is the same as the input layer, and thus the propagation from the hidden layer to the output layer is similar to the reconstruction operation of RBM. Assuming that the weights between the input layer and hidden layer are stored in a matrix $\mathbf{w} \in \mathbb{R}^{D\times k}$ and the weights between the hidden layer and output layer are stored in a matrix $\mathbf{w}' \in \mathbb{R}^{k\times D}$, the hidden layer $\mathbf{h} \in \mathbb{R}^{1\times k}$ can be calculated based on an input vector $\mathbf{x} \in \mathbb{R}^{1\times D}$:

$$\mathbf{h} = \sigma(\mathbf{a} + \mathbf{xw}), \tag{4.9}$$

where $\mathbf{a} \in \mathbb{R}^{1\times k}$ denotes the weights of the bias. Then, the output layer $\mathbf{y} \in \mathbb{R}^{1\times D}$ can be calculated based on $\mathbf{h}$:

$$\mathbf{y} = \sigma(\mathbf{a}' + \mathbf{hw}'), \tag{4.10}$$

where $\mathbf{a}' \in \mathbb{R}^{1\times D}$ denotes the weights of the bias. Given a set of training samples as input vectors, an auto-encoder can be trained by using gradient descent [17] for minimizing the difference between each pair of $\mathbf{x}$ and $\mathbf{y}$ in an unsupervised manner. Thus, the hidden layer $\mathbf{h}$ can be used as a reduced representation of $\mathbf{x}$ for $k < D$. In MOEA/PSL, the DAE rather than the original auto-encoder is used, which is a popular variant that randomly sets the elements in each input vector to zero [18].

4.3.2 Pareto Optimal Subspace Learning in MOEA/PSL

As illustrated in Fig. 4.2, since RBM can learn the probability distribution of the input obeying a binomial distribution, it is desirable to learn a sparse distribution from the binary vectors used in bi-level encoding. Analogously, since DAE can learn a compact representation of the input in a continuous space, it is desirable to reduce the dimensionality of the real vectors used in bi-level encoding. By using the promising solutions found so far as training samples, the reduced binary vectors and real vectors form a promising approximation of the Pareto optimal subspace. Therefore, the efficiency of finding sparse optimal solutions can be improved by evolving a population in the learnt subspace, even if the search strategies in general MOEAs are employed.

In comparison to other dimensionality reduction methods such as principal component analysis, the advantages of RBM and DAE lie in the fact that they can learn nonlinear interactions between real variables and interactions between binary variables. More importantly, the reduced binary vectors and real vectors can be easily recovered as shown in Fig. 4.2, which is necessary for the function evaluation of decision vectors.

The pseudocode of the Pareto optimal subspace learning-based offspring generation is presented in Algorithm 4.4. First of all, the binary vectors of the

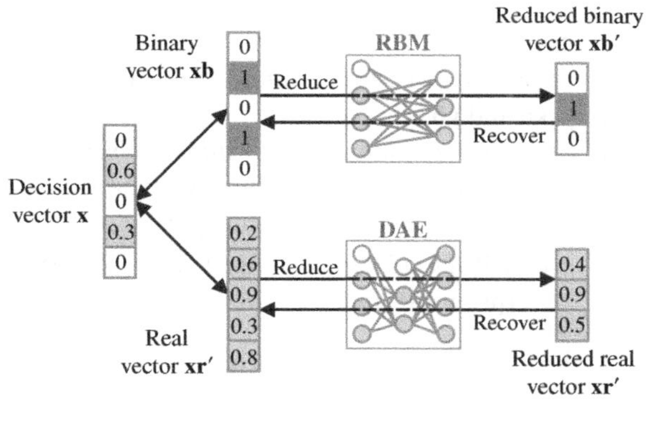

Figure 4.2 The unsupervised neural network-based dimensionality reduction strategy in MOEA/PSL.

Algorithm 4.4: *Variation*(P, P', ρ, k)

Input: P (current population), P' (a set of parents), ρ (ratio of solutions generated in the Pareto optimal subspace), k (size of hidden layers)

Output: O (offspring population)

```
//Model training
NB ← Binary vectors of all the non-dominated solutions in P;
Train an RBM with k hidden neurons based on NB;
if it is a continuous optimization problem then
    NR ← Real vectors of all the non-dominated solutions in P;
    Train a DAE with k hidden neurons based on NR;
//Offspring generation
O ← ∅;
while P' ≠ ∅ do
    [{xb, xr}, {yb, yr}] ← Randomly select two parents from P';
    P' ← P' \ {{xb, xr}, {yb, yr}};
    if ρ > rand then
        [xb', yb'] ← Reduce xb and yb by Eqs. (4.5) and (4.6);
        [pb', qb'] ← Perform uniform crossover and bitwise mutation based on
            xb' and yb';
        [pb, qb] ← Recover pb' and qb' by Eqs. (4.7) and (4.8);
        if it is a continuous optimization problem then
            [xr', yr'] ← Reduce xr and yr by Eq. (4.9);
            [pr', qr'] ← Perform simulated binary crossover and polynomial
                mutation based on xr' and yr';
            [pr, qr] ← Recover pr' and qr' by Eq. (4.10);
        else if it is a binary optimization problem then
            [pr, qr] ← Two 1 × D vectors of ones;
    else
        [pb, qb] ← Perform uniform crossover and bitwise mutation based on
            xb and yb;
        if it is a continuous optimization problem then
            [pr, qr] ← Perform simulated binary crossover and polynomial
                mutation based on xr and yr;
        else if it is a binary optimization problem then
            [pr, qr] ← Two 1 × D vectors of ones;
    O ← O ∪ {{pb, pr}, {qb, qr}};
return O;
```

non-dominated solutions in the current population are used to train an RBM (Lines 1 and 2), and the real vectors of the non-dominated solutions in the current population are used to train a DAE (Lines 4 and 5). Note that the training of DAE can be omitted when solving binary sparse LSMOPs. Afterwards, the binary vectors $\mathbf{xb}, \mathbf{yb}$ of two parents are used to generate the binary vectors $\mathbf{pb}, \mathbf{qb}$ of two offspring solutions each time, and the real vectors $\mathbf{xr}, \mathbf{yr}$ of the same parents are used to generate the real vectors $\mathbf{pr}, \mathbf{qr}$ of the same offspring solutions (Line 8). The generation of binary vectors is implemented by uniform crossover and bit-wise mutation, and the generation of real vectors is implemented by simulated binary crossover [10] and polynomial mutation [11]. Besides, the probability of generating offspring in the learnt subspace is controlled by a parameter ρ.

When generating two offspring solutions, if $\rho > rand$, the two $1 \times D$ binary vectors $\mathbf{xb}, \mathbf{yb}$ are first reduced to two $1 \times k$ binary vectors $\mathbf{xb}', \mathbf{yb}'$ via RBM (Line 11), then two $1 \times k$ binary vectors $\mathbf{pb}', \mathbf{qb}'$ are generated by uniform crossover and bitwise mutation (Line 12), and they are further recovered via RBM to obtain two $1 \times D$ binary vectors $\mathbf{pb}, \mathbf{qb}$ (Line 13). Also, when solving continuous sparse LSMOPs, the two $1 \times D$ real vectors $\mathbf{xr}, \mathbf{yr}$ are first reduced to two $1 \times k$ real vectors $\mathbf{xr}', \mathbf{yr}'$ via DAE (Line 15), then two $1 \times k$ real vectors $\mathbf{pr}', \mathbf{qr}'$ are generated by simulated binary crossover and polynomial mutation (Line 16), and they are further recovered via DAE to obtain two $1 \times D$ real vectors $\mathbf{pr}, \mathbf{qr}$ (Line 17). When solving binary sparse LSMOPs, the real vectors $\mathbf{pr}, \mathbf{qr}$ of offspring solutions are directly set to $1 \times D$ vectors of ones (Line 19). By contrast, if $\rho > rand$, the binary vectors $\mathbf{xb}, \mathbf{yb}$ are directly used to generate $\mathbf{pb}, \mathbf{qb}$ (Line 21), and the real vectors $\mathbf{xr}, \mathbf{yr}$ are directly used to generate $\mathbf{pr}, \mathbf{qr}$ (Line 23) when solving continuous sparse LSMOPs.

4.3.3 Parameter Adaptation in MOEA/PSL

As can be observed from the above procedure, the probability of generating offspring in the learnt subspace ρ and the hidden layer size k of the neural networks are given in advance. Intuitively, a larger ρ generates more offspring solutions in the learnt subspace, but also leads to poorer population diversity. A smaller k learns a smaller subspace for higher convergence speed, but also increases the possibility of trapping into local optimums. In order to strike a good balance between greed and randomness, the two parameters should be carefully set. Instead of predefining the parameters by users, MOEA/PSL suggests a parameter adaptation strategy to dynamically estimate the values of ρ and k according to the current population.

The pseudocode of the parameter adaptation in MOEA/PSL is described in Algorithm 4.5. At the end of each generation, the number of offspring solutions generated in the learnt subspace that successfully survives to the next generation s_1 and the number of offspring solutions generated in the original decision space

Algorithm 4.5: *ParameterAdaptation*(P, ρ)

Input: P (current population), ρ (ratio of offspring solutions generated in the Pareto optimal subspace)

Output: ρ (ratio of offspring solutions generated in the Pareto optimal subspace), k (size of hidden layers)

`//Update the parameter` ρ

1 $s_1 \leftarrow$ Number of successful offspring solutions generated in the learnt subspace at the current generation;

2 $s_2 \leftarrow$ Number of successful offspring solutions generated in the original decision space at the current generation;

3 Update ρ by Eq. (4.11);

`//Determine the parameter` k

4 $NP \leftarrow$ All the non-dominated solutions in P;

5 Calculate **dec** by Eq. (4.12);

6 Calculate k by Eq. (4.14);

7 **return** ρ and k;

---

that successfully survives to the next generation s_2 are determined (Lines 1 and 2), and the parameter $\rho \in [0, 1]$ is updated (Line 3):

$$\rho = 0.5 \times \rho + 0.5 \times \frac{s_1}{s_1 + s_2}, \tag{4.11}$$

where the initial value of ρ is set to 0.5. Obviously, a larger s_1 indicates higher performance of generating offspring in the learnt subspace, hence ρ is increased to generate more offspring solutions in the learnt subspace at the next generation. By contrast, a larger s_2 indicates higher performance of generating offspring in the original decision space, hence ρ is decreased to generate more offspring solutions in the original decision space at the next generation. Then, the non-dominated solution set NP in the current population is identified (Line 4), and the probability that the decision vector **dec** being one can be estimated based on NP (Line 5):

$$p(\mathbf{dec} = \mathbf{1}|NP) = \frac{1}{|NP|} \sum_{\mathbf{x} \in NP} |\text{sign}(\mathbf{x})|, \tag{4.12}$$

where $|\text{sign}(x_i)| = 0$ if $x_i = 0$ and 1 otherwise. Afterwards, the value of each element dec_i can be sampled by comparing $p(dec_i = 1|NP)$ with a uniformly distributed random number within $[0, 1]$:

$$dec_i = p(dec_i = 1|NP) > rand, \quad i = 1, \dots, D, \tag{4.13}$$

and the parameter k can be calculated (Line 6):

$$k = \sum_{i=1}^{D} dec_i. \tag{4.14}$$

This way, k approximates the number of nonzero variables in the decision vector, which is reasonable to be used as the hidden layer size to eliminate the zero variables from the Pareto optimal subspace.

4.3.4 Procedure of MOEA/PSL

With the Pareto optimal subspace learning strategy and parameter adaptation strategy, the procedure of MOEA/PSL is described in Algorithm 4.6. Firstly, a population P with N solutions is randomly initialized (Line 1), and the non-dominated front number and crowding distance of each solution in P are determined (Lines 2 and 3). Then, the parameters ρ and k are initialized (Lines 4 and 5). At each generation, the mating pool P' is constructed by selecting N parents from the population P via binary tournament selection according to the non-dominated front numbers and crowding distances of solutions (Line 7). After that, N offspring solutions are generated by the unsupervised neural network-based Pareto optimal subspace learning strategy (Line 8). To select N solutions from the combination of P and O, MOEA/PSL adopts the same environmental selection strategy as SparseEA, where N solutions with better non-dominated front numbers and crowding distances survive to the next generation (Line 9). Furthermore, the parameters ρ and k are updated by the parameter adaptation strategy (Line 10). Finally, the population P is returned as the output of the algorithm (Line 11).

Algorithm 4.6: Procedure of MOEA/PSL

Input: N (population size)
Output: P (final population)

1 $P \leftarrow$ Randomly generate N solutions;
2 $Front \leftarrow NondominatedSorting(P)$; `//Algorithm 1.2`
3 $Crowd \leftarrow CrowdingDistance(P, Front)$; `//Algorithm 1.3`
4 $\rho \leftarrow 0.5$;
5 $k \leftarrow N$;
6 **while** termination criterion not fulfilled **do**
7     $P' \leftarrow TournamentSelection(P, N, Front, -Crowd)$; `//Algorithm 1.4`
8     $O \leftarrow Variation(P, P', \rho, k)$; `//Algorithm 4.4`
9     $[P, Front, Crowd] \leftarrow EnvironmentalSelection(P \cup O, N)$; `//Algorithm 1.6`
10     $[\rho, K] \leftarrow ParameterAdaptation(P, \rho)$; `//Algorithm 4.5`
11 **return** P;

At each generation of MOEA/PSL, the time complexity is $O(NEDk)$ for neural network training, $O(NDk)$ for offspring generation, $O(MN^2)$ for environmental selection, and $O(ND)$ for parameter adaptation, where E denotes the number of epochs for training, D denotes the number of decision variables, and M denotes the number of objectives. Therefore, the time complexity of one generation of MOEA/PSL is $O(MN^2 + NEDk)$.

4.4 Data Mining-Assisted Evolutionary Algorithm

The pattern mining-based multi-objective evolutionary algorithm (PM-MOEA) also focuses on the dimensionality reduction of the decision spaces of sparse LSMOPs [19]. In contrast to MOEA/PSL using unsupervised neural networks, PM-MOEA suggests a novel evolutionary frequent pattern mining method to mine the sparse distribution from the binary vectors, which is quite different and more flexible than machine learning methods. Furthermore, PM-MOEA suggests a binary crossover operator and a binary mutation operator for ensuring the sparsity of offspring solutions. In comparison to the binary operators of SparseEA flipping a single variable, the expectation of the number of flipped variables is controllable in PM-MOEA and thus is more flexible to be used.

4.4.1 Evolutionary Frequent Pattern Mining

As introduced in Section 4.1.1, frequent pattern mining aims to find the optimal patterns from a transaction dataset [3]. In addition to considering it as a sparse LSMOP, it is also an effective tool for dimensionality reduction. In particular for solving sparse LSMOPs, the current population can be regarded as a transaction dataset with each decision vector being a transaction, and the set of nonzero variables is the optimal pattern to be mined. This way, the set of nonzero variables can be estimated through frequent pattern mining techniques to guide the offspring generation in later generations, and the dimensionality of the binary search space is naturally reduced.

According to Eq. (2.28), the optimal patterns are the most frequent and complete item sets of the transaction dataset. For solving sparse LSMOPs, the optimal patterns are the minimum estimations of the nonzero variable set, in which each binary variable is probably set to one. In addition, PM-MOEA also aims to obtain the maximum estimations of the nonzero variable set, in which each binary variable is possibly set to one. More specifically, when generating offspring solutions, the binary variables out of the maximum estimation are always set to zero, the binary variables in the maximum estimation are determined by parents, and the binary variables in the minimum estimation are always set to one. For this aim, PM-MOEA suggests two objectives for mining the maximum estimations:

$$\text{Maximize} f_1(\mathbf{xb}) = \frac{|NB'_{\mathbf{xb}}|}{|NB|}$$
$$f_2(\mathbf{xb}) = \frac{1}{|NB'_{\mathbf{xb}}|} \sum_{\mathbf{b} \in NB'_{\mathbf{xb}}} \frac{\sum \mathbf{b}}{\sum \mathbf{xb}} \tag{4.15}$$

and two objectives for mining the minimum estimations:

$$\text{Maximize} f_3(\mathbf{xb}) = \frac{|NB_{\mathbf{xb}}|}{|NB|},$$
$$f_4(\mathbf{xb}) = \frac{1}{|NB_{\mathbf{xb}}|} \sum_{\mathbf{b} \in NB_{\mathbf{xb}}} \frac{\sum \mathbf{xb}}{\sum \mathbf{b}}, \tag{4.16}$$

where $\mathbf{xb} \in \{0, 1\}^D$ is an estimation of the nonzero variable set, *NB* is the set of binary vectors of all the non-dominated solutions in the current population, $NB'_{\mathbf{xb}} = \{\mathbf{b} \in NB | b_i \leq xb_i, i = 1, \dots, D\}$ denotes the set of binary vectors whose nonzero variables form a subset of the nonzero variables in **xb**, and $NB_{\mathbf{xb}} = \{\mathbf{b} \in NB | b_i \geq xb_i, i = 1, \dots, D\}$ denotes the set of binary vectors whose nonzero variables form a superset of the nonzero variables in **xb**. Table 4.1 presents an example to illustrate the calculation of the four objectives.

It can be found that the objectives f_3 and f_4 are the same as f_1 and f_2 in Eq. (2.28), where f_3 aims to find a minimum candidate set being an intersection set of the nonzero variables in all the binary vectors, and f_4 makes the minimum candidate set as large as possible. On the other hand, the objectives f_1 are f_2 are novel, where f_1 aims to find a maximum candidate set being a union set of the nonzero variables in all the binary vectors, and f_2 makes the maximum candidate set as small as possible. At the beginning of the optimization procedure, the maximum candidate

Table 4.1 Example of Calculating the Four Objectives in the Evolutionary Frequent Pattern Mining Method.

$NB = \{\mathbf{b}_1, \mathbf{b}_2, \mathbf{b}_3, \mathbf{b}_4, \mathbf{b}_5, \mathbf{b}_6\}$
$\mathbf{b}_1 = \{1111110\}, \mathbf{b}_2 = \{1110001\}$
$\mathbf{b}_3 = \{0100001\}, \mathbf{b}_4 = \{1110000\}$
$\mathbf{b}_5 = \{1101110\}, \mathbf{b}_6 = \{1110000\}$
$\mathbf{xb} = \{1110001\}, NB'_{\mathbf{xb}} = \{\mathbf{b}_2, \mathbf{b}_3, \mathbf{b}_4, \mathbf{b}_6\}$
$f_1(\mathbf{xb}) = \frac{4}{6}, f_2(\mathbf{xb}) = \frac{1}{4} \times (\frac{4}{4} + \frac{2}{4} + \frac{3}{4} + \frac{3}{4})$
$\mathbf{yb} = \{1110000\}, NB_{\mathbf{yb}} = \{\mathbf{b}_1, \mathbf{b}_2, \mathbf{b}_4, \mathbf{b}_6\}$
$f_3(\mathbf{yb}) = \frac{4}{6}, f_4(\mathbf{yb}) = \frac{1}{4} \times (\frac{3}{6} + \frac{3}{4} + \frac{3}{3} + \frac{3}{3})$

set is likely to contain most variables and the minimum candidate set is likely to contain few variables. With the evolution of the population, the maximum candidate set will be shortened and the minimum candidate set will be enlarged, and the nonzero variable set can be exactly found when the maximum candidate set equals the minimum candidate set. In comparison to machine learning-based dimensionality reduction methods, frequent pattern mining has three main advantages: first, it mines the maximum and minimum candidate sets of nonzero variables rather than an exact estimation of nonzero variables, where the former are less greedy and can better tackle the inaccuracy of the binary vectors in the current population (since the current population is not optimal). Second, it provides multiple maximum candidate sets and minimum candidate sets, which can make diverse trade-offs between f_1 and f_2 or f_3 and f_4, thus preserving the population diversity in the decision space. Third, it does not contain any hyperparameter that needs to be tuned (e.g., the hidden layer size of RBM and DAE), which is flexible to be used since the ratio of nonzero variables in the optimal solutions is unknown.

It is obvious that the mining of maximum candidate sets and the mining of minimum candidate sets are also sparse LSMOPs, which have the same number of binary variables as the original sparse LSMOP. Nevertheless, the former are computationally cheap since they are pursued based on the binary vectors of the current population, regardless of the objectives or dataset of the original sparse LSMOP. While many practical frequent pattern mining methods have been proposed [3, 5, 20], they are inefficient for finding diverse patterns. Thus, PM-MOEA suggests a fast evolutionary frequent pattern mining method to mine the maximum and minimum candidate sets of nonzero variables, which is actually the same as PM-MOEA itself. Although the evolutionary frequent pattern mining method is adopted twice at each generation, it does not consume many computational resources since it uses a few function evaluations to evolve the population in an incremental manner, i.e., the final population in the previous mining operation is used as the initial population in the next mining operation.

4.4.2 Mining Candidate Sets in PM-MOEA

As illustrated in Fig. 4.3, the whole binary vector is divided into four parts. First, the variables being zero in all the binary vectors are directly fixed to zero and ignored in the mining operation. After obtaining the maximum and minimum candidate sets, the variables out of the maximum candidate set are also fixed to zero, the variables in the maximum candidate set may be one or zero, and the variables in the minimum candidate set are fixed to one. When generating an offspring solution, the variables in the maximum candidate set but out of the minimum candidate set are determined by its two parents.

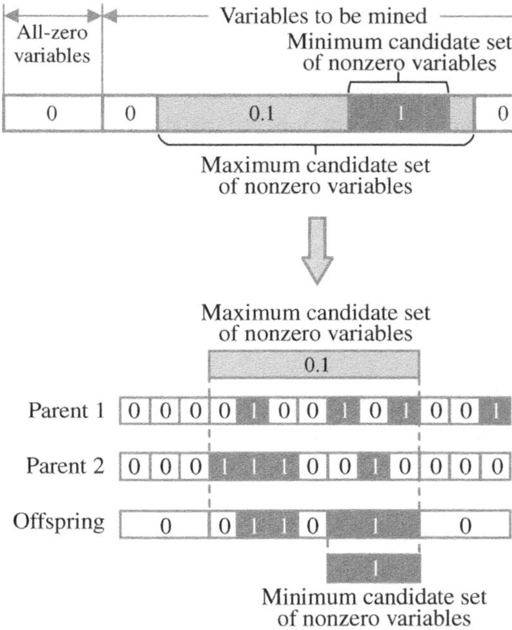

Figure 4.3 The evolutionary frequent pattern mining method in PM-MOEA.

The pseudocode of mining maximum candidate sets is presented in Algorithm 4.7, where the maximum candidate sets Q_{max} are initialized to those obtained in the previous mining operation. First of all, the binary vectors of the non-dominated solutions in the current population are stored in NB (Line 1), and the dimensions that are zero in all the binary vectors in NB are eliminated from NB (Lines 2 and 3) and Q_{max} (Lines 4 and 5). To initialize N' more maximum candidate sets according to the current NB, half the binary vectors are selected from NB (Line 7) each time, and the dimensions that are nonzero in at least one of the selected binary vectors constitute a new initial maximum candidate set (Lines 8 and 9). After evaluating all the maximum candidate sets by Eq. (4.15) based on NB (Line 10), the evolutionary frequent pattern mining method iteratively evolves them by using the selection strategies of NSGA-II (Lines 13 and 16) and the binary genetic operators suggested in PM-MOEA (Line 14). For the sake of efficiency, the number of candidate sets N' is set to 20 and the number of iterations is set to 10.

The pseudocode of mining minimum candidate sets is presented in Algorithm 4.8, which is very similar to the pseudocode of mining maximum candidate sets. First of all, the binary vectors of the non-dominated solutions in

Algorithm 4.7: *MiningMaxSets*(P, Q_{max}, N')

Input: P (current population), Q_{max} (maximum candidate sets of nonzero variables), N' (population size for evolutionary pattern mining)
Output: Q_{max} (new maximum candidate sets of nonzero variables)

1 $NB \leftarrow$ Binary vectors of all the non-dominated solutions in P;
2 $Zero \leftarrow$ Set of dimensions that are zero in all the binary vectors in NB;
3 $NB \leftarrow$ Delete the dimensions in $Zero$ from NB;
4 **if** $Q_{max} \neq \emptyset$ **then**
5 　$Q_{max} \leftarrow$ Delete the dimensions in $Zero$ from Q_{max};
6 **for** $i = 1, \dots, N'$ **do**
7 　$NB' \leftarrow$ Randomly select half the binary vectors from NB;
8 　$\mathbf{p} \leftarrow$ Set of dimensions that are nonzero in at least one binary vector in NB';
9 　$Q_{max} \leftarrow Q_{max} \cup \{\mathbf{p}\}$;
10 Evaluate Q_{max} by Eq. (4.15) based on NB;
11 $Front \leftarrow$ *NondominatedSorting*(Q_{max}); `//Algorithm 1.2`
12 **for** $generation = 1, \dots, 10$ **do**
13 　$Q'_{max} \leftarrow$ *TournamentSelection*($Q_{max}, 2N', Front, \emptyset$); `//Algorithm 1.4`
14 　$Q''_{max} \leftarrow$ *BinaryOperators*(Q'_{max}); `//Algorithm 4.10`
15 　Evaluate Q''_{max} by Eq. (4.15) based on NB;
16 　$[Q_{max}, Front] \leftarrow$ *EnvironmentalSelection*($Q_{max} \cup Q''_{max}, N'$); `//Algorithm 1.6`
17 $Q_{max} \leftarrow$ Add the dimensions in $Zero$ to Q_{max};
18 **return** Q_{max};

the current population are stored in *NB* (Line 1), and the dimensions that are zero in all the binary vectors in *NB* are eliminated from *NB* (Lines 2 and 3) and Q_{min} (Lines 4 and 5). To initialize N' more minimum candidate sets according to the current *NB*, half the binary vectors are selected from *NB* (Line 7) each time, and the dimensions that are nonzero in all the selected binary vectors constitute a new initial minimum candidate set (Lines 8 and 9). After evaluating all the minimum candidate sets by Eq. (4.16) based on *NB* (Line 10), the evolutionary frequent pattern mining method iteratively evolves them by using the selection strategies of NSGA-II (Lines 13 and 16) and the binary genetic operators suggested in PM-MOEA (Line 14).

Algorithm 4.8: *MiningMinSets*(P, Q_{min}, N')

Input: P (current population), Q_{min} (minimum candidate sets of nonzero variables), N' (population size for evolutionary pattern mining)
Output: Q_{min} (new minimum candidate sets of nonzero variables)

$NB \leftarrow$ Binary vectors of all the non-dominated solutions in P;
$Zero \leftarrow$ Set of dimensions that are zero in all the binary vectors in NB;
$NB \leftarrow$ Delete the dimensions in $Zero$ from NB;
if $Q_{min} \neq \emptyset$ **then**
  $Q_{min} \leftarrow$ Delete the dimensions in $Zero$ from Q_{min};
for $i = 1, \dots, N'$ **do**
  $NB' \leftarrow$ Randomly select half the binary vectors from NB;
  $\mathbf{q} \leftarrow$ Set of dimensions that are nonzero in all the binary vectors in NB';
  $Q_{min} \leftarrow Q_{min} \cup \{\mathbf{q}\}$;
Evaluate Q_{min} by Eq. (4.16) based on NB;
$Front \leftarrow NondominatedSorting(Q_{min})$; `//Algorithm 1.2`
for $generation = 1, \dots, 10$ **do**
  $Q'_{min} \leftarrow TournamentSelection(Q_{min}, 2N', Front, \emptyset)$; `//Algorithm 1.4`
  $Q''_{min} \leftarrow BinaryOperators(Q'_{min})$; `//Algorithm 4.10`
  Evaluate Q''_{min} by Eq. (4.16) based on NB;
  $[Q_{min}, Front] \leftarrow EnvironmentalSelection(Q_{min} \cup Q''_{min}, N')$; `//Algorithm 1.6`
$Q_{min} \leftarrow$ Add the dimensions in $Zero$ to Q_{min};
return Q_{min};

4.4.3 Binary Genetic Operators of PM-MOEA

After obtaining the maximum and minimum candidate sets by the evolutionary frequent pattern mining method, PM-MOEA can generate the binary vectors of offspring solutions in the reduced binary search spaces. Different from MOEA/PSL that generates parts of the offspring solutions in the original decision space for alleviating prematurity, PM-MOEA generates all the offspring solutions within the dimensions determined by the maximum candidate sets, since the evolutionary frequent pattern mining method provides multiple maximum and minimum candidate sets with good diversity, and one maximum candidate set and one minimum candidate set are randomly selected each time.

Algorithm 4.9: *Variation*(P', Q_{max}, Q_{min})

Input: P' (a set of parents), Q_{max} (maximum candidate sets of nonzero variables), Q_{min} (minimum candidate sets of nonzero variables)
Output: O (offspring population)

1 $O \leftarrow \emptyset$;
2 **while** $P' \neq \emptyset$ **do**
    `//Randomly select a maximum set and a minimum set`
3   **if** *no set in* Q_{min} *is the subset of any set in* Q_{max} **then**
4         $\mathbf{p} \leftarrow$ Randomly select a set from Q_{max};
5         $\mathbf{q} \leftarrow \emptyset$;
6   **else**
7         $Q'_{min} = \emptyset$;
8   **while** $Q'_{min} = \emptyset$ **do**
9             $\mathbf{p} \leftarrow$ Randomly select a set from Q_{max};
10             $Q'_{min} \leftarrow \{\mathbf{q} \in Q_{min} | \mathbf{q} \subseteq \mathbf{p}\}$;
11         $\mathbf{q} \leftarrow$ Randomly select a set from Q'_{min};
    `//Offspring generation`
12     $[\{\mathbf{xb}, \mathbf{xr}\}, \{\mathbf{yb}, \mathbf{yr}\}] \leftarrow$ Randomly select two parents from P';
13     $P' \leftarrow P' \setminus \{\{\mathbf{xb}, \mathbf{xr}\}, \{\mathbf{yb}, \mathbf{yr}\}\}$;
14     $\mathbf{ob} \leftarrow$ A $1 \times D$ vector of zeros;
15     $\mathbf{ob}_{\mathbf{p}} \leftarrow$ *BinaryOperators*($\{\mathbf{xb}_{\mathbf{p}}, \mathbf{yb}_{\mathbf{p}}\}$); `//Algorithm 4.10`
16     $\mathbf{ob}_{\mathbf{q}} \leftarrow 1$; `//`$\mathbf{ob}_{\mathbf{q}}$ `denotes the binary variables in dimensions` $\mathbf{q}$ `of` $\mathbf{ob}$
17   **if** *it is a continuous optimization problem* **then**
18         $\mathbf{or} \leftarrow$ Perform simulated binary crossover and polynomial mutation based on $\mathbf{xr}$ and $\mathbf{yr}$;
19   **else if** *it is a binary optimization problem* **then**
20         $\mathbf{or} \leftarrow$ A $1 \times D$ vector of ones;
21     $O \leftarrow O \cup \{\{\mathbf{ob}, \mathbf{or}\}\}$;
22 **return** O;

The pseudocode of offspring generation in PM-MOEA is described in Algorithm 4.9, where two parents are used to generate one offspring solution each time. Before offspring generation, a maximum candidate set $\mathbf{p}$ is randomly selected from Q_{max} (Line 9) and a minimum candidate set $\mathbf{q}$ is randomly selected from Q_{min} (Line 11), where $\mathbf{q}$ should be a subset of $\mathbf{p}$ (Line 10). While if there does not exist any $\mathbf{q} \in Q_{min}$ being a subset of any $\mathbf{p} \in Q_{max}$ (Line 3), $\mathbf{p}$ is randomly selected (Line 4) while $\mathbf{q}$ is set to empty (Line 5). Then, the binary vectors $\mathbf{xb}, \mathbf{yb}$ of two parents are used to generate the binary vector $\mathbf{ob}$ of one offspring solution, and the real vectors $\mathbf{xr}, \mathbf{yr}$ of two parents are used to generate the real vector $\mathbf{or}$ of one offspring solution (Line 12). The generation of binary vectors is implemented by the binary genetic operators suggested in PM-MOEA (Line 15), and the generation of real vectors is implemented by simulated binary crossover [10] and polynomial mutation [11] (Line 18). Besides, the binary variables out of $\mathbf{p}$ are fixed to zero (Line 14) and the binary variables in $\mathbf{q}$ are fixed to one (Line 16).

The binary genetic operators of PM-MOEA aim to **(a)** equate the expectation of the number of flipped nonzero variables with the expectation of the number of flipped zero variables, and **(b)** keep the expectation of the total number of flipped variables unchanged. By doing so, the generated binary vectors are not expected to have the same number of zeros and ones, and the sparsity of offspring solutions can be ensured. Assuming that a binary vector having d nonzero variables and $(D-d)$ zero variables is flipped with a probability of p on each dimension, then $p_1 n = p_0(D-n)$ holds for condition **(a)** and $p_1 n + p_0(D-n) = pD$ holds for condition **(b)**, where p_1 denotes the probability of flipping each nonzero variable and p_0 denotes the probability of flipping each zero variable. Thus,

$$
\begin{aligned}
p_1 &= \frac{pD}{2n}, \\
p_0 &= \frac{pD}{2(D-n)}.
\end{aligned}
\tag{4.17}
$$

Besides, p is set to $2n/D$ if it is larger than this value, otherwise p_1 will be larger than 1. The pseudocode of the binary genetic operators is described in Algorithm 4.10. For the crossover operator, the offspring $\mathbf{o}$ is first set to the same as one parent $\mathbf{x}$ (Line 5), and the dimensions that are different in two parents $\mathbf{x}, \mathbf{y}$ are stored in *diff* (Line 6). Then, the probability of flipping each variable in dimensions *diff* is calculated by Eq. (4.17) (Line 7), and the binary variables in dimensions *diff* of $\mathbf{o}$ are flipped accordingly (Line 8). Here, the number of variables D in Eq. (4.17) equals $|diff|$, n counts the number of nonzero variables in dimensions *diff* of $\mathbf{o}$, and p is set to 0.5. For the mutation operator, the probability of flipping each variable is also calculated by Eq. (4.17) (Line 9), and the binary variables of $\mathbf{o}$ are flipped accordingly (Line 10). Here, the number of variables D in Eq. (4.17) equals the total number of binary variables, n counts the total number of nonzero variables of $\mathbf{o}$, and p is usually set to $1/D$.

4.4.4 Procedure of PM-MOEA

With the evolutionary frequent pattern mining method and the binary genetic operators, the procedure of PM-MOEA is presented in Algorithm 4.11. First, a

Algorithm 4.10: *BinaryOperators*(P')

Input: P' (a set of parents)
Output: O (offspring population)

```
O ← ∅;
while P′ ≠ ∅ do
    [x, y] ← Randomly select two parents from P′;
    P′ ← P′ \ {x, y};

    //Crossover
    o ← x;
    diff ← xor(x, y); //The dimensions that are different in
        x and y
    prob_diff ← Calculate the probability to flip each variable in dimensions
    diff of o by Eq. (4.17);
    Flip each variable in dimensions diff of o with the probability prob_diff;
    //Mutation
    prob ← Calculate the probability to flip each variable of o by Eq. (4.17);
    Flip each variable of o with the probability prob;
    O ← O ∪ {o};
return O;
```

Algorithm 4.11: Procedure of PM-MOEA

Input: N (population size), N' (population size for evolutionary pattern mining)
Output: P (final population)

```
P ← Randomly generate N solutions;
Front ← NondominatedSorting(P); //Algorithm 1.2
Q_max ← ∅; //Maximum candidate sets of nonzero variables
Q_min ← ∅; //Minimum candidate sets of nonzero variables
while termination criterion not fulfilled do
    Q_max ← MiningMaxSets(P, Q_max, N′); //Algorithm 4.7
    Q_min ← MiningMinSets(P, Q_min, N′); //Algorithm 4.8
    P′ ← TournamentSelection(P, 2N, Front, ∅); //Algorithm 1.4
    O ← Variation(P′, Q_max, Q_min); //Algorithm 4.9
    [P, Front] ← EnvironmentalSelection(P ∪ O, N); //Algorithm 1.6
return P;
```

population P with N solutions is randomly initialized (Line 1), and the maximum candidate sets Q_{max} and minimum candidate sets Q_{min} are initialized to empty (Lines 3 and 4). At each generation, Q_{max} and Q_{min} are updated through the evolutionary frequent pattern mining method based on the current population (Lines 6 and 7), and the mating pool P' is constructed by selecting $2N$ parents from the population P via binary tournament selection according to the non-dominated front numbers (Line 8). After that, N offspring solutions are generated with the assistance of Q_{max}, Q_{min}, and the new binary genetic operators (Line 9). To select N solutions from the combination of P and O, PM-MOEA adopts the same environmental selection strategy as SparseEA and MOEA/PSL, where N solutions with better non-dominated front numbers and crowding distances survive to the next generation (Line 10). Finally, the population P is returned as the output of the algorithm (Line 11).

At each generation of PM-MOEA, the time complexity is $O(G'(N'^2 + N'D))$ for frequent pattern mining, $O(ND)$ for offspring generation, and $O(MN^2)$ for environmental selection, where G' denotes the number of iterations for frequent pattern mining, N' denotes the maximum number of maximum and minimum candidate sets, D denotes the number of decision variables, and M denotes the number of objectives. Therefore, the time complexity of one generation of PM-MOEA is $O(G'(N'^2 + N'D) + ND + MN^2)$.

4.5 Experimental Comparisons

To investigate the performance of SparseEA, MOEA/PSL, and PM-MOEA, they are compared with several representative MOEAs on benchmark and real-world sparse LSMOPs. The compared MOEAs include a classical evolutionary algorithm NSGA-II [6], two decision variable grouping-based evolutionary algorithms MOEA/DVA [21] and LMEA [22], two decision space reduction-based evolutionary algorithms WOF-SMPSO [23] and LSMOF-NSGA-II [24], and a novel search strategy-based evolutionary algorithm LMOCSO [25]. The tested sparse LSMOPs include the benchmark problems SMOP1–SMOP8 [7], the neural network training problem (including four test instances NN1–NN4 with different datasets) introduced in Eq. (2.23), and the frequent pattern mining problem (including four test instances PM1–PM4 with different datasets) introduced in Eq. (2.28). The population size is set to 100 and the number of function evaluations is set to $100 \times D$, where D is the number of decision variables ranging from 100 to 6241. The populations obtained on benchmark problems are assessed by IGD [26], and the populations obtained on real-world problems are assessed by HV [27].

Table 4.2 lists the experimental results averaged over 30 runs, where the best results are shown in bold and the statistically similar ones are highlighted.

Table 4.2 IGD Values Obtained by Nine MOEAs on Eight Benchmark Sparse LSMOPs and HV Values Obtained by Nine MOEAs on Two Real-World Sparse LSMOPs.

Problem (*D*)	NSGA-II	MOEA/DVA	LMEA	WOF-SMPSO	LSMOF-NSGA-II	LMOCSO	SparseEA	MOEA/PSL	PM-MOEA
SMOP1 (100)	1.3881e−1	1.1699e+0	7.5073e−1	2.7445e−1	5.4056e−2	4.5991e−1	9.6493e−3	**7.5149e−3**	1.0619e−2
SMOP2 (100)	5.9326e−1	1.9698e+0	1.6498e+0	9.4523e−1	1.4431e−1	1.6623e+0	3.7669e−2	**1.8031e−2**	2.4476e−2
SMOP3 (100)	8.9196e−1	2.3901e+0	1.9913e+0	7.2635e−1	7.1078e−1	1.6021e+0	1.6880e−2	**1.2013e−2**	1.3663e−2
SMOP4 (100)	1.8464e−1	1.0420e+0	8.0055e−1	5.2006e−1	6.4901e−3	8.0353e−1	5.0302e−3	4.6541e−3	**4.1171e−3**
SMOP5 (100)	3.7593e−1	8.2484e−1	5.6683e−1	3.7125e−1	3.6417e−1	4.1629e−1	6.0333e−3	6.7538e−3	**5.8328e−3**
SMOP6 (100)	5.1558e−2	3.7933e−1	2.2736e−1	8.3967e−2	8.2906e−3	1.5128e−1	8.3891e−3	**8.0869e−3**	9.3224e−3
SMOP7 (100)	3.7936e−1	1.9933e+0	1.5488e+0	1.7419e−1	2.2770e−1	5.4583e−1	3.7442e−2	**2.8884e−2**	4.2378e−2
SMOP8 (100)	1.6116e+0	3.3817e+0	3.2100e+0	8.1791e−1	7.0334e−1	2.4326e+0	1.6878e−1	**1.5041e−1**	1.7536e−1
SMOP1 (1000)	2.3802e−1	1.4011e+0	6.3298e−1	2.5378e−1	5.8896e−2	5.5384e−1	2.5844e−2	**1.5394e−2**	2.1628e−2
SMOP2 (1000)	9.5843e−1	2.1029e+0	1.4979e+0	4.4390e−1	1.3484e−1	1.8253e+0	7.1310e−2	**4.1015e−2**	5.2607e−2
SMOP3 (1000)	1.2041e+0	2.4833e+0	1.8615e+0	7.0493e−1	6.8275e−1	1.6860e+0	2.8652e−2	**1.4260e−2**	2.4684e−2
SMOP4 (1000)	4.2098e−1	1.0962e+0	7.0554e−1	5.2047e−2	4.8741e−3	8.8435e−1	4.9091e−3	4.6623e−3	**4.1184e−3**
SMOP5 (1000)	3.8407e−1	9.5534e−1	5.0427e−1	3.5629e−1	3.6945e−1	4.3506e−1	5.7937e−3	7.2164e−3	**4.7038e−3**
SMOP6 (1000)	6.4852e−2	4.2930e−1	1.9409e−1	8.2004e−2	6.1533e−3	1.8522e−1	7.5569e−3	7.2044e−3	**4.9861e−3**
SMOP7 (1000)	4.7519e−1	2.4557e+0	1.1352e+0	9.3192e−2	1.8547e−1	6.1863e−1	8.2947e−2	**5.2677e−2**	7.3822e−2
SMOP8 (1000)	1.9015e+0	3.5972e+0	2.8570e+0	5.6434e−1	7.1966e−1	2.3568e+0	2.5142e−1	**1.7740e−1**	1.8902e−1
NN1 (301)	4.0033e−1	8.2934e−2	4.1292e−1	3.9293e−1	3.9402e−1	3.9862e−1	9.9175e−1	**9.9425e−1**	9.7726e−1
NN2 (521)	3.0025e−1	5.9979e−2	3.0456e−1	3.0817e−1	3.0323e−1	2.9797e−1	7.9986e−1	**8.0317e−1**	7.7881e−1
NN3 (1241)	3.4234e−1	6.8343e−2	3.4964e−1	3.6260e−1	3.4953e−1	3.5647e−1	**8.9675e−1**	8.6952e−1	8.6733e−1
NN4 (6241)	3.5159e−1	6.7343e−2	3.4481e−1	3.3685e−1	3.4354e−1	3.6199e−1	9.2786e−1	9.3084e−1	**9.3878e−1**
PM1 (100)	2.8312e−1	8.2645e−3	5.7655e−2	2.1942e−1	2.1178e−1	9.5224e−2	3.3445e−1	3.0572e−1	**3.5000e−1**
PM2 (500)	1.7204e−1	8.2645e−3	9.0380e−2	1.2756e−1	1.2305e−1	9.7368e−2	**1.9429e−1**	1.6867e−1	1.8008e−1
PM3 (1000)	1.2228e−1	8.2645e−3	7.8952e−2	1.0825e−1	1.0755e−1	4.5595e−2	1.4634e−1	1.3407e−1	**2.0506e−1**
PM4 (5000)	8.7578e−2	8.2645e−3	6.6640e−2	9.6512e−2	9.5333e−2	6.2657e−2	1.1165e−1	1.0248e−1	**1.8328e−1**

It can be observed that the decision variable grouping-based and novel search strategy-based evolutionary algorithms (i.e., MOEA/DVA, LMEA, and LMOCSO) perform worse than the classical NSGA-II, which indicates that these MOEAs are ineffective for solving sparse LSMOPs, though they are tailored for solving general LSMOPs. The decision space reduction-based evolutionary algorithms (i.e., WOF-SMPSO and LSMOF-NSGA-II) perform better, as they can converge to quasi-optimal regions very quickly. The three sparse evolutionary algorithms (i.e., SparseEA, MOEA/PSL, and PM-MOEA) exhibit the best performance, which verifies the effectiveness of the strategies for finding sparse Pareto optimal solutions. More specifically, MOEA/PSL has the best overall performance on real-coded problems (i.e., SMOP1–SMOP8 and NN1–NN4) since it uses RBM and DAE to reduce the dimensionality of both the binary vectors and real vectors, while PM-MOEA has the best overall performance on binary-coded problems (i.e., PM1–PM4), since it uses an evolutionary frequent pattern mining method to reduce the dimensionality of the binary vectors.

Moreover, Fig. 4.4 plots the populations with the median HV values obtained by the nine MOEAs on a neural network training problem, where the left axis shows the network complexity and training loss and the right axis shows the network complexity and test loss. Obviously, the solutions obtained by SparseEA, MOEA/PSL, and PM-MOEA have worse training loss but similar test loss to the solutions obtained by the other MOEAs, which indicates that the solutions obtained by the other MOEAs overfit the training samples due to their high complexities. As a consequence, it is reasonable to use sparse MOEAs to minimize both the network complexity and loss in training neural networks.

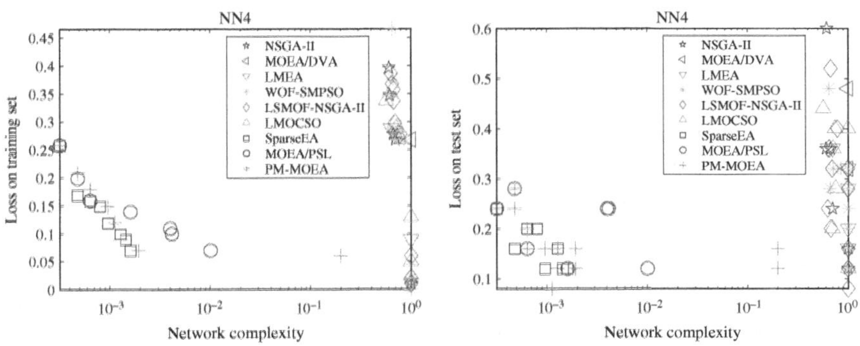

Figure 4.4 Solutions obtained by nine MOEAs on a neural network training problem.

4.6 Summary

This chapter presents the first three MOEAs for sparse large-scale multi-objective optimization. The most significant feature of these MOEAs is that they search for the sparse Pareto optimal solutions of specific LSMOPs by customizing effective strategies, such as the population initialization strategy in SparseEA, the Pareto optimal subspace learning strategy in MOEA/PSL, and the evolutionary frequent pattern mining method in PM-MOEA. Due to the wide application scenarios of sparse large-scale multi-objective optimization, it appears as an emerging topic in evolutionary multi-objective optimization, and an increasing number of new methods are being proposed [28–37].

In the following chapters, a number of large-scale MOEAs tailored for the applications in several representative areas are presented, where each MOEA customizes novel strategies to tackle the difficulties of a specific application.

References

1 S. S. Petrova and A. D. Solov'ev, "The origin of the method of steepest descent," *Historia Mathematica*, vol. 24, no. 4, pp. 361–375, 1997.

2 S. Yang, Y. Tian, C. He, X. Zhang, K. C. Tan, and Y. Jin, "A gradient guided evolutionary approach to training deep neural networks," *IEEE Transactions on Neural Networks and Learning Systems*, vol. 33, no. 9, pp. 4861–4875, 2021.

3 L. Tang, L. Zhang, P. Luo, and M. Wang, "Incorporating occupancy into frequent pattern mining for high quality pattern recommendation," in *Proceedings of the 21th ACM International Conference on Information and Knowledge Management*, 2012, pp. 75–84.

4 Y. Sun, G. G. Yen, and Y. Zhang, "Evolving unsupervised deep neural networks for learning meaningful representations," *IEEE Transactions on Evolutionary Computation*, vol. 23, no. 1, pp. 89–103, 2018.

5 Y. Tian, S. Yang, L. Zhang, F. Duan, and X. Zhang, "A surrogate-assisted multiobjective evolutionary algorithm for large-scale task-oriented pattern mining," *IEEE Transactions on Emerging Topics in Computational Intelligence*, vol. 3, no. 2, pp. 106–116, 2018.

6 K. Deb, A. Pratap, S. Agarwal, and T. Meyarivan, "A fast and elitist multiobjective genetic algorithm: NSGA-II," *IEEE Transactions on Evolutionary Computation*, vol. 6, no. 2, pp. 182–197, 2002.

7 Y. Tian, X. Zhang, C. Wang, and Y. Jin, "An evolutionary algorithm for large-scale sparse multi-objective optimization problems," *IEEE Transactions on Evolutionary Computation*, vol. 24, no. 2, pp. 380–393, 2019.

8 Y. Jin and B. Sendhoff, "Pareto-based multiobjective machine learning: an overview and case studies," *IEEE Transactions on Systems, Man, and Cybernetics, Part C (Applications and Reviews)*, vol. 38, no. 3, pp. 397–415, 2008.

9 F. Streichert, H. Ulmer, and A. Zell, "Evaluating a hybrid encoding and three crossover operators on the constrained portfolio selection problem," in *Proceedings of the 2004 IEEE Congress on Evolutionary Computation*, vol. 1, 2004, pp. 932–939.

10 K. Deb and R. B. Agrawal, "Simulated binary crossover for continuous search space," *Complex Systems*, vol. 9, no. 2, pp. 115–148, 1995.

11 K. Deb and M. Goyal, "A combined genetic adaptive search (GeneAS) for engineering design," *Computer Science and Informatics*, vol. 26, no. 4, pp. 30–45, 1996.

12 X. Zhang, Y. Tian, R. Cheng, and Y. Jin, "An efficient approach to non-dominated sorting for evolutionary multi-objective optimization," *IEEE Transactions on Evolutionary Computation*, vol. 19, no. 2, pp. 201–213, 2014.

13 Y. Tian, C. Lu, X. Zhang, K. C. Tan, and Y. Jin, "Solving large-scale multiobjective optimization problems with sparse optimal solutions via unsupervised neural networks," *IEEE Transactions on Cybernetics*, vol. 51, no. 6, pp. 3115–3128, 2020.

14 A. Fischer and C. Igel, "An introduction to restricted Boltzmann machines," in *Proceedings of the Iberoamerican Congress on Pattern Recognition*, 2012, pp. 14–36.

15 G. E. Hinton, "Training products of experts by minimizing contrastive divergence," *Neural Computation*, vol. 14, no. 8, pp. 1771–1800, 2002.

16 J. Zhai, S. Zhang, J. Chen, and Q. He, "Autoencoder and its various variants," in *Proceedings of the 2018 IEEE International Conference on Systems, Man, and Cybernetics*, 2018, pp. 415–419.

17 Y. LeCun, Y. Bengio, and G. Hinton, "Deep learning," *Nature*, vol. 521, no. 7553, pp. 436–444, 2015.

18 P. Vincent, H. Larochelle, Y. Bengio, and P. Manzagol, "Extracting and composing robust features with denoising autoencoders," in *Proceedings of the 25th International Conference on Machine Learning*, 2008, pp. 1096–1103.

19 Y. Tian, C. Lu, X. Zhang, F. Cheng, and Y. Jin, "A pattern mining based evolutionary algorithm for large-scale sparse multi-objective optimization problems," *IEEE Transactions on Cybernetics*, vol. 52, no. 7, pp. 6784–6797, 2020.

20 X. Zhang, F. Duan, L. Zhang, F. Cheng, Y. Jin, and K. Tang, "Pattern recommendation in task-oriented applications: a multi-objective perspective [application notes]," *IEEE Computational Intelligence Magazine*, vol. 12, no. 3, pp. 43–53, 2017.

21 X. Ma, F. Liu, Y. Qi, X. Wang, L. Li, L. Jiao, M. Yin, and M. Gong, "A multiobjective evolutionary algorithm based on decision variable analyses for multiobjective optimization problems with large-scale variables," *IEEE Transactions on Evolutionary Computation*, vol. 20, no. 2, pp. 275–298, 2015.

22 X. Zhang, Y. Tian, R. Cheng, and Y. Jin, "A decision variable clustering-based evolutionary algorithm for large-scale many-objective optimization," *IEEE Transactions on Evolutionary Computation*, vol. 22, no. 1, pp. 97–112, 2016.

23 H. Zille, H. Ishibuchi, S. Mostaghim, and Y. Nojima, "A framework for large-scale multiobjective optimization based on problem transformation," *IEEE Transactions on Evolutionary Computation*, vol. 22, no. 2, pp. 260–275, 2017.

24 C. He, L. Li, Y. Tian, X. Zhang, R. Cheng, Y. Jin, and X. Yao, "Accelerating large-scale multiobjective optimization via problem reformulation," *IEEE Transactions on Evolutionary Computation*, vol. 23, no. 6, pp. 949–961, 2019.

25 Y. Tian, X. Zheng, X. Zhang, and Y. Jin, "Efficient large-scale multiobjective optimization based on a competitive swarm optimizer," *IEEE Transactions on Cybernetics*, vol. 50, no. 8, pp. 3696–3708, 2019.

26 C. A. Coello Coello and N. C. Cortés, "Solving multiobjective optimization problems using an artificial immune system," *Genetic Programming and Evolvable Machines*, vol. 6, pp. 163–190, 2005.

27 L. While, P. Hingston, L. Barone, and S. Huband, "A faster algorithm for calculating hypervolume," *IEEE Transactions on Evolutionary Computation*, vol. 10, no. 1, pp. 29–38, 2006.

28 G. Peng, H. Wu, H. Wu, and K. Wolter, "Evolutionary large-scale sparse multi-objective optimization for collaborative edge-cloud computation offloading," in *Proceedings of the 12th International Joint Conference on Computational Intelligence*, 2020, pp. 100–111.

29 I. Kropp, A. P. Nejadhashemi, and K. Deb, "Benefits of sparse population sampling in multi-objective evolutionary computing for large-scale sparse optimization problems," *Swarm and Evolutionary Computation*, vol. 69, no. 3, p. 101025, 2022.

30 Z. Tan and H. Wang, "A kriging-assisted evolutionary algorithm using feature selection for expensive sparse multi-objective optimization," in *Proceedings of the 2020 IEEE Congress on Evolutionary Computation*, 2020, pp. 1–8.

31 Z. Tan, H. Wang, and S. Liu, "Multi-stage dimension reduction for expensive sparse multi-objective optimization problems," *Neurocomputing*, vol. 440, pp. 159–174, 2021.

32 Y. Zhang, Y. Tian, and X. Zhang, "A comparison study of evolutionary algorithms on large-scale sparse multi-objective optimization problems," in *Proceedings of the 2021 International Conference on Evolutionary Multi-Criterion Optimization*, 2021, pp. 424–437.

33 Y. Zhang, Y. Tian, and X. Zhang, “Improved SparseEA for sparse large-scale multi-objective optimization problems,” *Complex & Intelligent Systems*, vol. 9, pp. 1127–1142, 2021.

34 G. Y. H. Woo, H. S. Sim, Y. K. Goh, and W. J. Leong, “Spectral proximal method for solving large scale sparse optimization,” in *Proceedings of ITM Web of Conferences*, vol. 36, p. 04007, 2021.

35 X. Wang, K. Zhang, J. Wang, and Y. Jin, “An enhanced competitive swarm optimizer with strongly convex sparse operator for large-scale multi-objective optimization,” *IEEE Transactions on Evolutionary Computation*, vol. 26, no. 5, pp. 859–871, 2021.

36 Y. Su, Z. Jin, Y. Tian, X. Zhang, and K. C. Tan, “Comparing the performance of evolutionary algorithms for sparse multi-objective optimization via a comprehensive indicator,” *IEEE Computational Intelligence Magazine*, vol. 17, no. 3, pp. 34–53, 2022.

37 Y. Tian, Y. Feng, X. Zhang, and C. Sun, “A fast clustering based evolutionary algorithm for super-large-scale sparse multi-objective optimization,” *IEEE/CAA Journal of Automatica Sinica*, vol. 10, no. 4, pp. 1048–1063, 2022.

5

Evolutionary Large-Scale Multi-Objective Optimization for Community Detection in Complex Networks

5.1 Introduction

The complex network is composed of a set of nodes and edges, where a node represents an item and an edge indicates that there is a relationship between the nodes connected by it. Fig. 5.1 gives an illustrative example of complex network. In the real world, there is a wide variety of complex systems that can be modeled as complex networks, such as social platforms, power systems, and World Wide Web. The community detection is to divide the nodes in a complex network into different groups, so that the nodes densely connect to the nodes in the same group (called community or cluster), but sparsely connect to the nodes in the other groups. In Fig. 5.1, the complex network can be divided into two communities. One is made up of nodes $\{1, 2, 3, 4, 5\}$ and the other is made up of nodes $\{6, 7, 8, 9\}$. The community detection is a vital analysis technique for revealing useful information hidden in complex networks. For example, in social networks, the communities are helpful to find a group of people with similar interests [1]. In protein–protein interaction networks, the communities can be used to identify the functional modules [2, 3]. In recent years, a large number of community detection algorithms have been proposed based on different ideas [4–6]. Generally, the existing algorithms tailored for community detection can be divided into three groups [7]. The first group adopts traditional methods in data clustering to merge densely connected nodes, such as hierarchical clustering [8] and spectral clustering [9]. The second group removes the edges that connect different communities to partition networks [10]. As for the third group, it resorts to optimization algorithms including mathematical programming [11] and heuristics [12], where the communities are determined by optimizing the metrics for the quality of communities.

As a meta-heuristic algorithm, evolutionary algorithms have been successfully applied for community detection in complex networks, which can be divided into

Evolutionary Large-Scale Multi-Objective Optimization and Applications: Integrating Evolutionary Computation, Machine Learning, and Data Science,
First Edition. Xingyi Zhang, Ran Cheng, Ye Tian, and Yaochu Jin.

two categories according to the number of objectives to be optimized. In the first category, the communities are obtained by evolutionary algorithms to optimize a single objective, such as the modularity and edge betweenness. In the second category, the evolutionary algorithms simultaneously optimize multiple conflicting objectives to divide the nodes into different communities. Compared to single-objective evolutionary algorithms, multi-objective evolutionary algorithms (MOEAs) have attracted more attention in complex networks for community detection. The main reasons are twofold. On the one hand, there exist some potential disadvantages in community detection by optimizing a single objective, such as the limited resolution of modularity. As pointed out in [13, 14], only optimizing the modularity might overlook the communities with small size, and empirical results indicate that multi-objective optimization is an effective way to tackle this issue. On the other hand, different from single-objective optimization, the MOEAs can find a set of Pareto optimal solutions rather than a single optimal solution in one run. Since each solution in the Pareto optimal set represents a trade-off between the multiple objectives, the MOEA-based community detection methods can provide a set of network partitions with different hierarchical levels, which means that this kind of methods is a rich framework for finding the hierarchical community structure of complex networks [15, 16].

Despite the promising performance achieved by MOEAs on community detection, most of the existing algorithms suffer from the scalability issue due to the curse of dimensionality. As analyzed in [17, 18], the search space of MOEAs increases exponentially with the increase of network size in community detection. This fact can naturally lead to most of the existing MOEA-based algorithms being only suitable for networks having a relatively small number of nodes. To enhance the scalability of MOEAs in large-scale complex networks, a number of ideas and/or strategies have been proposed for MOEA-based community detection algorithms in recent years. In the following sections, two representative ideas tailored for MOEAs to detect communities in large-scale complex networks will be presented, namely, the network reduction for multi-objective evolutionary algorithms (RMOEAs) and the parallel framework for multi-objective evolutionary algorithms (PMOEAs).

5.2 Network Reduction-Based Multi-Objective Evolutionary Algorithm for Community Detection

As discussed above, the main bottleneck of MOEAs for community detection in large-scale complex networks lies in the curse of dimensionality. An intuitive idea is whether we can reduce the tremendous search space of MOEAs by utilizing the problem-specific knowledge and information, i.e., those related to community detection and complex networks. Inspired by this idea, the network reduction-based multi-objective evolutionary algorithm (RMOEA) was first suggested in [19], where two kinds of strategies are proposed to reduce the size of complex networks, namely, prereduction and reduction in evolution. These two strategies are based on the following observation. For a given complex network, there usually exist some nodes in the network that can easily be checked for belonging to the same community. The prereduction reduces the size of complex networks by simply checking the local topology structures, and the reduction in evolution achieves further reduction of networks by finding the local communities found during the evolution of MOEAs. The proposed RMOEA considers the local communities obtained by prereduction and reduction in evolution as a whole in later evolution of the algorithm. By doing so, the size of complex networks being handled is reduced; hence, the scalability of RMOEA to large-scale complex networks can be considerately enhanced.

5.2.1 Formulation of Multi-Objective Community Detection Problem

In the proposed RMOEA, the community detection is formulated as the following bi-objective optimization problem by minimizing two common metrics, the kernel k-means (**KKM**) [14] and the ratio cut (**RC**) [20].

$$\begin{aligned} &\text{Minimize}\, F(C) = (\mathbf{KKM}(C),\ \mathbf{RC}(C)), \\ &\mathbf{KKM}(C) = 2(n-k) - \sum_{i=1}^{k} \frac{L(V_i, V_i)}{|V_i|}, \\ &\mathbf{RC}(C) = \sum_{i=1}^{k} \frac{L(V_i, \bar{V}_i)}{|V_i|}, \end{aligned} \tag{5.1}$$

where $C = \{V_1, V_2, \dots, V_k\}$ is a partition of a complex network with k communities $V_1, V_2, \dots, V_k$, $L(V_i, \bar{V}_i) = \sum_{i\in V_i, j\in \bar{V}_i} A_{ij}$, and $\bar{V}_i = V - V_i$. Here, V is the set of nodes in the complex network and A is the adjacent matrix of the complex network. If there is an edge between nodes i and j, then $A_{ij} = 1$; otherwise, $A_{ij} = 0$.

From Eq. (5.1), it can be found that the **KKM** measures the link density of intracommunities and the **RC** evaluates the link density of intercommunities. The **KKM** decreases with the number of communities and the **RC** owns an opposite trend. This means that the **KKM** and **RC** are two conflicting objectives, and simultaneously minimizing them can ensure that the nodes belonging to the same community are densely connected and the nodes belonging to different communities are sparely connected, which is coincidence with the requirements of community detection.

5.2.2 Prereduction Based on Local Network Topology Structure

As discussed above, some local communities can be easily identified by only checking the local topology structure of networks. Hence, the proposed RMOEA first reduces the size of complex networks according to the local topology structure of networks, before the evolution of population starts. The pseudocode of prereduction in RMOEA is given in Algorithm 5.1. To be specific, the prereduction is performed as follows.

First, the RMOEA randomly selects a node s from the complex network G, and determines a subgraph G^{LS} of s (Lines 4–7). The subgraph G^{LS} consists of the nodes s, s^c (a neighboring node of s which has the largest degree), s^{cc} (a neighboring node of s^c which shares the largest number of neighbors with s^c), and all common

---

Algorithm 5.1: *Prereduction(G)*

Input: G (complex network)
Output: G_R (the reduced network)

1 $G_R \leftarrow G$;
2 *Nodes* $\leftarrow$ nodes in G;
3 **while** *Nodes* $\neq \emptyset$ **do**
4     $s \leftarrow$ randomly select one node from *Nodes*;
5     $s^c \leftarrow$ find the node in the neighborhood of s which has the largest degree;
6     $s^{cc} \leftarrow$ find the node in the neighborhood of s^c which has the largest number of common neighboring nodes with s^c;
7     $G^{LS} \leftarrow$ get the subgraph of s;
    `//contain` s, s^c, s^{cc}, `and all common neighbors of` s^c `and` s^{cc}
8     $NB \leftarrow$ find the neighbors which are not in G^{LS}, for each node in G^{LS};
9     $NC \leftarrow \emptyset$;
10   **while** $NB \neq NC$ **do**
11         $NC \leftarrow NB$;
12         $t \leftarrow$ randomly select one node from NB;
13   **if** *the number of links between t and* G^{LS} *is larger than one half of degree of t* **then**
14             $G^{LS} \leftarrow \cup\{t\}$;
15             $NB \leftarrow$ find the neighbors which are not in G^{LS}, for each node in G^{LS};
16     $G_R \leftarrow$ merge all nodes in G^{LS} into one node in G_R;
17   *Nodes* $\leftarrow$ remove nodes in G^{LS} from Nodes;
18 **return** G_R;

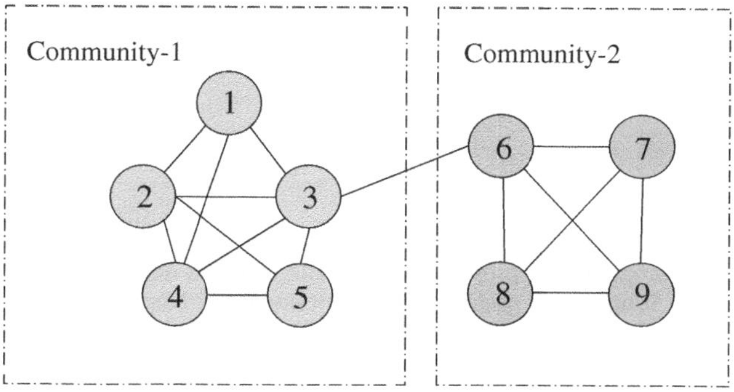

Figure 5.1 An example of complex network, where the network is divided into two communities. One consists of nodes {1, 2, 3, 4, 5} and the other contains nodes {6, 7, 8, 9}.

neighbors of s^c and s^{cc}. Then, the RMOEA extends the subgraph G^{LS} by adding some neighbors of G^{LS} (Lines 8–15). To ensure that these nodes belong to the same community, the RMOEA only adds each of the neighbors of G^{LS} whose number of links with G^{LS} is larger than one half of its degree. After that, the local communities of s is determined, and the RMOEA merges all nodes in G^{LS} into one node to reduce the size of network being handled (Lines 16 and 17). The RMOEA will repeat to find the other local communities for the remaining nodes in the network and this process stops until all nodes of the network have been assigned to local communities.

Fig. 5.2 gives an illustrative example of prereduction developed in the RMOEA, where a network with seven nodes is considered, and we assume that the node 1 is selected to identify the corresponding local community. First, the subgraph {1,2,3,5} plotted in Fig. 5.2 (b) is obtained by the RMOEA, given that the node 3 is the neighbor of node 1 that owns the largest degree, and nodes 2 and 5 share the largest number of neighbors with node 3. Second, as depicted in Fig. 5.2, since all the links of node 4 are connected to subgraph {1, 2, 3, 5}, whereas for nodes 6 and 7 only half of links are connected to subgraph {1, 2, 3, 5}, the node 4 is added to the local community. Third, the RMOEA merges the local community {1, 2, 3, 4, 5} into one node, so that the network to be handled is reduced to a network with only three nodes.

5.2.3 Network Reduction in Evolution Based on Elite Individuals

In addition to the local topology structure, some local communities can also be obtained in evolution by using the elite individuals searched by MOEAs. In the proposed RMOEA, two metrics are employed to determine the elite individuals

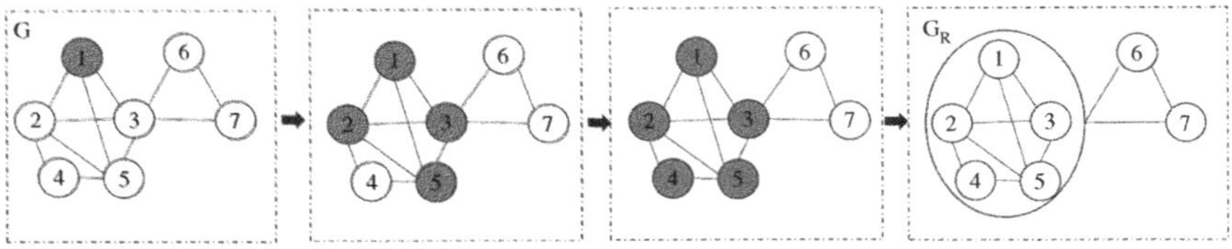

(a) (b) (c) (d)

Figure 5.2 An illustrative example of the prereduction. (a) The complex network consisting of seven nodes. (b) Subgraph of node 1, which contains four nodes, namely, nodes 1, 2, 3, and 5. (c) The local community of node 1 by adding node 4 to the subgraph. (d) The reduction network based on the found local community of $\{1, 2, 3, 4, 5\}$.

in population. One is the widely used metric for measuring the quality of found communities, namely, modularity Q. The larger the modularity Q is, the better quality the found communities have. The other is the number of communities C_{num}. Considering that the a large C_{num} implies that the size of each found community is small, it is helpful to reduce the probability of incorrectly identified local communities. The procedure of elite individual selection is presented in Algorithm 5.2. For a population P, the modularity Q and the number of communities C_{num} corresponding to each individual are first calculated, and the non-dominated individuals are selected as the elite individuals for network reduction.

The pseudocode of network reduction based on elite individuals is presented in Algorithm 5.3. First, the elite individuals P' are selected from current population P by using Algorithm 5.2. Then, the individual *ind* with the largest number of communities is selected from P'. Third, the individual *ind* is used to identify the local communities based on whether all the nodes in a community of *ind* belong to the same community in the remaining elite individuals. Specifically, for a community C_i in *ind*, if all nodes in C_i are identified as in one community by

---

Algorithm 5.2: *EliteSelection*(G, P)

Input: G (complex network); P (population)
Output: P' (the set of elite individuals)

pop $\leftarrow$ the number of individuals in P;
for $i = 1 \leftarrow pop$ **do**
  $Q(i) \leftarrow$ calculate the modularity Q of partition corresponding to individual p_i;
  $C_{num}(i) \leftarrow$ calculate the number of communities in the partition corresponding to individual p_i;
$P' \leftarrow$ obtain the non-dominated solutions in P by using Q and C_{num};
return P';

Algorithm 5.3: *ReductionElite*(G, P)

Input: G (complex network); P (population)

Output: G_R (the reduced network); P_R (the population for the reduced network)

$G_R \leftarrow G$;

$P_R \leftarrow P$;

$P' \leftarrow$ EliteSelection(G_R, P);

`//Algorithm 5.2`

$ind \leftarrow$ find the individual in P' with the largest number of communities;

$l \leftarrow$ the number of communities in *ind*;

$\mathbf{C} \leftarrow$ the communities $C_1, C_2, \ldots, C_l$ corresponding to *ind*;

for $i = 1 \leftarrow l$ **do**

  $C_i^s \leftarrow C_i$;

  for *each* $ind' \in P'$ *and* $ind' \neq ind$ **do**

    $\mathbf{C}' \leftarrow$ the communities $C_1', C_2', \ldots, C_l'$ corresponding to ind';

    $C_j' \leftarrow$ find one community $C_j', j \in \{1, 2, \ldots, t\}$ satisfying that $C_i \cap C_j'$ has the most number of nodes;

    $C_i^s \leftarrow C_i^s \cap C_j'$;

  $G_R \leftarrow$ merge C_i^s into one node in G_R;

  $P_R \leftarrow$ merge the genes of nodes of C_i^s into one gene for each individual in P_R;

return G_R and P_R;

---

the remaining elite individuals, then C_i is considered as a local community. Otherwise, the maximal subset of C_i which are identified in one community by the remaining elite individuals is considered as a local community. The above operations repeat until all communities in *ind* are checked. Lastly, a reduced network G_R is obtained by merging all nodes in each local community into one node in G, and the population P_R for the reduced network is obtained by merging the genes of nodes of each local community into one gene for all individuals in P. For each individual, an index of the local community is assigned to the merged gene, and the community information on the rest genes in the individual keeps unchanged.

Fig. 5.3 plots an illustrative example of elite individual-based network reduction. Fig. 5.3(a) displays the original complex network with seven nodes. In Fig. 5.3(b), it is assumed that the two elite individuals *ind*1 and *ind*2 have been selected from population P for determining the local communities. Here, the individuals are represented by the locus-based representation scheme, where each gene of the individual represents the index of a node that belongs to the

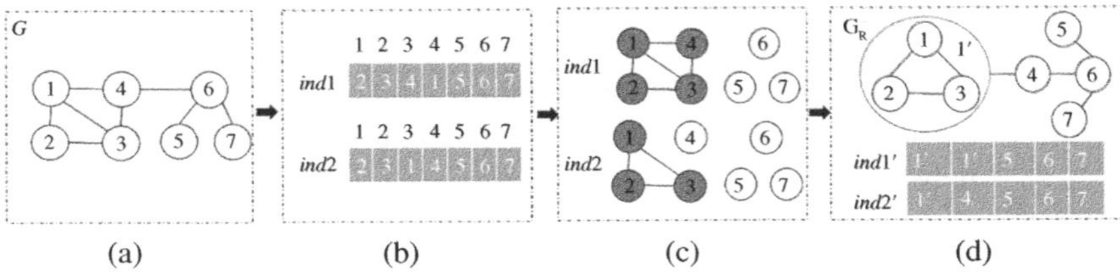

Figure 5.3 An illustrative example of elite individual-based network reduction. (a) Complex network G. (b) Two individuals *ind*1 and *ind*2 having been found for network G, where the locus-based individual representation scheme is adopted [21]. (c) Two divisions corresponding to individuals *ind*1 and *ind*2. The division of *ind*1 consists of four communities $\{1, 2, 3, 4\}$, $\{5\}$, $\{6\}$, and $\{7\}$, and the division of *ind*2 is $\{1, 2, 3\}$, $\{4\}$, $\{5\}$, $\{6\}$, and $\{7\}$. (d) Reduced network G_R based on the local communities found by individuals *ind*1 and *ind*2, where the local community $\{1, 2, 3\}$ is identified as a node in the reduced network since both individuals recognize nodes 1, 2, and 3 as in one community. *ind*1′ and *ind*2′ are the new individuals of *ind*1 and *ind*2 for the reduced network.

same community. The partitions corresponding to *ind*1 and *ind*2 are shown in Fig. 5.3(c), where *ind*1 has four communities including $\{1, 2, 3, 4\}$, $\{5\}$, $\{6\}$, and $\{7\}$, and the partition of *ind*2 is $\{1, 2, 3\}$, $\{4\}$, $\{5\}$, $\{6\}$, and $\{7\}$. Since the number of communities in *ind*2 is larger than that in *ind*1, the *ind*2 is used to obtain the local communities. For community $\{1, 2, 3\}$, it is identified as a local community since all of its nodes are identified as in one community by individual *ind*1. For community $\{4\}$, it is also considered as a local community due to the fact that only one node is contained in the community. This is also the case for communities $\{5\}$, $\{6\}$, and $\{7\}$. In this way, five local communities $\{1, 2, 3\}$, $\{4\}$, $\{5\}$, $\{6\}$, and $\{7\}$ are obtained from the elite individuals *ind*1 and *ind*2, and the original complex network is reduced to a network with five nodes $\{1', 4, 5, 6, 7\}$ as displayed in Fig. 5.3(d). Here, the local community $\{1, 2, 3\}$ is merged into one node $1'$. Accordingly, the *ind*1 and *ind*2 are changed to $\{1', 1', 5, 6, 7\}$ and $\{1', 4, 5, 6, 7\}$ as shown in Fig. 5.3(d), where $1'$ is assigned to the merged gene and the information on the rest genes keeps unchanged.

5.2.4 Procedure of RMOEA

The procedure of RMOEA is described in Algorithm 5.4, where the framework of MOEA/D [22] is adopted to evolve the population. First, the RMOEA conducts the prereduction for complex network G by using the local topology structure of network (Algorithm 5.1). Second, based on the network G_R obtained by the pre-reduction, the RMOEA initializes a population P and a reference point z^*. Here, the locus-based adjacency encoding schema is used to generate the population P, where each gene of individuals is first randomly assigned to an

Algorithm 5.4: Procedure of RMOEA

Input: G (complex network); *maxgen* (maximum number of generations); *pop* (population size); $\nabla = \{\nabla_1, \nabla_2, \dots, \nabla_{pop}\}$ (weight vectors); *ns* (the size of neighborhood); p_c (crossover probability); p_m (mutation probability); T (the number of reductions in evolution)

Output: The final population

$G_R \leftarrow$ Prereduction(G);
`//Algorithm 5.1`
Initialize the population $P = \{p_1, p_2, \dots, p_{pop}\}$;
Initialize reference point z^* ;
for $i = 1 \leftarrow pop$ **do**
    $N_i \leftarrow$ find the *ns* individuals from P with the nearest Euclidean distance to the weight vector ∇_i;
$P_R \leftarrow P$;
for $i = 1 \leftarrow maxgen$ **do**
    if $i\%(\lceil (maxgen + 1)/(T + 1) \rceil) == 0$ **then**
        $[G_R, P_R] \leftarrow$ ReductionElite(G_R, P_R);
        `//Algorithm 5.3`
        $[G_R, P_R] \leftarrow$ LocalCommunityRepairing(G_R, P_R);
        `//Algorithm 5.5`
        Update P_R and reference point z^* based on G_R;
    for $j = 1 \leftarrow pop$ **do**
        Randomly select one individual from N_j;
        Generated an offspring by crossover and mutation operators;
        Update individuals in N_j;
        Update reference point z^*;
return The final population ;

index of a neighbor of the node corresponding to this gene or index of the node itself, and then the genes in the individuals are iteratively replaced by the index shared by most of its neighbors, until the genes keep unchanged. For reference point z^*, it is initialized as the best values of **KKM** and **RC** in the population. Third, the RMOEA assigns *ns* individuals to each weight vector according to the Euclidean distance between the individuals and weight vectors. Fourth, the RMOEA further reduces the size of network by using elite individuals based on the network reduction in evolution. Specifically, for each generation i satisfying $i\%(\lceil (maxgen + 1)/(T + 1) \rceil) == 0$, the reduced network G_R is reduced by using elite individuals found during evolution (Algorithm 5.3). Here, % is the modular operation and T is a parameter controlling the number of reductions during evolution. Moreover, for each individual j, the RMOEA randomly selects an individual from N_j (a set of ns individuals assigned to the j weight vector) to generate an offspring by crossover and mutation operators in each generation. The N_j and reference point z^* are updated according to the Tchebycheff value of the offspring. The details of crossover and mutation operations are given in Fig. 5.4. The RMOEA will repeat to evolve the population until the maximum number of generations is reached.

It is worth noting that the network reduction method might find incorrect local communities, which means that the local communities identified by the reduction in evolution may contain some nodes belonging to different communities. Hence, the RMOEA integrates a local community repairing strategy to correct the misidentified nodes in the reduced network. The pseudocode of the local community repairing strategy is elaborated in Algorithm 5.5. First, the set of elite individuals P' are selected from the population P (Line 1). Second, for each elite individual p'_i in P', the misidentified nodes are detected by checking whether the communities of p'_i contain some nodes belonging to different local communities. More specifically, a node is considered as misidentified one if a better modularity Q can be achieved by moving it to the communities of its neighbors. Last, the misidentified nodes are removed from the merged nodes of reduced network G_R as a separate node, and the genes of the misidentified nodes are also added to each individual in population P_R.

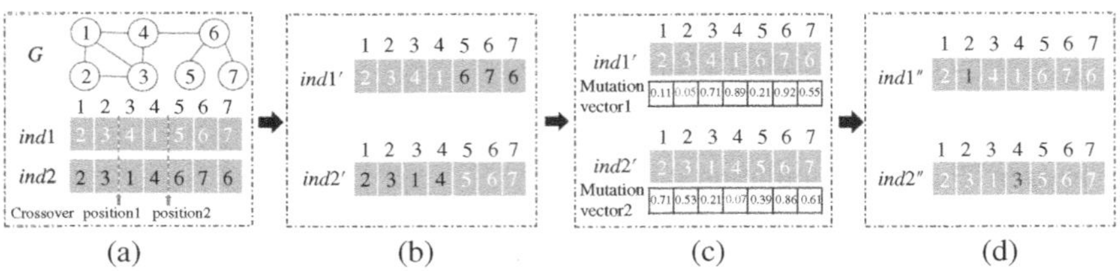

Figure 5.4 An illustrative example of crossover and mutation operators. (a) Complex network G and two individuals *ind*1 and *ind*2 with crossover positions *position*1 and *position*2 that are generated randomly. In crossover process, a randomly number *rand* is generated from {1, 2, 3}, which means that the parts before *position*1, between *position*1 and *position*2, and after *position*2 of *ind*1 and *ind*2 are mutually exchanged, respectively. (b) Two generated offspring *ind*1′ and *ind*2′ under crossover operator with *rand* = 3. (c) Two mutation vectors for *ind*1′ and *ind*2′, where the value of each node is randomly generated from [0, 1]. If the value of the node in mutation vector is smaller than mutation probability, the corresponding gene of the node will be changed to an index in its neighboring nodes. (d) Two offspring *ind*1″ and *ind*2″, which are generated by mutation operation with the mutation probability set as 0.1.

Algorithm 5.5: *LocalCommunityRepairing*(G_R, P_R)

Input: G_R (the reduced network); P_R (population for reduced network)
Output: G_R^r (the repaired network); P_R^r (population for the repaired network)
$P' \leftarrow$ EliteSelection(G_R, P);
//Algorithm 5.2
for $i = 1 \leftarrow |P'|$ **do**
  $k \leftarrow$ the number of communities corresponding to p_i' in P';
  $\mathbf{C} \leftarrow$ the communities $C_1, C_2, \ldots, C_k$ corresponding to p_i';
  $Q_i' \leftarrow$ *Modularity*(p_i', G_R);
  for $j = 1 \leftarrow k$ **do**
    Nodes $\leftarrow$ select nodes in C_j if there exist one neighbor not in C_j;
    while *Nodes* $\neq \emptyset$ **do**
      $s \leftarrow$ randomly select one node from *Nodes*;
      $p_i'' \leftarrow$ move s to the communities of its neighbors;
      $Q_i'' \leftarrow$ *Modularity*(p_i'', G_R);
      if $Q_i'' > Q_i'$ **then**
        Fault_Nodes $\leftarrow$ *Fault_Nodes* $\cup \{s\}$;
      Nodes $\leftarrow$ remove s from *Nodes*;
$G_R^r \leftarrow$ remove nodes in *Fault_Nodes* from merged nodes in G_R as a separate node;
$P_R^r \leftarrow$ add the genes of nodes of *Fault_Nodes* for each individual in P_R;
return G_R^r and P_R^r;

Fig. 5.5 shows an illustrative example of the local community repairing strategy. As plotted in Fig. 5.5(a), a local community $\{3, 5, 6, 7, 8\}$ is identified by the elite individuals in the population, and only node 3 has at least one neighbor that does not belong to the local community. Furthermore, the modularity Q increases from 0.2219 to 0.3517, when node 3 is removed from the local community $\{3, 5, 6, 7, 8\}$. Hence, the local community is repaired as $\{5, 6, 7, 8\}$, by which the quality of local communities found by elite individuals can be considerably improved.

For a complex network with n nodes and m edges, the RMOEA has a time complexity of $O(maxgen * pop * (m' + n'))$, where n' and m' are the number of nodes and edges of the reduced network, respectively. As presented in Algorithm 5.4, the RMOEA contains four main steps, network reduction based on the local topology, population initialization, network reduction based on the elite individuals, and population evolution. For the network reduction based on local topology, it holds a time complexity of $O(n * D * D)$ since in the worst case, it needs the time complexity of $O(D * D)$ to identify the local community for a given node.

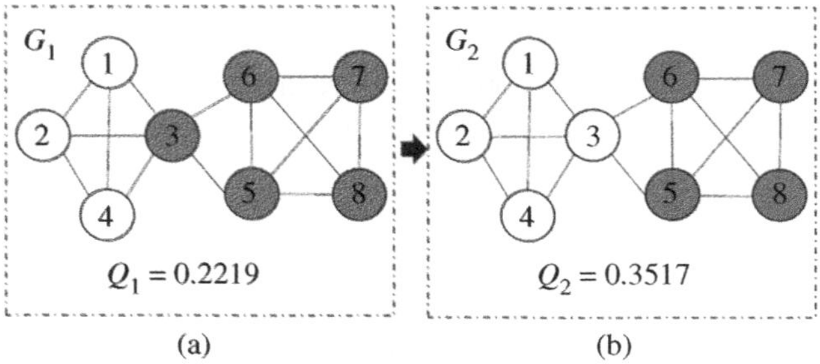

Figure 5.5 An illustrative example of local community repairing strategy. (a) The local community $\{3, 5, 6, 7, 8\}$ found according to elite individuals during evolution. (b) Repaired local community $\{5, 6, 7, 8\}$ by moving node 3 to the community $\{1, 2, 4\}$.

Here, D is the maximum degree of nodes in the complex network. The population initialization takes the time complexity of $O(pop * n * D)$, due to the fact that for each gene of individuals in the population, the maximum number of checking neighbors is D. The time complexity of the network reduction based on elite individuals is $O(T * pop * m)$. It is because that the most time-consuming operation in this step is the calculation of modularity, which holds the time complexity of $O(m)$ for each individual. As for the population evolution, its time complexity is $O(maxgen * pop * (n' + m'))$, considering that in each generation, the RMOEA needs to calculate the values of the two objectives, **KKM** and **RC**, which takes the time complexity of $O(n' + m')$. Since $maxgen * (n' + m')$ is larger than $T * m + D * n$, the RMOEA holds a time complexity of $O(maxgen * pop * (m' + n'))$.

5.2.5 Experimental Comparisons

To verify the performance of RMOEA, it is compared with several representative community detection algorithms. The compared algorithms include five popular evolutionary algorithm-based community detection algorithms, namely, CoCoMi [23], MOCD [13], MOGA-net [16], MODTLBO/D [24], and MODPSO [14], where the CoCoMi is a single-objective evolutionary algorithm, while the MOCD, MOGA-net, MODTLBO/D, and MODPSO are MOEAs. Moreover, two non-EA-based community detection algorithms, namely, Louvain [25] and SSCF [9], are also employed as compared algorithms. The parameters of the compared algorithms are set to the values recommended in [14], which is also listed in Table 5.1. For the RMOEA, the number of reductions in evolution T is set to 2 in the experiments.

The test datasets include synthetic benchmark networks and real-world networks. The synthetic benchmark networks are LFR networks [26], which are widely used to verify the performance of community detection algorithms [6, 27]. The real-world networks include *net-science* and *blogs* networks that are used

Table 5.1 Parameter Setting of the Compared Algorithms, Where *pop* is the Population Size, *maxgen* is the Maximum Generations of the Algorithm, *pc* is the Probability of Crossover, *pm* is the Probability of Mutation, and *ns* is the Neighborhood Size.

Algorithm	*pop*	*maxgen*	*pc*	*pm*	*ns*	References
CoCoMi	100	100	0.9	0.1	—	[23]
MOCD	100	100	0.9	0.1	—	[13]
MOGA-net	100	100	0.9	0.1	—	[16]
MODTLBO/D	100	100	0.9	0.1	40	[24]
MODPSO	100	100	—	0.1	40	[14]
Louvain	—	—	—	—	—	[25]
SSCF	—	—	—	—	—	[9]
RMOEA	100	100	—	0.1	40	[19]

Table 5.2 Details of Seven Real-World Networks Used in Experiments.

Network	Node number	Link number	Average degree
net-science	1,589	2,742	3.46
blogs	3,982	6,803	3.42
ca-GrQc	5,242	14,496	5.53
ca-GrQc	9,877	25,998	5.26
ca-Hepth	12,008	118,521	19.74
ca-AstroPh	18,772	198,080	21.10
ca-CondMat	23,133	93,468	8.08

in [28], and *ca-GrQc*, *ca-GrQc*, *ca-Hepth*, *ca-AstroPh* and *ca-CondMat* that are collected by the Stanford Network Analysis Project [29]. The details of the seven real-world networks are given in Table 5.2.

The evaluation metrics used to measure the quality of detected communities are normalized mutual information (*NMI*) [30] and modularity (*Q*) [31], which can be calculated as follows:

$$NMI(A,B) = \frac{-2\sum_{i=1}^{C_A}\sum_{j=1}^{C_B} C_{i,j}\log\left(\frac{C_{i,j}N}{C_{i.}C_{.j}}\right)}{\sum_{i=1}^{C_A} C_{i.}\log\left(\frac{C_{i.}}{N}\right) + \sum_{j=1}^{C_B} C_{.j}\log\left(\frac{C_{.j}}{N}\right)}. \tag{5.2}$$

$$Q = \sum_{s=1}^{k}\left[\frac{l_s}{M} - \left(\frac{d_s}{2M}\right)^2\right]. \tag{5.3}$$

Table 5.3 The *NMI* Values and Runtime(s) of the Eight Algorithms Averaging Over 20 Runs on Larger LFR Benchmark Networks.

Network	Measure	CoCoMi	MOCD	MOGA-net	MODTLBO/D	MODPSO	SSCF	Louvain	RMOEA
LFR_10000	*NMI_avg*	/	0.532 ± 0.019(−)	0.524 ± 0.004(−)	/	0.974 ± 0.001(≈)	**0.996** ± 0.002(+)	0.857 ± 0.005(−)	0.986 ± 0.003
	Runtime	/	6,879	28,533	/	32,561	1,666	63	1,811
LFR_20000	*NMI_avg*	/	0.588 ± 0.013(−)	/	/	/	**0.995** ± 0.001(≈)	0.871 ± 0.003(−)	0.994 ± 0.004
	Runtime	/	20,638	/	/	/	10,430	224	4,326
LFR_30000	*NMI_avg*	/	0.605 ± 0.018(−)	/	/	/	/	0.882 ± 0.003(−)	**0.994** ± 0.004
	Runtime	/	39,252	/	/	/	/	476	9,490
LFR_40000	*NMI_avg*	/	/	/	/	/	/	0.889 ± 0.002(−)	**0.995** ± 0.004
	Runtime	/	/	/	/	/	/	868	19,331
LFR_50000	*NMI_avg*	/	/	/	/	/	/	0.894 ± 0.002(−)	**0.985** ± 0.004
	Runtime	/	/	/	/	/	/	1,372	39,356
+/−/≈		0/5/0	0/5/0	0/5/0	0/5/0	0/4/1	1/3/1	0/5/0	-

Note that "/" means that *NMI* values and runtime are not provided here since these results cannot be obtained within 12 hours for one run. The symbols "+," "−," and "≈" indicate that the performance is significantly better, worse, and similar to that of RMOEA, respectively.

In Eq. (5.2), C_A and C_B are the numbers of communities in divisions A and B and C is a confusion matrix, where $C_{i,j}$ is the number of common nodes that belong to community i in division A and community j in division B, and $C_{i.}(C_{.j})$ is the sum of the elements in the ith row (jth column) of the confusion matrix. In Eq. (5.3), M is the number of edges in the network, k is the number of detected communities, l_s is the number of edges contained in community s, and d_s is the sum of degrees of nodes in s. For both *NMI* and Q, a larger value of them represents a higher quality of community division. Due to the fact that the ground truth of synthetic networks is known, the *NMI* is used to measure the performance of algorithms on synthetic networks. As for real-world networks, the Q is employed to evaluate the performance since the ground truth of real-world networks is unknown.

Table 5.3 lists the experimental results of the RMOEA and seven compared algorithms, averaged over 20 runs on LFR benchmark networks with the nodes ranging from 10,000 to 50,000. The Wilcoxon rank sum test results with a significance level of 0.05 are also given at the bottom of the table. From Table 5.3, it can be found that the RMOEA owns better scalability than the other five evolutionary algorithm-based algorithms. Compared to non-EA-based algorithms, although the efficiency of RMOEA is worse than Louvain, the detection performance of RMOEA is significantly better than that of Louvain. Despite that the SSCF shows a better detection performance than RMOEA on the LRF networks with 10,000 and 20,000 nodes, it cannot obtain results within 12 hours on the other networks. Different from the SSCF, the RMOEA can always obtain community results within reasonable runtime on all sizes of LFR networks, and achieve competitive detection performance.

Table 5.4 gives the experimental results of the community detection algorithms averaged over 20 runs on seven real-world networks. In Table 5.4, the Louvain achieves the best computational efficiency on all seven real-world networks. The RMOEA also holds a competitive efficiency, which is much better than the other five evolutionary algorithm-based algorithms. In terms of detection performance, the RMOEA owns the best overall performance. It achieves the best Q on five of seven networks. Based on the above experimental comparison, it can be concluded that the RMOEA is a promising community detection algorithm for large-scale networks, in terms of detection performance and runtime.

5.3 Parallel Multi-Objective Evolutionary Algorithm for Community Detection

Parallelization is a widely adopted technique to enhance the scalability of evolutionary algorithms in tackling large-scale optimization problems. The idea of parallelization has also adopted for community detection in large-scale complex

Table 5.4 The Q Values and Runtime(s) of the Eight Algorithms Averaging Over 20 Runs on Seven Real-World Networks.

Network	Measure	CoCoMi	MOCD	MOGA-net	MODTLBO/D	MODPSO	SSCF	Louvain	RMOEA
net-science	*Q_avg*	$0.905 \pm 0.007(-)$	$0.902 \pm 0.011(-)$	$0.847 \pm 0.008(-)$	$0.907 \pm 0.005(-)$	$0.947 \pm 0.003(\approx)$	$0.830 \pm 0.012(-)$	$0.894 \pm 0.008(-)$	0.952 ± 0.004
	Runtime	1,318	379	546	786	995	16	2	321
blogs	*Q_avg*	$0.758 \pm 0.006(-)$	$0.742 \pm 0.008(-)$	$0.694 \pm 0.012(-)$	$0.718 \pm 0.004(-)$	$0.788 \pm 0.005(-)$	$0.711 \pm 0.005(-)$	$0.729 \pm 0.007(-)$	0.806 ± 0.003
	Runtime	25,657	1,028	2,156	7,965	3,215	3,215	4	628
ca-GrQc	*Q_avg*	/	$0.703 \pm 0.007(-)$	$0.688 \pm 0.008(-)$	$0.698 \pm 0.002(-)$	$0.793 \pm 0.015(\approx)$	$0.734 \pm 0.004(-)$	$0.750 \pm 0.003(-)$	0.796 ± 0.007
	Runtime	/	2,726	3,492	9,682	5,784	369	6	1,272
ca-GrQc	*Q_avg*	/	$0.460 \pm 0.001(\approx)$	$0.486 \pm 0.009(+)$	/	$0.462 \pm 0.016(\approx)$	$0.478 \pm 0.001(-)$	$0.352 \pm 0.001(-)$	0.453 ± 0.005
	Runtime	/	3,835	9,853	/	18,235	836	10	2,465
ca-Hepth	*Q_avg*	/	$0.405 \pm 0.012(-)$	$0.563 \pm 0(-)$	/	$0.566 \pm 0.014(\approx)$	/	$0.583 \pm 0.005(+)$	0.576 ± 0.004
	Runtime	/	9,323	27,864	/	43,168	/	86	3,130
ca-AstroPh	*Q_avg*	/	$0.276 \pm 0.009(-)$	/	/	/	/	$0.530 \pm 0.004(-)$	0.535 ± 0.005
	Runtime	/	21,237	/	/	/	/	305	8,129
ca-CondMat	*Q_avg*	/	/	/	/	/	/	$0.578 \pm 0.002(-)$	0.619 ± 0.001
	Runtime	/	/	/	/	/	/	116	12,917
$+/-/\approx$		0/7/0	0/6/1	1/6/0	0/7/0	0/3/4	0/7/0	1/6/0	–

Note that "/" means that Q values and runtime are not provided here since these results cannot be obtained within 12 hours for one run. The symbols "+," "−," and "≈" indicate that the performance is significantly better, worse, and similar to that of RMOEA, respectively.

networks. Based on this idea, a parallel multi-objective evolutionary algorithm called PMOEA is reported in [27]. In the PMOEA, several key nodes of a complex network are first identified, and then the communities associated with these key nodes are parallel detected by MOEAs. After that, a single-objective evolutionary algorithm is employed to integrate the communities associated with the key nodes, by which the communities of the entire network are obtained. Since the communities associated with the key nodes are detected in parallel, the efficiency of PMOEA in handling large-scale complex networks can be significantly improved. The framework of PMOEA is presented in Fig. 5.6 and the pseudocode is given in Algorithm 5.6.

Algorithm 5.6: Procedure of PMOEA

Input: G (complex network)
Output: $\mathscr{C}$ (the community partition of entire network)

$KV \leftarrow$ Find the key nodes in G by using Algorithm 5.7;
for *all key node* $kv \in KV$ **do**
  $Cset \leftarrow$ Detect communities associated with each key node kv in parallel by using Algorithm 5.8;
  $\mathscr{C}Set \leftarrow \mathscr{C}Set \cup Cset$;
$\mathscr{C} \leftarrow$ Identify the community partition of entire network from $\mathscr{C}Set$ by using Algorithm 5.9;
return $\mathscr{C}$;

5.3.1 Key Node Identification

The pseudocode of identifying key nodes is given in Algorithm 5.7, which consists of the following steps. First, the average degree of all nodes in G is calculated. Next, the nodes that own degree larger than the average degree are determined and added into the set *Lset*. All nodes with the maximum degree are considered as key nodes in G. Then, the nodes with the maximum degree and all other nodes that have at least one edge with these nodes are removed from the *Lset*. Among the remaining nodes in *Lset*, those with the maximum degree are also considered as key nodes. This process is repeated until no node remains in set *Lset*. It is necessary to stress that the key nodes of a network can also be identified by other effective strategies.

5.3.2 Multi-Objective Evolutionary Algorithm for Detecting the Communities Associated with a Key Node

After identifying key nodes of a network, the PMOEA will detect the communities associated with these key nodes in parallel. To this end, an MOEA is devised based

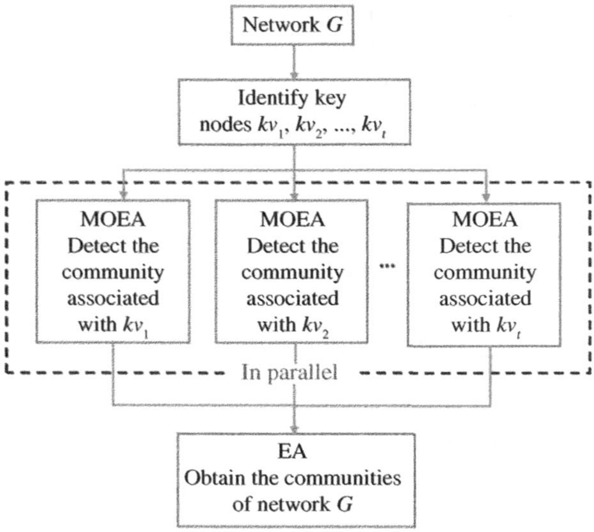

Figure 5.6 Framework of the PMOEA.

Algorithm 5.7: *KeyNodesIdentification*(G)

Input: G (a complex network)
Output: KV (the set of key nodes)

$KV \leftarrow \emptyset$ $d_{abg} \leftarrow$ Calculate the average degree of nodes in G;
$Lset \leftarrow$ Find all the nodes in G that owns degree larger than d_{abg};
while $Lset \neq \emptyset$ **do**
  $kv_i \leftarrow$ Find the node with the largest degree in $Lset$;
  $KV \leftarrow KV \cup kv_i$;
  $Neig(kv_i) \leftarrow$ Find all neighbors of node kv_i from $Lset$;
  $Lset$ Remove the node kv_i and all nodes in $Neig(kv_i)$ from $Lset$;
return The key nodes set KV;

---

on the decomposition-based MOEA (MOEA/D) [22], which detects a community C_i of a key node by simultaneously minimizing the following two objectives.

$$
\begin{aligned}
Cond(C_i) &= \frac{E(C_i, V \setminus C_i)}{min(E(C_i, V), E(V \setminus C_i, V))}, \\
Num(C_i) &= |C_i \cap KV|,
\end{aligned}
\tag{5.4}
$$

where $Cond(C_i)$ is the conductance of community C_i, $E(A, B)$ is the number of edges between nodes in sets A and B, V denotes the set of nodes in G, and $V \setminus C_i$

represents the complement of C_i. If the number of edges in C_i is less than that in $V \setminus C_i$, namely, $E(C_i, V) \leq E(V \setminus C_i, V)$, the $Cond(C_i)$ is calculated by the ratio of number of edges leaving community C_i to that in C_i. If $E(C_i, V) > E(V \setminus C_i, V)$, the $Cond(C_i)$ is determined by the ratio of number of edges leaving community C_i to that in $V \setminus C_i$. The $Cond(C_i)$ has been widely used to detect local communities, and achieved promising results on detecting a single community near a seed point or set [32, 33]. The $Num(C_i)$ is defined as the number of key nodes in community C_i, and minimizing it enables the algorithm to focus on detecting the community associated with a single key node.

The pseudocode of the MOEA developed for detecting community of a key node is presented in Algorithm 5.8. The procedure of MOEA is similar to most of existing MOEA/D variants. Here, the MOEA uses the following Tchebycheff approach to decompose the multi-objective optimization problem into a number of single-objective optimization problems.

$$Minimize \quad g^{te}(x|\lambda_i, z^*) = max_{j=1}^{2}\{\lambda_i^j * (|F_j(x) - z^*|)\}, \tag{5.5}$$

where $\lambda_i = \langle \lambda_i^1, \lambda_i^2 \rangle (\lambda_i^1, \lambda_i^2 \in [0, 1], \lambda_i^1 + \lambda_i^2 = 1)$ is the weight vector, $F_j(x)$ is the jth optimization objective value, and $z^* = \langle z_1^*, z_2^* \rangle$ is the reference point with z_j^*

Algorithm 5.8: *CommunityDetectionMOEA(G, kv)*

Input: G (a complex network); kv (a key node)
Output: $Cset$ (the communities associated with kv)
Initialize the population $P = \{p_1, p_2, \cdots, p_{pop}\}$;
Initialize the reference point z^*;
$W \leftarrow$ Generate pop well-distributed weight vectors $\lambda_1, \lambda_2, \ldots, \lambda_{pop}$;
for *each weight vectors* λ_i **do**
    $N_i \leftarrow$ Find T individuals from P with the smallest Euclidean distance to the weight vectors λ_i;
for $gen = 1$ *to maximum number of generations* **do**
    for $i = 1$ *to pop* **do**
        Randomly select two individuals from N_i;
        Generate an offspring *offs* by crossover and mutation operators;
        Evaluate *offs* by the two objectives defined in Eq. (5.4);
        Replace each individual in N_i with offspring *offs* if the Tchebycheff value of *offs* is better than the individual in N_i;
        Update reference point z^* if the offspring *offs* owns better value than z^* on one or two objectives;
$Cset \leftarrow$ Select individual with the best Tchebycheff value in each N_i;
return *Cset*;

$(1 \leq j \leq 2)$ being the minimal value of jth optimization objective for all individuals in population.

Specifically, the procedure of Algorithm 5.8 for detecting communities associated with a key node consists of four main steps. The first step is the population initialization. For a given complex network G, the individual in the population is encoded as $\langle x_{kv}, x_1, x_2, \dots, x_i \rangle, 1 \leq i \leq N_{kv}$, where N_{kv} is the number of neighbors of key node kv, x_{kv} is the label of key node kv, and $x_1, x_2, \dots, x_i$ are the labels of nodes that belong to the same community of key node kv. For initializing the population, *pop* individuals are generated by using the labels of key node kv and a subset of neighbors of kv. The second step is the initialization of reference point z^* and a set of well-distributed weight vectors $\lambda_1, \lambda_2, \dots, \lambda_{pop}$, where *pop* is a predefined parameter. The reference point z^* is initialized by the best values of *Cond* and *Num* of individuals in the initial population, and the weight vectors $\lambda_1, \lambda_2, \dots, \lambda_{pop}$ are uniformly generated. The third step is the assignment of individuals to weight vectors. For each weight vectors λ_i, it is assigned T individuals, denoted by N_i according to the Euclidean distance between individuals and λ_i in the objective space. The N_i is called the neighborhood of λ_i and all individuals in N_i are called neighbors. The fourth step is the evolution of population. For each weight vector λ_i, two individuals are randomly selected from its neighborhood N_i, and an offspring *offs* is generated based on these two individuals by using the crossover and mutation operators described in Sections 5.3.2.1 and 5.3.2.2. The offspring *offs* will replace each individual in N_i if the *offs* owns better Tchebycheff value than the individual. Moreover, the reference point z^* also needs to be updated if *offs* has better one or two objective values than that in z^*. The fourth step will be repeated until the maximum number of generations is reached.

Using the MOEA above, a set of communities associated with a given key node can be detected, and the detection of communities for different key nodes can be easily parallelized. Since the MOEA only detects the communities associated with a key node instead of all communities in the complex network, the dedicated crossover and mutation operators need to be devised to guide the evolution of population, which are described as follows.

5.3.2.1 Crossover Operator

For any two individuals $p_1 = \langle x_1, x_2, \dots, x_i \rangle$ and $p_2 = \langle y_1, y_2, \dots, y_j \rangle$ in the neighborhood of a weight vector, the labels of nodes in p_2 but not in p_1 are first found. Let us assume that $y_{k_1}, y_{k_2}, \dots, y_{k_l}$ are these labels. Then, for each y_{k_s}, $1 \leq s \leq l$, we calculate the interest of node y_{k_s} joining the community encoded by individual p_1 as follows:

$$J(y_{k_s}, p_1) = \begin{cases} E_{in}/E_{out}, & E_{out} \neq 0, \\ 1, & E_{out} = 0, \end{cases} \tag{5.6}$$

where the E_{in} is the number of edges between the community encoded by individual p_1 and node y_{k_s}, and the E_{out} denotes the number of edges between nodes outside the community of individual p_1 and node y_{k_s}. The larger the value of $J(y_{k_s}, p_1)$ is, the higher the probability of node y_{k_s} belonging to the community of p_1 is.

After calculating the interest of all found nodes, a number *rand* is randomly generated in the range of $[0, 1]$. If the values of $J(y_{k_s}, p_1)$ is larger than *rand*, then the label of node y_{k_s} will be added to individual p_1. Otherwise, it will not be added. Assuming that the added labels are $y_{k_1}, y_{k_2}, \dots, y_{k_m}$, an offspring $offs = \langle x_1, x_2, \dots, x_i, y_{k_1}, y_{k_2}, \dots, y_{k_m} \rangle$ can be generated from the crossover between individuals p_1 and p_2.

5.3.2.2 Mutation Operator

For an offspring $offs = \langle z_1, z_2, \dots, z_t \rangle$ generated by the crossover operator, the mutation is conducted as follows. First, a number is randomly generated from the range of $[0, 1]$ to decide whether to add or delete some nodes from *offs*. If the random number is less than 0.5, then the removal operator is conducted. Otherwise, the addition operator is performed.

If the removal operator is selected, another random number is generated from the range of $[0, 1]$ for each label in *offs* and the label will be removed from *offs* if the random number is less than 0.5. For the addition operator, all the neighbors of the nodes in community of *offs* are first found, and then the interest of each node joining the community of *offs* is calculated according to Eq. (5.6). The label of a neighbor will be added to *offs* if a random number generated from the range of $[0, 1]$ is less than the corresponding value of interest. By using the removal and addition operators, a new individual can be generated by the mutation from offspring *offs*.

Fig. 5.7 plots an illustrative example of crossover and mutation operators. As shown in Fig. 5.7(a), the complex network to be detected owns 10 nodes labeled as $1, 2, 3, \dots, 10$, and the node 1 is the key node. Fig. 5.7(b) gives two individuals $p1$ and $p2$ of a weight vector, which encode two communities associated with the key node 1 $\langle 1, 2, 4 \rangle$ and $\langle 1, 4, 5, 6, 7 \rangle$, respectively. From Fig. 5.7(b), it can be found that the labels 5, 6, and 7 in $p2$ do not appear in $p1$. According to Eq. (5.6), the interest of nodes 5, 6, and 7 added to the community of $p1$ are 0.5, 2, and 0, respectively. Assume that the random numbers generated for nodes 5, 6, and 7 are 0.6, 0.8, and 0.2, respectively. This means that only the label of node 6 is added to the community of $p1$, and thus an offspring $p' = \langle 1, 2, 4, 6 \rangle$ is generated by the crossover operator as shown in Fig. 5.7(c).

In Fig. 5.7(d), the mutation operator is applied to $p' = \langle 1, 2, 4, 6 \rangle$. Let us assume that the addition operator is conducted. In Fig. 5.7(d), there are two neighbors of nodes in the community of p', namely, nodes 3 and 5. The interests of them joining

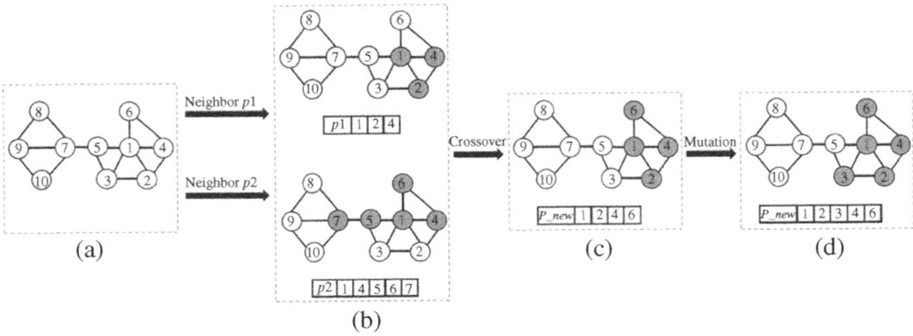

Figure 5.7 An illustrative example of crossover and mutation operators. (a) A network G with 10 nodes, where node 1 is assumed to be a key node. (b) Two individuals $p1$ and $p2$ of a weight vector, which encode two communities associated with key node 1 $\{1, 2, 4\}$ and $\{1, 4, 5, 6, 7\}$, respectively. (c) An offspring p' generated by crossover operator from individuals $p1$ and $p2$. (d) An individual p_{new} generated by the mutation operator from offspring *offs*.

the community of p' are 2 and 0.5, respectively. If the randomly number generated for nodes 3 and 5 are smaller than 2 and larger than 0.5, only the label of node 3 is added to p', and, hence, a new individual $p_{new} = \langle 1, 2, 3, 4, 6\rangle$ is obtained by the mutation operator as displayed in Fig. 5.7(d).

5.3.3 Evolutionary Algorithm for Determining the Communities of the Entire Network

After detecting the communities associated with all key nodes, the PMOEA will obtain the communities of the entire network based on these detected communities. Due to the fact that the MOEA returns a set of communities for each key node, the PMOEA employs an evolutionary algorithm to select a suitable community from the set of communities detected for each key node and then constitutes the community partition of the entire network by using the selected communities. The pseudocode of evolutionary algorithm for determining the community partition of the entire network is presented in Algorithm 5.9.

Algorithm 5.9 consists of four main steps. First, a population P' is randomly generated, where the individual p' is encoded as $\langle x'_1, x'_2, \dots, x'_t\rangle$, $x'_i \in \{1, 2, \dots, |\mathscr{C}_i|\}$ represents the community selected from the set of detected communities $\mathscr{C}_i$ of ith key node, and t is the number of identified key nodes in network G. Second, two individuals are randomly selected from the population P' and then two offsprings are generated by using the one-point crossover and mutation operators. The mutation operator is conducted by changing each x'_i in individual p' with a probability m_p, $i \in \{1, 2, \dots, t\}$. Third, the offsprings will replace their parents if they have better modularity than their parents. The second and third steps will

Algorithm 5.9: *CommunityDeterminationEA*(G, kv)

Input: $\mathscr{C}$ *Set* $= \{\mathscr{C}_1, \mathscr{C}_2, \ldots, \mathscr{C}_t\}$ ($\mathscr{C}_i, i \leq i \leq t$, denotes the set of communities detected by the developed MOEA for the ith key node, where t is the number of identified key nodes in network G)

Output: $\mathscr{C}$ (the community partition of entire network)

```
Initialize the population P' = {p'_1, p'_2, ..., p'_pop'};
for gen = 1 to maximum number of generations do
    Randomly select two individuals from P';
    Generate two offspring by one-point crossover and mutation operators;
    Evaluate offspring according to modularity Q;
    Replace parents with offspring if the offspring owns better modularity;
𝒞 ← Find communities represented by the best solution in P;
𝒞 ← Merge the communities in 𝒞 with overlapping rate overlap larger than ℵ;
Rest ← Find the nodes that are not assigned to any community;
𝒞 ← Assign the nodes in Rest to the communities that contain the most
    neighbors of this node;
return 𝒞;
```

be repeated until the maximum number of generations is reached. Fourth, the community partition $\mathscr{C}$ of the entire network is improved by using the following strategy. If the similarity between two communities C_i and C_j in $\mathscr{C}$ is larger than a predefined threshold $\aleph \in [0, 1]$, then they are merged into one community. The similarity can be calculated as follows:

$$Similarity(C_i, C_j) = \frac{|C_i \cap C_j|}{min(|C_i|, |C_j|)}, \tag{5.7}$$

where $|C_i|$ represents the number of nodes in community C_i. Moreover, considering that some nodes might not be assigned to any community, we also need to assign each of these nodes to a community to which most of its neighbors have been assigned.

5.3.4 Time Complexity of PMOEA

For a given complex network with n nodes and m edges, the PMOEA owns the time complexity of $O(maxgen * pop * m)$. Specifically, the key node identification step of PMOEA has the time complexity of $O(n' * D)$, where D is the largest degree of nodes in the complex network and n' is the number of nodes that have larger degree than the average degree. The time complexity of detecting the communities associated with key nodes is $O(maxgen_1 * pop_1 * (n_c *$

$D + (n_c + m_c))) \approx O(maxgen_1 * pop_1 * n_c * D)$. It is because that the crossover and mutation operators check at most $O(n_c * D)$ neighboring nodes for an individual which encodes a community with n_c nodes and m_c edges, and the complexity of evaluating such individual is $O(n_c + m_c)$. Since the population has pop_1 individuals and the maximum number of generations is $maxgen_1$, the time complexity of the second step is $O(maxgen_1 * pop_1 * (n_c * D + (n_c + m_c)))$. As for the third step, the calculation of modularity for each individual consumes the most computational time, of which time complexity is $O(maxgen_2 * pop_2 * m)$ for an evolutionary algorithm with $maxgen_2$ generations and pop_2 individuals. Assuming that $maxgen = max(maxgen_1, maxgen_2)$ and $pop = max(pop_1, pop_2)$, the PMOEA holds a complexity of $O(maxgen * pop * m)$, given that m is usually larger than $n' * D$ and $n_c * D$.

5.3.5 Experimental Comparisons

To verify the performance of PMOEA, it is compared to eight algorithms tailored for community detection. For these compared algorithms, six of them are evolutionary algorithm-based algorithms, namely, CoCoMi [23], MOCD [13], MOGA-net [16], MODPSO [14], QDM-PSO [34], and PaGeCoDe [35], and two of them are non-EA-based algorithms including LET [36] and FOCS [37]. The parameters of these algorithms are set to the values recommended in the original references, which are also listed in Table 5.5. It is worth noting that in the MOEA of PMOEA, the crossover and mutation probabilities are set to 0.9 and 0.6,

Table 5.5 Parameter Setting of the Compared Algorithms, Where *pop* is the Population Size, *maxgen* is the Maximum Generations of the Algorithm, c_p is the Probability of Crossover, m_p is the Probability of Mutation, and T is the Neighborhood Size.

Algorithm	*pop*	*maxgen*	c_p	m_p	T	References
CoCoMi	100	100	0.9	0.1	—	[23]
MOCD	100	100	0.9	0.1	—	[13]
QDM-PSO	100	100	0.9	—	—	[34]
MOGA-net	100	100	0.9	0.1	—	[16]
PaGeCoDe	100	100	0.8	0.01	—	[35]
MODPSO	100	100	—	0.1	40	[14]
LET	—	—	—	—	—	[36]
FOCS	—	—	—	—	—	[37]
PMOEA	100	50	0.9	0.6	40	[19]

respectively, and in the EA of PMOEA, the crossover and mutation probabilities are set to 0.6 and 0.01, respectively. Moreover, the parameter α in PMOEA for merging two communities is set to 0.4 in the experiments.

Two groups of complex networks are used to verify the detection performance of different algorithms. The first group is the synthetic networks which are generated by using the LFR model [26] with nodes ranging from 50,000 to 200,000. The second group is the real-world networks, including five networks from the Stanford large network dataset [29], namely, *ca-HepTh2*, *ca-AstroPh*, *ca-CondMat*, *loc-Brightkite*, and *loc-Gowalla*. The details of these networks are given in Table 5.6. The evaluation metric used for synthetic networks is the *NMI* since the ground truth of these networks is known, and the modularity (Q) is adopted for real-world networks due to the fact that most of them do not have a known ground truth.

Table 5.7 presents the detection performance of the PMOEA and the eight compared algorithms on LFR benchmark networks with a size ranging from 50,000 to 200,000. The Wilcoxon rank-sum test results with a significance level of 0.05 are also given at the bottom of the tables. From Table 5.7, it can be found that the PMOEA achieves the best detection quality on all LFR networks. On the contrary, most of the evolutionary algorithm-based compared algorithms cannot obtain results within 17 hours. Although the FOCS owns higher computational efficiency than PMOEA, the detection quality is much worse than PMOEA. Table 5.8 lists the detection quality of different algorithms on five real-world networks, where the PMOEA also holds the best value of Q. As for the other evolutionary algorithm-based community detection algorithms, although some of them can complete within 17 hours, the Q value of their detection results is worse than that of PMOEA. The non-EA-based algorithm FOCS consumes less computational time than PMOEA, but its detection quality is inferior to that of the PMOEA. Hence, it can be concluded that the PMOEA owns a good scalability for large-scale complex networks in terms of both efficiency and detection performance.

Table 5.6 Details of Five Real-World Networks Used in Experiments.

Network	Node number	Link number	Average degree
ca-Hepth	12,008	118,521	19.74
ca-AstroPh	18,772	198,080	21.10
ca-CondMat	23,133	93,468	8.08
loc-Brightkite	58,228	214,078	7.35
loc-Gowalla	196,591	950,327	9.67

Table 5.7 The *NMI* Values and Runtime(s) of the Nine Algorithms Averaging Over 20 Runs on Larger LFR Benchmark Networks.

Network	Measure	CoCoMi	QDM-PSO	MOGA-net	MODPSO	MOCD	PaGeCoDe	LET	FOCS	PMOEA
LFR_50000	*NMI_avg*	/	/	/	/	/	0.630 ± 0.007(−)	0.988 ± 0.001(−)	0.716 ± 0(−)	**0.999** ± 0.001
	Runtime	/	/	/	/	/	30,716	21,235	89	2,923
LFR_100000	*NMI_avg*	/	/	/	/	/	/	0.981 ± 0.001(−)	0.721 ± 0(−)	**0.998** ± 0.002
	Runtime	/	/	/	/	/	/	51,712,224	229	11,950
LFR_150000	*NMI_avg*	/	/	/	/	/	/	/	0.721 ± 0(−)	**0.999** ± 0.001
	Runtime	/	/	/	/	/	/	/	1,979	25,576
LFR_200000	*NMI_avg*	/	/	/	/	/	/	/	0.720 ± 0 (−)	**0.999** ± 0.002
	Runtime	/	/	/	/	/	/	/	1,979	59,424
+/−/≈		0/4/0	0/4/0	0/4/0	0/4/0	0/4/0	0/4/0	0/4/0	0/4/0	—

Note that "/" means that *NMI* values and runtime are not provided here since these results cannot be obtained within 17 hours for one run. The symbols "+," "−," and "≈" indicate that the performance is significantly better, worse, and similar to that of PMOEA, respectively.

Table 5.8 The Q Values and Runtime(s) of the Nine Algorithms Averaging Over 20 Runs on Five Real-World Networks.

Network	Measure	CoCoMi	QDM-PSO	MOGA-net	MODPSO	MOCD	PaGeCoDe	LET	FOCS	PMOEA
ca-Hepth	*Q_avg*	/	/	$0.563 \pm 0(-)$	$0.566 \pm 0.014(-)$	$0.405 \pm 0.012(-)$	$0.361 \pm 0.001(-)$	$0.579 \pm 0.001(-)$	$0.361 \pm 0(-)$	**0.596** $\pm$ 0.009
	Runtime	/	/	27,864	43,168	9,323	4,863	1,954	10	110
ca-AstroPh	*Q_avg*	/	/	/	/	$0.276 \pm 0.009(-)$	$0.238 \pm 0.005(-)$	$0.483 \pm 0.002(-)$	$0.252 \pm 0(-)$	**0.579** $\pm$ 0.001
	Runtime	/	/	/	/	21,237	10,015	10,581	18	324
ca-CondMat	*Q_avg*	/	/	/	/	$0.451 \pm 0.002(-)$	$0.433 \pm 0.002(-)$	$0.537 \pm 0.001(-)$	$0.294 \pm 0(-)$	**0.655** $\pm$ 0.001
	Runtime	/	/	/	/	45,118	16,157	16,157	87	613
lo-Brightkite	*Q_avg*	/	/	/	/	/	$0.349 \pm 0.006(-)$	$0.487 \pm 0.002(-)$	$0.090 \pm 0(-)$	**0.575** $\pm$ 0.005
	Runtime	/	/	/	/	/	54,571	35,079	435	3,308
loc-Gowalla	*Q_avg*	/	/	/	/	/	/	/	$0.112 \pm 0(-)$	**0.338** $\pm$ 0.003
	Runtime	/	/	/	/	/	/	/	2,921	40,194
+/−/≈		0/5/0	0/5/0	0/5/0	0/5/0	0/5/0	0/5/0	0/5/0	0/5/0	—

Note that "/" means that *NMI* values and runtime are not provided here since these results cannot be obtained within 17 hours for one run. The symbols "+," "−," and "≈" indicate that the performance is significantly better, worse, and similar to that of PMOEA, respectively.

5.4 Summary

This chapter presents two ideas of enhancing the scalability of MOEAs in solving the community detection problems for large-scale complex networks, and two MOEAs, namely, RMOEA and PMOEA, are developed based on the ideas. The RMOEA utilizes the local communities detected during the evolution to reduce the size of complex networks, so that the scalability of MOEA to large-scale complex networks can be considerately enhanced. The PMOEA is based on the idea of parallelization to tackle large-scale complex networks, where the community detection of the entire network is decomposed into detecting the communities associated with a number of key nodes performed in parallel. Although the reduction and parallelization are two widely used technologies for large-scale optimization, there are many other technologies that can be used to deal with large-scale optimization, such as decomposition [38] and search enhancement [39, 40]. In the following chapters, a number of MOEAs with different strategies to tackle the large-scale optimization in other applications are presented.

References

1 G. Appel, L. Grewal, R. Hadi, and A. T. Stephen, "The future of social media in marketing," *Journal of the Academy of Marketing Science*, vol. 48, no. 1, pp. 79–95, 2020.

2 G. C. Koh, P. Porras, B. Aranda, H. Hermjakob, and S. E. Orchard, "Analyzing protein–protein interaction networks," *Journal of Proteome Research*, vol. 11, no. 4, pp. 2014–2031, 2012.

3 H. Jiang, F. Zhan, C. Wang, J. Qiu, Y. Su, C. Zheng, X. Zhang, and X. Zeng, "A robust algorithm based on link label propagation for identifying functional modules from protein–protein interaction networks," *IEEE/ACM Transactions on Computational Biology and Bioinformatics*, vol. 19, no. 3, pp. 1435–1448, 2020.

4 G. Rossetti and R. Cazabet, "Community discovery in dynamic networks: a survey," *ACM Computing Surveys*, vol. 51, no. 2, pp. 1–37, 2018.

5 W. Luo, D. Zhang, H. Jiang, L. Ni, and Y. Hu, "Local community detection with the dynamic membership function," *IEEE Transactions on Fuzzy Systems*, vol. 26, no. 5, pp. 3136–3150, 2018.

6 H. Jiang, Z. Liu, C. Liu, Y. Su, and X. Zhang, "Community detection in complex networks with an ambiguous structure using central node based link prediction," *Knowledge-Based Systems*, vol. 195, p. 105626, 2020.

7 Y. Su, C. Liu, Y. Niu, F. Cheng, and X. Zhang, "A community structure enhancement-based community detection algorithm for complex networks," *IEEE Transactions on Systems, Man, and Cybernetics: Systems*, vol. 51, no. 5, pp. 2833–2846, 2019.

8 M. E. Newman and M. Girvan, "Finding and evaluating community structure in networks," *Physical Review E*, vol. 69, no. 2, p. 026113, 2004.

9 A. Mahmood and M. Small, "Subspace based network community detection using sparse linear coding," *IEEE Transactions on Knowledge and Data Engineering*, vol. 28, no. 3, pp. 801–812, 2015.

10 A. Pothen, "Graph partitioning algorithms with applications to scientific computing," in Proceedings of the 1997 *Parallel Numerical Algorithms*, 1997, pp. 323–368.

11 S. Srinivas and C. Rajendran, "Community detection and influential node identification in complex networks using mathematical programming," *Expert Systems with Applications*, vol. 135, pp. 296–312, 2019.

12 C.-H. Mu, J. Xie, Y. Liu, F. Chen, Y. Liu, and L.-C. Jiao, "Memetic algorithm with simulated annealing strategy and tightness greedy optimization for community detection in networks," *Applied Soft Computing*, vol. 34, pp. 485–501, 2015.

13 C. Shi, Z. Yan, Y. Cai, and B. Wu, "Multi-objective community detection in complex networks," *Applied Soft Computing*, vol. 12, no. 2, pp. 850–859, 2012.

14 M. Gong, Q. Cai, X. Chen, and L. Ma, "Complex network clustering by multiobjective discrete particle swarm optimization based on decomposition," *IEEE Transactions on Evolutionary Computation*, vol. 18, no. 1, pp. 82–97, 2013.

15 A. Lancichinetti, S. Fortunato, and J. Kertész, "Detecting the overlapping and hierarchical community structure in complex networks," *New Journal of Physics*, vol. 11, no. 3, p. 033015, 2009.

16 C. Pizzuti, "A multiobjective genetic algorithm to find communities in complex networks," *IEEE Transactions on Evolutionary Computation*, vol. 16, no. 3, pp. 418–430, 2011.

17 X. Ma, F. Liu, Y. Qi, X. Wang, L. Li, L. Jiao, M. Yin, and M. Gong, "A multiobjective evolutionary algorithm based on decision variable analyses for multiobjective optimization problems with large-scale variables," *IEEE Transactions on Evolutionary Computation*, vol. 20, no. 2, pp. 275–298, 2015.

18 X. Zhang, Y. Tian, R. Cheng, and Y. Jin, "A decision variable clustering-based evolutionary algorithm for large-scale many-objective optimization," *IEEE Transactions on Evolutionary Computation*, vol. 22, no. 1, pp. 97–112, 2016.

19 X. Zhang, K. Zhou, H. Pan, L. Zhang, X. Zeng, and Y. Jin, "A network reduction-based multiobjective evolutionary algorithm for community detection in large-scale complex networks," *IEEE Transactions on Cybernetics*, vol. 50, no. 2, pp. 703–716, 2018.

20 L. Zhang, H. Pan, Y. Su, X. Zhang, and Y. Niu, "A mixed representation-based multiobjective evolutionary algorithm for overlapping community detection," *IEEE Transactions on Cybernetics*, vol. 47, no. 9, pp. 2703–2716, 2017.

21 Y. Park and M. Song, "A genetic algorithm for clustering problems," in *Proceedings of the 3rd Annual Conference on Genetic Programming*, vol. 1998, 1998, pp. 568–575.

22 Q. Zhang and H. Li, "MOEA/D: A multiobjective evolutionary algorithm based on decomposition," *IEEE Transactions on Evolutionary Computation*, vol. 11, no. 6, pp. 712–731, 2007.

23 S. He, G. Jia, Z. Zhu, D. A. Tennant, Q. Huang, K. Tang, J. Liu, M. Musolesi, J. K. Heath, and X. Yao, "Cooperative co-evolutionary module identification with application to cancer disease module discovery," *IEEE Transactions on Evolutionary Computation*, vol. 20, no. 6, pp. 874–891, 2016.

24 D. Chen, F. Zou, R. Lu, L. Yu, Z. Li, and J. Wang, "Multi-objective optimization of community detection using discrete teaching–learning-based optimization with decomposition," *Information Sciences*, vol. 369, pp. 402–418, 2016.

25 V. D. Blondel, J.-L. Guillaume, R. Lambiotte, and E. Lefebvre, "Fast unfolding of communities in large networks," *Journal of Statistical Mechanics: Theory and Experiment*, vol. 2008, no. 10, p. P10008, 2008.

26 A. Lancichinetti, S. Fortunato, and F. Radicchi, "Benchmark graphs for testing community detection algorithms," *Physical Review E*, vol. 78, no. 4, p. 046110, 2008.

27 Y. Su, K. Zhou, X. Zhang, R. Cheng, and C. Zheng, "A parallel multi-objective evolutionary algorithm for community detection in large-scale complex networks," *Information Sciences*, vol. 576, pp. 374–392, 2021.

28 S. Gregory, "Finding overlapping communities in networks by label propagation," *New Journal of Physics*, vol. 12, no. 10, p. 103018, 2010.

29 J. Leskovec and A. Krevl, "Snap datasets: stanford large network dataset collection," *Available*: http://snap.stanford.edu/data, 2014.

30 L. Danon, A. Diaz-Guilera, J. Duch, and A. Arenas, "Comparing community structure identification," *Journal of Statistical Mechanics: Theory and Experiment*, vol. 2005, no. 09, p. P09008, 2005.

31 M. E. Newman, "Fast algorithm for detecting community structure in networks," *Physical Review E*, vol. 69, no. 6, p. 066133, 2004.

32 J. J. Whang, D. F. Gleich, and I. S. Dhillon, "Overlapping community detection using neighborhood-inflated seed expansion," *IEEE Transactions on Knowledge and Data Engineering*, vol. 28, no. 5, pp. 1272–1284, 2016.

33 H. Yin, A. R. Benson, J. Leskovec, and D. F. Gleich, "Local higher-order graph clustering," in *Proceedings of the 23rd ACM SIGKDD International Conference on Knowledge Discovery and Data Mining*, 2017, pp. 555–564.

34 L. Li, L. Jiao, J. Zhao, R. Shang, and M. Gong, "Quantum-behaved discrete multi-objective particle swarm optimization for complex network clustering," *Pattern Recognition*, vol. 63, pp. 1–14, 2017.

35 M. Joldos and C. C. Technical, "A parallel evolutionary approach to community detection in complex networks," in *Proceedings of the 13th IEEE International Conference on Intelligent Computer Communication and Processing*, 2017, pp. 247–254.

36 J. Huang, H. Sun, Y. Liu, Q. Song, and T. Weninger, "Towards online multiresolution community detection in large-scale networks," *PLoS One*, vol. 6, no. 8, p. e23829, 2011.

37 S. Bandyopadhyay, G. Chowdhary, and D. Sengupta, "FOCS: Fast overlapped community search," *IEEE Transactions on Knowledge and Data Engineering*, vol. 27, no. 11, pp. 2974–2985, 2015.

38 J. Xiao, T. Zhang, J. Du, and X. Zhang, "An evolutionary multiobjective route grouping-based heuristic algorithm for large-scale capacitated vehicle routing problems," *IEEE Transactions on Cybernetics*, vol. 51, no. 8, pp. 4173–4186, 2019.

39 Y. Tian, Y. Feng, C. Wang, R. Cao, X. Zhang, X. Pei, K. C. Tan, and Y. Jin, "A large-scale combinatorial many-objective evolutionary algorithm for intensity-modulated radiotherapy planning," *IEEE Transactions on Evolutionary Computation*, vol. 26, no. 6, pp. 1511–1525, 2022.

40 Y. Tian, L. Si, X. Zhang, R. Cheng, C. He, K. C. Tan, and Y. Jin, "Evolutionary large-scale multi-objective optimization: a survey," *ACM Computing Surveys*, vol. 54, no. 8, pp. 1–34, 2021.

6

Evolutionary Large-Scale Multi-Objective Optimization in Logistics Scheduling

6.1 Introduction

Logistics has been known as an important foundation of worldwide economic development and thus widely exists in social life [1]. A good logistics scheduling plan guarantees that the materials flow fluently in a logistics system and significantly contributes to economic growth [2]. Hence, logistics scheduling remains a hot research field over decades, which involves a large variety of optimization problems: facility location, vehicle routing, and container loading, to name a few [3–5]. A large number of algorithms have been developed to address these optimization problems, which are roughly categorized into two groups: exact algorithms and heuristic ones. Exact algorithms often formulate logistics scheduling problems as mathematical programming models that can be iteratively computed, such as linear programming models or dynamic programming models [6]. Heuristic algorithms iteratively seek optimal solutions in the search space of logistics scheduling problems in a flexible way [7]. Exact algorithms are often able to obtain the optimal solution of a small-scale single-objective logistics scheduling problem at an acceptable computational cost. Nevertheless, they can barely obtain high-quality solutions for a large-scale multi-objective logistics scheduling problem, which typically formulates a multi-objective problem as a single-objective optimization model by a weighted method and neglects a potential conflict between multiple objectives. Thereafter, heuristic algorithms are developed to solve multi-objective logistics scheduling problems due to their capability of simultaneously optimizing multiple objectives, which, however, still have difficulty in addressing large-scale multi-objective problems. Due to the NP-hard characteristic of many logistics scheduling problems, a large-scale multi-objective logistics scheduling problem generally has a huge solution search space, which hinders many existing heuristic algorithms from obtaining satisfying solutions at an acceptable computational cost.

Evolutionary Large-Scale Multi-Objective Optimization and Applications: Integrating Evolutionary Computation, Machine Learning, and Data Science,
First Edition. Xingyi Zhang, Ran Cheng, Ye Tian, and Yaochu Jin.

Evolutionary algorithms (EAs) are a type of popular metaheuristic algorithms and have exhibited good performance in addressing multi-objective logistics scheduling problems [8, 9]. Due to the fact that EAs are population-based searching methods, they are able to simultaneously optimize multiple objectives by searching in multiple directions in the decision space. Nevertheless, it is still difficult for EAs to handle large-scale multi-objective logistics scheduling problems, since the performance of an EA considerably deteriorates with the rapid growth of problem size. This motivates many researchers to improve EAs for large-scale multi-objective logistics scheduling problems, and this chapter presents two state-of-the-art EAs for two typical large-scale multi-objective logistics scheduling problems: an evolutionary multi-objective route grouping-based heuristic algorithm (EMRG-HA) for the vehicle routing problem (VRP) [10] and a clustering-based surrogate-assisted multi-objective EA with reference point adaptation (AR-MOEA+SA) for the facility location problem (FLP) [4]. The VRP presented in this chapter is the classic and basic capacitated vehicle routing problem (CVRP), which serves as a reliable testbed for many VRP optimization algorithms. The FLP presented in this chapter is a real-world application in emergency management: the uncertain shelter location problem (USLP).

6.2 Evolutionary Multi-Objective Route Grouping-Based Heuristic Algorithm for Large-Scale Capacitated Vehicle Routing Problems

The task of a typical CVRP is to plan the optimal traveling routes with the lowest traveling cost for a fleet of vehicles so as to serve a set of customers with known demands, where all the vehicles are assumed to be homogeneous and thus have the same capacity. The large-scale capacitated vehicle routing problem (LSCVRP) refers to the CVRP with thousands of customers to be served, which is challenging and draws much attention from CVRP researchers [11–14]. Particularly, the idea of divide-and-conquer has been widely used in developing large-scale optimization algorithms by decomposing a large-scale optimization problem into some small sub-components that can be easily and independently solved [15, 16]. In the field of large-scale logistics scheduling, the idea of divide-and-conquer is mainly used in two ways: to geographically decompose customers into groups of customers [17, 18] or to decompose routes into groups of routes for service of different vehicles [19, 20]. Regardless of the studies that use divide-and-conquer for LSCVRPs, many of these studies fail to produce a good decomposition for either customers or routes [18, 21]. The customer-decomposition strategies could easily result in that customers served by the same vehicles in the optimal routes cannot be divided into a group, since the customers are divided into groups based on their geographical locations. For the route-decomposition strategies, they are often hard to achieve a good decomposition since the clustering of routes is more difficult than the clustering of customers, especially for the definition of intraclustering and interclustering distances for routes.

Thereafter, this section presents a divide-and-conquer-based heuristic algorithm EMRG-HA for solving LSCVRPs, which reports an evolutionary multi-objective route grouping method to achieve promising decomposition for LSCVRPs [22]. By exploiting the advantage of evolutionary multi-objective optimization, the route grouping method mathematically formulates a typical CVRP as a tri-objective model by simultaneously optimizing the intra-group similarity, inter-group similarity and inter-group balance. The three objectives guide the optimization of LSCVRPs to find high-quality route groupings with low inter-group similarity, high intra-group similarity and inter-group balance. Because of the fact that route groupings have a great influence on the quality of routes, the high-quality route groupings thus can lead to high-quality routes in the optimization of LSCVRPs.

In the following, we first present the formal definition of CVRP, then the EMRG-HA for solving LSCVRPs is elaborated. Finally, experimental comparisons between the EMRG-HA and state-of-the-art approaches for CVRPs are presented and discussed.

6.2.1 Problem Formulation

The CVRP is defined on a undirected graph $G = (V, E)$, where $V = \{v_i | i = 0, 1, \dots, N\}$ is a set of vertices and $E = \{e_{ij} | i, j = 0, 1, \dots, N\}$ is a set of edges between these vertices. Vertex v_0 denotes the depot which all vehicles should start from and return back to when the vehicles finish serving customers. Vertices v_1 to v_N represent N customers to be served by the vehicles, and each v_i has a predefined demand d_i. All the vehicles have the same capacity Q. A vehicle can successfully execute a planned route only when the total demand of customers in this route does not exceed the vehicle capacity. The notations used in the CVRP formulation are presented in Table 6.1. The objective of CVRP is to minimize the total traveling cost of vehicles [23], which is presented as follows:

$$\min \sum_{i=0}^{N} \sum_{j=0}^{N} c_{ij} x_{ij}, \tag{6.1}$$

where $c_{ij} \geq 0\ (0 \leq i, j \leq N)$ denotes the traveling cost of vehicles from vertices v_i to v_j, $x_{ij} \in \{0, 1\}\ (0 \leq i, j \leq N)$ denotes whether the route contains the route segment from vertices v_i to v_j, $x_{ij} = 1$ indicates that the route does, and $x_{ij} = 0$ otherwise.

Table 6.1 Notations in CVRP Formulation.

Notation	Definition
x_{ij}	Whether the route contains the travel from vertices v_i to v_j, $x_{ij} \in \{0, 1\}$, $x_{ij} = 1$ indicates that one vehicle travels from vertices v_i to v_j, and $x_{ij} = 0$ otherwise
c_{ij}	The traveling cost of vehicles from vertices v_i to v_j
u_i	The load of the vehicle after visiting customer at vertex v_i
K	The maximum number of vehicles which are allowed to be used

The CVRP should satisfy the following constraints:

$$\begin{cases} \sum_{i=0}^{N} x_{ij} = 1, \ \forall j = 1, 2, \dots, N, \\ \sum_{j=0}^{N} x_{ij} = 1, \ \forall i = 1, 2, \dots, N, \\ \sum_{i=0}^{N} x_{i0} = K, \\ \sum_{j=0}^{N} x_{0j} = K, \\ u_i - u_j + Qx_{ij} \leq Q - d_j, \ \forall i, j = 1, 2, \dots, N, i \neq j, \ d_i + d_j \leq Q, \\ d_i \leq u_i \leq Q, \ \forall i = 1, 2, \dots, N, \\ x_{ij} \in \{0, 1\}, \ \forall i, j = 1, 2, \dots, N. \end{cases} \tag{6.2}$$

In the above constraints, the first two constraints ensure that each customer is served exactly once, the third and the fourth constraints indicate that the number of dispatched vehicles does not exceed the total number of available vehicles. The fifth constraint is the subtour elimination constraint, where u_i represents the remaining capacity of the vehicle after visiting customer v_i. The sixth constraint is the capacity constraint of a vehicle that already serves a customer.

6.2.2 Evolutionary Multi-Objective Route Grouping Method

To obtain promising route groupings, EMRG-HA uses an evolutionary tri-objective optimization-based route grouping method, termed EMRG, in the optimization of LSCVRPs. The three optimization objectives, namely, the intra-group similarity, the inter-group similarity and inter-group balance, are adopted to guarantee that the EMRG finds promising route groupings in the optimization of LSCVRPs. Specifically, minimizing the intra-group similarity ensures that the routes in the same group are similar enough to be served by the same vehicle, while maximizing the inter-group similarity enables that the routes in different groups are different enough from each other to be served by different vehicles. Minimizing the inter-group balance makes sure that EMRG achieves balanced route groupings so as to adapt the use of vehicles. Formally, the three optimization objectives are defined as follows.

Assume that $p = \langle 0, v_{1_1}, \dots, v_{1_{i_1}}, 0, v_{2_1}, \dots, v_{2_{i_2}}, 0, \dots, 0, v_{k_1}, \dots, v_{k_{i_k}}, 0\rangle$ is a solution for a LSCVRP to be solved and EMRG-HA performs EMRG for the routes in solution p. There are k routes $\langle v_{j_1}, \dots, v_{j_{i_j}}\rangle$ $(1 \leq j \leq k)$ in solution p, which are denoted as $R_1, \dots, R_k$. The EMRG produces the grouping of routes $R_1, \dots, R_k$ by simultaneously optimizing three objectives:

$$\min \begin{cases} f_1 = -\sum_{i \neq j} Dis(C_i, C_j), \\ f_2 = \sum_{i=1}^{T} Dis(C_i, C_i), \\ f_3 = \frac{1}{T} \sum_{i \neq j} |T_{C_i} - T_{C_j}|, \end{cases} \tag{6.3}$$

where $C = \{C_1, \dots, C_T\}$ denotes a grouping of routes $R_1, \dots, R_k$, C_i, $1 \leq i \leq T$, is a group in the grouping C, $\bigcup_{i=1}^{T} C_i = \{R_1, \dots, R_k\}$, $C_i \cap C_j = \emptyset$ $(i \neq j, 1 \leq i,j \leq T)$, $Dis(C_i, C_j)$ denotes the distance between routes in groups C_i and C_j, T_{C_i} denotes the number of routes in group C_i, and T is the number of groups in C.

The first objective maximizes the inter-group distance, i.e., the distance between routes in different groups of grouping C, while the second objective minimizes the intra-group distance, i.e., the distance between routes in the same group. Formally, the distance between routes in groups C_i and C_j is computed as follows:

$$Dis(C_i, C_j) = \frac{\sum_{R_u \in C_i} \sum_{R_v \in C_j} Dis(R_u, R_v)}{T_{C_i} \times T_{C_j}}, \tag{6.4}$$

where R_u (R_v) denotes a route in group C_i (C_j), T_{C_i} (T_{C_j}) denotes the total number of routes in group C_i (C_j), and $Dis(R_u, R_v)$ indicates the distance between routes R_u and R_v, which is formulated as:

$$Dis(R_u, R_v) = \frac{\sum_{u_i \in R_u} \sum_{v_j \in R_v} Dis(u_i, v_j)}{N_{R_u} \times N_{R_v}}, \tag{6.5}$$

the N_{R_u} (N_{R_v}) denotes the number of customers in route R_u (R_v), $Dis(u_i, v_j) = \sqrt{(x_{u_i} - x_{v_j})^2 + (y_{u_i} - y_{v_j})^2}$ denotes the distance between customers u_i and v_j, and $u_i = (x_{u_i}, y_{u_i})$ and $v_i = (x_{v_i}, y_{v_i})$ are the coordinates of u_i and v_j.

The third objective minimizes the inter-group balance in size, i.e., the difference of number of customers between groups. The third objective guarantees that each group in grouping C holds a similar number of customers so as to achieve a balanced grouping. For optimization of the above three objectives, EMRG-HA employs a classical multi-objective EA NSGA-II [24]. The pseudocode of the EMRG is presented in Algorithm 6.1.

Algorithm 6.1: *EMRG*(p)

Input: p (a solution of the LSCVRP)
Output: *RGSet* (a set of grouping for routes in p).
Dis ← Calculate the distance between any two routes in p by Eq. (6.5);
P ← Randomly generate a population with M solutions;
Obj ← Calculate the objective values of each solution in P by Eq. (6.3);
$[F_1, F_2, \dots]$ ← Do non-dominated sorting on P;
CrowdDis ← Calculate the crowding distance of each solution in the fronts;
while *termination criterion not fulfilled* **do**
    P_{parent} ← Select parents from P via binary tournament selection according to their non-dominated front number and crowding distance;
    P_{child} ← *Crossover* (P_{parent});
    $P \leftarrow P \cup P_{child}$;
    Obj ← Calculate the objective values of each solution in P by Eq. (6.3);
    $[F_1, F_2, \dots]$ ← Do non-dominant sorting on P;
    CrowdDis ← Calculate the crowding distance of each solution in the fronts;
    Delete $|F_1 \cup \cdots \cup F_k| - M$ solutions with the smallest crowding distance from F_k;
    /* k is the minimum number such that $|F_1 \cup \cdots \cup F_k| >= M$ /*;
    $P \leftarrow F_1 \cup \cdots \cup F_k$
return F_1;

For a grouping of k routes, EMRG-HA encodes it as a vector $\langle x_1, \dots, x_k, y_1, \dots, y_k \rangle$, where x_i $(1 \leq i \leq k)$ denotes a real number in $(0, 1)$ and y_i $(1 \leq i \leq k)$ denotes a nonnegative integer that is not larger than k. There are two components in the encoding: the first component $\langle x_1, \dots, x_k \rangle$ indicates the fitness value of each route in sorting and the second component $\langle y_1, \dots, y_k \rangle$ represents the grouping of the k routes. The x_i denotes the fitness value of the ith route in sorting. The y_i indicates that the routes whose fitness values ranked from the $(y_1 + \cdots + y_{i-1} + 1)$th to the $(y_1 + \cdots + y_i)$th are allocated in one group $1 \leq i \leq k$.

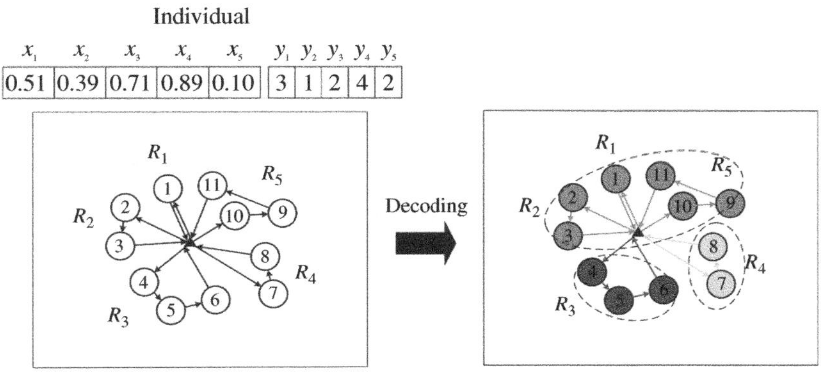

Figure 6.1 An illustrative example of the encoding scheme.

Figure 6.1 takes the individual $\langle 0.51, 0.39, 0.71, 0.89, 0.10, 3, 1, 2, 4, 2\rangle$ as an example, where a grouping of five routes denoted as $R_1, \ldots, R_5$ are encoded. Hence, the five routes are decomposed into three groups in this individual, i.e., the groups $\{R_1, R_2, R_5\}$, $\{R_3\}$, and $\{R_4\}$. The first group of routes $\{R_1, R_2, R_5\}$ is produced since $y_1 = 3$, indicating that the three routes with the first, second, and third minimum fitness values (namely, R_1, R_2, and R_5) are allocated in a group. The second group of routes $\{R_3\}$ is obtained due to $y_2 = 1$, and the route R_3 with the fourth minimum fitness value is allocated to one group. The third group of routes $\{R_4\}$ is obtained because $y_3 = 2$, which means that the routes with the fifth and sixth minimum fitness values are allocated in a group. Since the number of routes is five, the route R_4 with the fifth minimum fitness value is finally allocated in one group. The assignment of all five routes to the groups is then finished, and thus $y_4 = 4$ and $y_5 = 2$ is not considered in the individual. The encoding scheme can hinder the produced offspring to be trapped in local optima due to the single-point crossover adopted in the EMRG. Therefore, there is no need to adopt a mutation operator in EMRG.

6.2.3 Framework of EMRG-HA

Figure 6.2 illustrates the overall procedure of EMRG-HA that integrates the evolutionary multi-objective route grouping into the search process for solving LSCVRPs. As can be observed from the figure, integrating the route grouping into the search process guarantees that the route groupings can be flexibly adapted and updated based on route quality, due to the fact that the intra-group similarity, inter-group similarity, and inter-group balance are continuously optimized for subsequent promising groupings. Furthermore, the route quality is in return improved with the quality improvement of grouping during the optimization

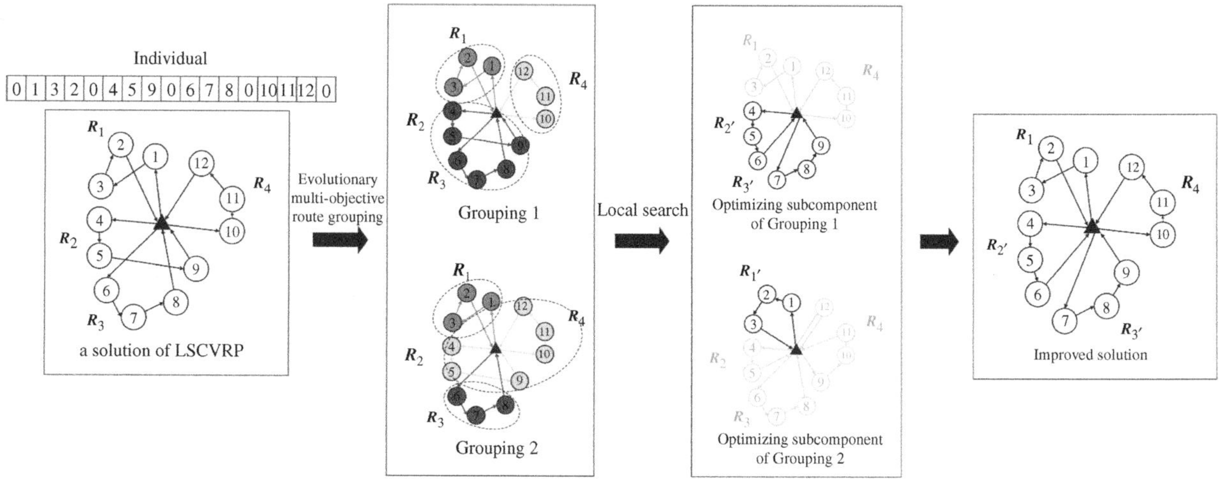

Figure 6.2 Illustration of the overall procedure of EMRG-HA.

process. More detailed steps of EMRG-HA are presented in Algorithm 6.3, which mainly consists of three steps.

First, EMRG-HA randomly generates a solution of the CVRP so that all the vehicles satisfy capacity constraint. For a CVRP with N customers, EMRG-HA encodes a solution p as $\langle 0, v_{1_1}, \dots, v_{1_{i_1}}, 0, v_{2_1}, \dots, v_{2_{i_2}}, 0, \dots, 0, v_{k_1}, \dots, v_{k_{i_k}}, 0\rangle$, where $\langle v_{j_1}, \dots, v_{j_{i_j}}\rangle$, $1 \leq j \leq k$, is a route for the jth vehicle, $v_{j_1}, \dots, v_{j_{i_j}}$ denote the customers served by the jth vehicle satisfying that $\sum_{j=1}^{k} i_j = N$, and k is the number of vehicles used in solution p. Second, EMRG-HA performs the evolutionary multi-objective optimization-based route grouping method EMRG to produce a set of groupings for routes of solution p. Third, EMRG-HA conducts the tabu search based local search in the produced route groupings so as to improve the quality of solution p. The same operations for the improved solution in the above two steps will repeat until the termination criterion of EMRG-HA is satisfied. Otherwise, the obtained solution will be output as the final solution for the LSCVRP. With the improvement of solution quality during the evolution, the EMRG-HA is able to dynamically adapt the ingredients of decomposed subcomponents in the search process via the route grouping.

For the route grouping in each solution, a tabu search-based local search (TS-LS) is employed to improve the route quality in the groups of the grouping, which is presented in Algorithm 6.2. First, the TS-LS randomly selects $\lfloor \delta \times |RGSet| \rfloor$ groupings from *RGSet*, where δ is used to control the number of groupings for reducing computational cost of TS-LS. Second, for each of the randomly selected groupings, the TS-LS selects one group via the roulette wheel method to apply local search to customers in the group. Third, TS-LS applies the tabu search to improve the route quality for each of the selected groups, where three operators: single insertion, swap, and 2-opt, are randomly selected for route quality improvement. The local search is performed for t times for one group of routes. Finally, the routes improved by tabu search replace those in the selected group and a new solution thus is obtained for each grouping.

6.2.4 Experimental Comparisons

This section verifies the competitiveness of EMRG-HA by comparing it with five state-of-the-art algorithms for addressing LSCVRPs in terms of both solution quality and computational efficiency.

The comparisons utilize two commonly used benchmark test sets for LSCVRPs: a set of 20 LSCVRP benchmark instances suggested by Golden *et al.* in [25] ranging from 240 to 483 customers (denoted as G1 to G20), and a set of 12 benchmark instances suggested by Li *et al.* in [26] ranging from 560 to 1200 customers (denoted as L1 to L12). Five existing heuristic algorithms for solving LSCVRPs are used as compared algorithms of EMRG-HA, namely, hybrid genetic search with

Algorithm 6.2: *TS_LS(RGSet, p)*

Input: *RGSet* (a set of groupings for routes in p), p (a solution of the LSCVRP).
Output: p_{new} (the improved solution).

RGTemp ← Randomly select $\lfloor \delta \times |RGSet| \rfloor$ groupings from *RGSet*;
for *each RG* $\in$ *RGTemp* **do**
  G ← Find one group from *RG* via roulette wheel selection based on the average serving cost of routes in each group;
  s^*, s_{inf} ← Initialize s^*, s_{inf} by using the subcomponent of p in group G;
  while *termination condition not fulfilled* **do**
    Execute a randomly selected local search for t times on s_{inf};
    Calculate the serving cost of vehicles in the generated solutions;
    Select a solution s with the least serving cost;
    if *s is non-tabu or s satisfies the aspiration criterion* **then**
      Replace s^* and s_{inf} with s;
    else
      Replace s_{inf} with the solution with the second least serving cost;
    Update the tabu list;
  p_{RG} ← Update p by replacing the subcomponent G of p with s^*;
p_{new} ← p_{RG} with the least serving cost;

Algorithm 6.3: General Framework of the Algorithm EMRG-HA

Input: G (the LSCVRP to be solved).
Output: p (the solution of the LSCVRP).

p ← Randomly generate a solution that satisfies the capacity constraint of vehicles;
while *termination criterion not fulfilled* **do**
  RGSet ← *EMRG(p)*/* Grouping the routes of solution p by the evolutionary multi-objective route grouping method */;
  p ← *TS_LS(RGSet, p)*/* Local search for improving the quality of solution p */;
return p;

adaptive diversity control (HGSADC) [27], cooperative parallel metaheuristic (CPM) [28], parallel random record-to-record (PRRTR) [29], fixed-J quantum annealing (FJ-QA) [30], and variable neighbourhood simulated annealing (VNSA) [31]. All the parameters of each of the compared algorithms are set to

Table 6.2 The Parameter Settings in the Proposed EMRG-HA.

Parameter	Setting
The maximum number of iterations	30
The δ for controlling the number of groupings	0.1
The maximum number of groups	k
The maximum number of iterations in EMRG	100
The population size of EMRG	50
The maximum number of iterations in TS_LS	$200\sqrt{NC}$
The t for controlling the times of local search	$15 + random[0, 10]$
The tabu tenure	8

k is the number of routes in the solution adopted for route grouping and NC is the number of customers in the group to be optimized.

the same values that are suggested in their original references. For EMRG-HA, the maximum number of iterations is set to 30. For EMRG in EMRG-HA, the population size is set to 50 and the maximum number of iterations is set to 100. Table 6.2 presents the parameter settings in EMRG-HA. For fair comparisons, all algorithms are programmed in Matlab and executed on the CPU Intel Core i5-7500 with 3.4 GHz and 16 GB memory.

Tables 6.3 and 6.4 present the best traveling cost achieved by EMRG-HA and 5 compared algorithms on 20 test instances of Golden *et al.* and 12 test instances of Li *et al.* over 20 independent runs, in which the best values achieved by the compared algorithms refer to their original references. The best-known traveling cost for each instance is given in Tables 6.3 and 6.4 and the result for each algorithm represents the gap between the result achieved by the algorithm and the best-known result. Hence, the result 0 means that the algorithm achieves the best-known result, and "+" ("−") indicates that the value is worse (better) than the best-known result.

It can be observed from Tables 6.3 and 6.4 that EMRG-HA can achieve the best-known traveling cost on most LSCVRP test instances. Specifically, EMRG-HA achieves the best-known traveling cost on 16 of 20 Golden test instances, and it achieves the best-known traveling cost on 7 instances of 12 Li test instances. Furthermore, EMRG-HA produces a new best result on some test instances in comparison with the current best-known result, i.e., the instance (G10) in Golden *et al.* benchmarks and the instance (L12) in Li *et al.* benchmarks. Thereafter, it can be concluded that the performance of EMRG-HA is overall superior over five compared algorithms on the 32 LSCVRP test instances. Both HGSADC and CPM also have a competitive performance, despite that they fail to find a low

Table 6.3 The Best Serving Cost Obtained by the Proposed EMRG-HA and Five Compared Algorithms on the 20 Test Instances of Golden *et al.* Benchmarks.

							EMRG-HA		
Problem (*N*, *K*, *Q*)	**Best-so-far**	**HGSADC**	**PRRTR**	**CPM**	**FJ-QA**	**VNSA**	**Best**	**Average**	**Time**
G1 (240, 10, 550)	5,623.47	0	0	0	0	+3.34	0	+4.15	11.33
G2 (320, 10, 700)	8,404.61	0	+43.31	+1.2	+43.31	+45.71	0	+12.16	12.06
G3 (400, 10, 900)	11,036.22	0	0	0	0	0	0	+6.58	14.37
G4 (480, 10, 1000)	13,590	+34.52	+34.52	0	+34.52	+34.6	0	+13.81	17.26
G5 (200, 5, 900)	6,460.98	0	0	0	0	0	0	+21.80	8.46
G6 (280, 8, 900)	8,400.33	+12.57	+12.57	0	+12.57	+12.57	+12.57	+22.85	11.64
G7 (360, 9, 900)	10,102.68	+0.02	+92.91	+4.81	+92.91	+92.91	0	+9.57	12.89
G8 (440, 11, 900)	11,635.3	0	+28.25	+0.04	+36.81	+28.25	+0.04	+4.88	21.86
G9 (255, 14, 1000)	579.71	0	0	0	0	+0.75	0	+0.20	11.58
G10 (323, 16, 1000)	735.66	+0.6	+1.62	0	+1.75	+3.37	−0.82	−0.37	12.15
G11 (399, 18, 1000)	912.03	+0.81	+1.32	0	+0.74	+4.32	0	+0.52	14.55
G12 (483, 19, 1000)	1,101.5	+1.19	+1.26	0	+0.62	+7.6	0	+0.95	18.44
G13 (252, 27, 1000)	857.19	0	0	0	0	+1.7	0	0	12.36
G14 (320, 30, 1000)	1,080.55	0	0	0	0	0	0	+1.13	14.87
G15 (396, 34, 1000)	1,337.87	+0.05	+0.32	0	+2.37	+7.24	0	+0.60	16.02
G16 (480, 38, 1000)	1,611.50	+1	+2.16	+0.06	0	+10.55	+0.06	+0.88	20.46
G17 (240, 22, 200)	707.76	0	0	0	0	+0.43	0	+1.30	13.04
G18 (300, 28, 200)	995.13	0	0	+2.45	+0.85	+5.45	0	0	14.10
G19 (360, 33, 200)	1,365.6	0	0	0	+0.99	+3.5	0	+0.40	15.59
G20 (420, 41, 200)	1,817.59	+0.73	+0.66	+0.3	+1.36	+6.16	0	+0.25	19.71

Table 6.4 The Best Serving Cost Obtained by the Proposed EMRG-HA and Five Compared Algorithms on the 12 Test Instances of Li *et al.* Benchmarks.

							EMRG-HA		
Problem (*N*, *K*, *Q*)	**Best-so-far**	**HGSADC**	**PRRTR**	**CPM**	**FJ-QA**	**VNSA**	**Best**	**Average**	**Time**
L1 (560, 10, 1200)	16,212.826	0	0	0	0	+1.20	0	+7.81	27.58
L2 (600, 15, 900)	14,499.04	+29.15	+85.38	+40.75	+87.07	+119.14	0	+2.35	28.38
L3 (640, 10, 1400)	18,801.13	0	0	0	0	+0.73	0	+6.31	30.04
L4 (750, 10, 1500)	21,389.43	0	0	0	0	+2.4	0	+14.72	32.01
L5 (760, 19, 900)	16,668.51	+36.60	+95.21	+40.93	+183.47	+344.35	0	+5.05	31.78
L6 (800, 10, 1700)	23,977.73	0	0	+2.39	0	+3.6	+1.52	+6.25	33.10
L7 (840, 20, 900)	17,343.38	+39.80	+90.31	0	+165.01	+89.47	+29.26	+34.51	33.86
L8 (880, 10, 1800)	26,566.03	+0.01	0	+1.20	+0.01	+1.21	0	+12.14	34.53
L9 (960, 10, 2000)	29,154.33	0	0	0	0	+2.4	0	+6.36	39.44
L10 (1040, 10, 2100)	31,742.64	0	0	0	0	0	+2.71	+7.73	42.91
L11 (1120, 10, 2300)	34,330.94	0	0	0	0	+1.2	+0.32	+15.23	44.86
L12 (1200, 11, 2500)	37,159.41	0	+26.44	+3.13	+171.70	+44.64	−7.27	−2.03	45.82

traveling cost in a few instances, such as instances L2 and L5. It can be also found in Tables 6.3 and 6.4 that EMRG-HA is computationally efficient in addressing LSCVRPs. To this end, EMRG-HA is competitive in solving LSCVRPs in terms of both solution quality and computational efficiency.

6.3 Clustering-Based Surrogate-Assisted Multi-Objective Evolutionary Algorithm for Shelter Location Problem Under Uncertainty of Road Networks

Over the years, the USLP has attracted many researchers from a variety of fields due to its importance in tackling the threat of natural disasters, such as earthquakes, hurricanes, and tsunamis [32, 33]. There are two tasks for a shelter location problem, namely, selecting sites for shelter construction and assigning evacuees to shelters [34, 35]. The two tasks are obviously in conflict with each other since it is expected that less construction cost can be used to achieve a smaller evacuation distance from residential communities to shelters. The large-scale uncertain shelter location problem (LSUSLP) refers to the USLP that contains hundreds of residential communities to be assigned, which is difficult to solve because it is computationally expensive to evaluate a solution for a USLP. One solution evaluation needs to compute the average traveling cost over a set of simulated damage scenarios. Since EAs are population-based optimization algorithms, it will cost much computational budget to evaluate solutions at each generation of an EA optimization. Regardless of the wide applications of EAs in solving FLPs, few studies have been reported for LSUSLPs by using EAs [36–38].

This section introduces a surrogate-assisted EA, termed AR-MOEA+SA, for efficiently addressing LSUSLPs by using a surrogate modeling strategy [39]. The surrogate modeling strategy helps create computationally cheap function evaluation that well approximates the computationally expensive exact function evaluation, to replace the exact evaluation with the approximate one. In this case, the computational efficiency of solving LSUSLPs can be significantly improved to obtain high-quality solutions for LSUSLPs at a low computational cost.

In this section, we first give the mathematical formulation of USLP, then the clustering-based surrogate modeling method and framework of ARMOEA+SA are detailed. Finally, experimental comparisons between AR-MOEA+SA and state-of-the-art approaches for USLPs are presented and comparison results are discussed.

6.3.1 Problem Formulation

The task of USLP is to select a set of shelters from candidate shelters for construction, to achieve that the traveling distance between evacuees and selected shelters are shortest in both undamage and damage road network scenarios, and the shelter construction cost is as low as possible. It is assumed that there are L candidate shelters, M residential communities and a defined road network between L candidate shelters and M communities. Each road in the defined road network holds a probability to be broken, and the probability for each road is generally obtained based on properties of the road, such as the collapse resistance and quality of buildings along the road, etc. The notations used in the USLP formulation are presented in Table 6.5, and the USLP is formulated as follows:

$$\min_{\substack{x_{ij} \\ 1 \le i \le M, 1 \le j \le L}} \begin{cases} f_1 = \sum_{j=1}^{L} (x_{1j} \vee x_{2j} \vee \cdots \vee x_{Mj}) \cdot c_j, \\ f_2 = \sum_{i=1}^{M} \sum_{j=1}^{L} x_{ij} \cdot d_{ij} \cdot w_i, \\ f_3 = \frac{1}{N_s} \sum_{k=1}^{N_s} \sum_{i=1}^{M} \sum_{j=1}^{L} y_{ij}^k \cdot \tau_{ij}^k \cdot w_i, \end{cases} \tag{6.6}$$

where the operator $\vee$ denotes the OR operator, $x_{ij}(y_{ij}^k) = 1$ indicates that the ith community is assigned to the jth candidate shelter in the undamage (or kth damage) scenario and 0 otherwise.

Table 6.5 Notations in the USLP Formulation.

Notation	Definition
$L(M)$	The number of candidate shelters (residential communities)
$x_{ij}(y_{ij}^k)$	Whether the ith residential community is allocated to the jth candidate shelter in the undamage (kth damage) scenario
$d_{ij}(\tau_{ij}^k)$	The shortest distance between the ith residential community and the jth candidate shelter in the undamage (kth damage) scenario
w_i	The population size in the ith residential community
N_s	The number of damage scenarios
c_j	The capacity of the jth candidate shelter
md_i	The maximum traveling distance for the ith residential community
K, G	The number of clusters and groups of shelters
NR	New representation after clustering
DB, B_{max}	Sample data and the maximum size of sample data
A	Archive of the individuals with exact evaluations
O	Locations of candidate shelters
r	The number of subintervals in the initialization

The three optimization objectives minimize the construction area for the selected shelters, the total evacuation distance in the undamage scenario, and that in the damage scenario, respectively. The objective f_3 is computationally expensive since the roads in the road network will be randomly damaged under their breaking probability in the damage scenario. Hence, a number of damage scenarios are necessary for estimating the evacuation distance in the damage scenario in one individual evaluation.

The constraints of USLP are listed as follows:

$$\begin{cases} \sum_{i=1}^{M} x_{ij} \cdot w_i < c_j, \sum_{i=1}^{M} y_{ij}^k \cdot w_i < c_j, \\ \forall j \in \{j | x_{1j} \vee x_{2j} \vee \cdots \vee x_{Mj} = 1\}, k = 1, 2, \ldots, N_s, \\ \sum_{j=1}^{L} d_{ij} \cdot x_{ij} \le md_i, \sum_{j=1}^{L} \tau_{ij}^k \cdot y_{ij}^k \le md_i, \\ \forall i = 1, 2, \ldots, M, \forall k = 1, 2, \ldots, N_s, \\ \sum_{j=1}^{L} x_{ij} = 1, \sum_{j=1}^{L} y_{ij}^k = 1, \\ \forall i = 1, 2, \ldots, M, \forall k = 1, 2, \ldots, N_s, \end{cases} \tag{6.7}$$

where md_i indicates the maximum traveling distance for the ith community. The first constraint ensures that the population assigned to each shelter does not exceed the capacity of the shelter, the second constraint guarantees that the traveling distance of each community is equal to or smaller than the maximum distance, and the third constraint represents that each community will be exactly assigned to one shelter.

6.3.2 Clustering-Based Surrogate Modeling Method

The AR-MOEA+SA utilizes a surrogate modeling method to conduct computationally cheap function evaluation which approximates the original computationally expensive exact one. Surrogates are a popular method to replace time-consuming exact function evaluations in an EA optimization and thus leads to significant efficiency improvement of EA optimization. Nevertheless, it is difficult for most existing surrogate-assisted EAs to accurately train a surrogate model with hundreds of dimensions, since training a high-dimensional high-accuracy surrogate model needs a very large amount of training data which are often hard to obtain in optimization process of EAs. Hence, AR-MOEA+SA adopts a clustering-based surrogate modeling method to transform the high-dimensional LSUSLP decision variables into low-dimensional ones before training a surrogate

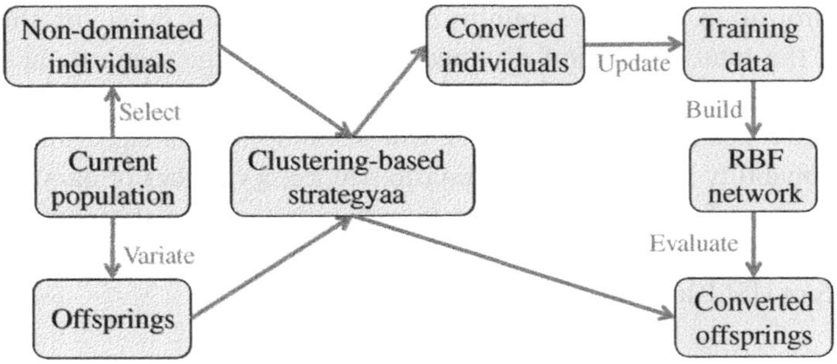

Figure 6.3 The main procedure of the clustering-based surrogate modeling method.

model, which effectively approximates original function evaluation and guarantees high efficiency in addressing LSUSLPs.

The procedure of the clustering-based surrogate modeling method is shown in Fig. 6.3. The surrogate modeling method first selects non-dominated individuals from current population via non-dominated sorting. Next, the surrogate modeling method uses a clustering-based strategy on the non-dominated individuals in order to convert them into data points of low dimensions. Then, the surrogate modeling method updates the training data using the converted data points and an RBF network model is built by training on data. Finally, the surrogate modeling method uses the built RBF network model to evaluate the offsprings reproduced from current parent population. Note that here the clustering-based strategy is also used to convert the offsprings into individuals of low dimensions, which is presented in Algorithm 6.4. Much sample data are necessary for effective surrogate modeling, which may be not affordable since the USLP generally holds hundreds of decision variables. Hence, a clustering-based strategy is suggested in AR-MOEA+SA to build the surrogate model at a low computational budget via converting the modeling of high dimensions into the one of low dimensions for

---

Algorithm 6.4: Converting (sp, G, D, W)

Input: sp (an individual), G (groups of shelters), D (distances between communities and shelters in the undamage scenario), W (population size in each community)

Output: NR (new representation)

$K \leftarrow$ the number of groups in G

for $i = 1$ *to* K **do**

  $s \leftarrow \{j | sp_j \in g_i\}$ /* g_i is the set of candidate shelters in the ith group, sp_j is the jth value in sp, d_{ij} is the value at the ith row and the jth column in D */

  $nr_i \leftarrow \sum_{j=1}^{|s|} d_{s_j sp_{s_j}} \cdot w_{s_j}$ /* s_j is the jth value in s */

Return NR

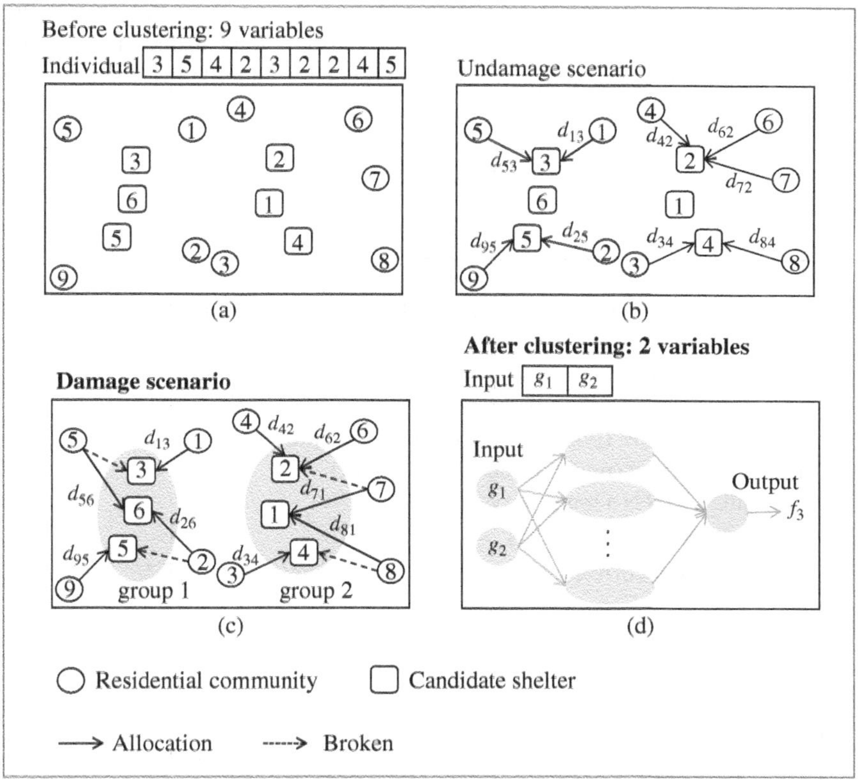

Figure 6.4 Illustration of the clustering-based strategy for building the surrogate model.

the USLP. Specifically, Fig. 6.4 illustrates the process of building a surrogate model by clustering. The candidate shelters are clustered into multiple groups based on the geography locations of shelters. For each individual found in AR-MOEA+SA optimization, the residential communities are covered by the groups according to the shelter coverage situation in the individual. This is because of the fact that the residential communities covered by one group are often assigned to another shelter in the same group if they cannot be evacuated to the original assigned shelter in a damage scenario. The assignment of communities in each group can thus be regarded as a whole that is represented by the total evacuation distance in this group.

The surrogate model built using the above method should be continuously updated, due to the fact that a fixed surrogate model often works only in the cases where the decision space has very low dimensions and the training data used in the surrogate modeling is quite sufficient. The surrogate update also refers to model management in surrogate-assisted EAs, which needs to handle two issues: when is the surrogate updated and which individuals in current population are selected for updating the surrogate. Specifically, Algorithm 6.5 presents the pseudocode of model management in the clustering-based surrogate modeling method.

Algorithm 6.5: Management($DB, A, B_{max}, P, F, G, D, W$)

Input: DB (sample data), A (archive of the individuals with exact evaluations), B_{max} (maximum size of sample data), P (current population), F (objective values of P), G (groups of shelters), D (distances between communities and shelters in the undamage scenario), W (population size in each community)

Output: DB, A, *net* (updated the RBF network)

$TB \leftarrow$ null /*TB is a temporary database*/
$I \leftarrow$ non-dominated sorting(F)
  /*I is the set of corresponding indexes of the Pareto front in F */
for $j \in I$ **do**
  if $p_j \notin A$ **then**
    $f_j \leftarrow$ calculate the real objectives of p_j and modify them in F /*p_j is the jth individual in P*/
    $A \leftarrow$ add p_j and f_j
    $NR \leftarrow$ Converting(p_j, G, D, W), $TB \leftarrow$ add NR and f_j
for $i = 1$ *to* $|TB|$ **do**
  if $|DB| < B_{max}$ **then**
    $DB \leftarrow$ add tb_i /*$|TB|$ is the number of data points in TB, tb_i is the ith data point in TB*/
  else
    $s_{far} \leftarrow$ find the data point farthest from P in DB based on decision variables after clustering
    if *tb_i is nearer to P than s_{far}* **then**
      $DB \leftarrow$ replace s_{far} with tb_i
net $\leftarrow$ trainRBF(DB)
Return $DB, A,$ *net*

6.3.3 Procedure of AR-MOEA+SA

Based on the clustering-based surrogate modeling method, AR-MOEA+SA uses a state-of-the-art multi-objective EA, termed AR-MOEA that is proposed in [40]. The AR-MOEA is an indicator-based multi-objective EA using a performance indicator of inverted generational distance with noncontributing solution detection IGD-NS [40]. Specifically, Algorithm 6.6 shows the procedure of AR-MOEA+SA,

---

Algorithm 6.6: AR-MOEA+SA (W, C, L, O, D)

Input: W (population size in each community), C (capacity of each candidate shelter), L (number of candidate shelters), O (locations of candidate shelters), D (distances between communities and shelters in the undamage scenario)

Output: PS (Pareto optimal solutions)

$P \leftarrow$ Initialization (N, W, L, C, r) /*N is the population size*/
$P \leftarrow$ handle constraints and for P
$F \leftarrow$ calculate the real objective values of individuals in P
$G \leftarrow Kmeans(O, K)$ /*K is the number of clusters*/
$NR \leftarrow$ Converting (P, G, D, W)
$DB, A \leftarrow$ initialize training data and archive by F and NR
$net \leftarrow$ RBFtrain (DB)
$ref \leftarrow$ initialize reference points
$t \leftarrow N$
while $t < FE$ **do**
  /*FE is the maximum number of function evaluations*/$Q \leftarrow$ create offsprings based on P and ref
  $Q \leftarrow$ handle constraints for Q
  $NR \leftarrow$ Converting (Q, G, D, W)
  $QF \leftarrow$ approximation (net, Q, NR)
  $ref \leftarrow$ update reference points
  $P, F \leftarrow$ environmental selection (P, F, Q, QF, ref)
  $DB, A, net \leftarrow$ Management $(DB, A, B_{max}, P, F, G, D, W)$
  $t \leftarrow$ the number of individuals in A
$F \leftarrow$ evaluate the individuals in A by using 100 randomly generated damage scenarios
$PS \leftarrow$ get Pareto optimal solutions from A
Return PS

---

which employs the radial basis function (RBF) as the surrogate model and adopts the Gaussian function as the basis function for RBF network. There are six main steps in the procedure of AR-MOEA+SA.

First, AR-MOEA+SA performs the initialization of population and surrogate model, which uses the clustering strategy to convert the initial population into initial training data for surrogate modeling. Based on the initial training data, a surrogate model is built and an archive is created to store all individuals under exact evaluations. Second, AR-MOEA+SA performs the uniform crossover and mutation to produce offsprings from current parent population. Third, AR-MOEA+SA uses the latest surrogate model to approximately evaluate each individual in the offspring population. Fourth, AR-MOEA+SA utilizes the environmental selection in AR-MOEA to select the parent population in next generation, which is based on the approximate evaluation results of individuals to be selected. Fifth, AR-MOEA+SA updates the surrogate model according to the model management method presented in Algorithm 6.5, and the training data and the archive are also updated. The second to the fifth steps repeat until the termination criterion is satisfied. Finally, AR-MOEA+SA averagely reevaluates the individuals from the archive on a set of 100 randomly generated damage scenarios so as to alleviate the influence of randomness. The non-dominated individuals that are identified based on the reevaluations are obtained from the archive and regarded as the final result achieved by AR-MOEA+SA.

6.3.4 Experimental Comparisons

In this section, we evaluate the performance of AR-MOEA+SA in terms of both solution quality and computational efficiency by comparing it with four state-of-the-art algorithms for USLPs.

A set of four USLP test instances are used as test instances for all the compared algorithms, which consist of three simulated numerical instances and one real case. Particularly, three simulated test instances are created on the basis of the parameter settings in Table 6.6, in which the community scale varies from small, medium to large. Four compared algorithms are utilized for comparisons with AR-MOEA+SA, namely, a mathematical programming method BD [33], an EA MNSGA-II [41], a surrogate-based optimization algorithm SA-COSO [42], and AR-MOEA without the surrogate modeling method [40].

The population size for AR-MOEA+SA is set to 100, while the two sub-populations in SA-COSO/MNSGA-II are set to hold 80 individuals and 20 ones, respectively. Each algorithm that uses uniform crossover and mutation operators holds a crossover probability of 0.8 and a mutation probability of $1/d$,

Table 6.6 Parameter Setting of Instances A, B, C, and the Real Case.

Problem	Residential communities	Candidate shelters	Intersections	Region length	Population in each community	The maximum distance	Capacity of each candidate shelter	Damage probability
A	100	50	[2,5]	100	[100, 300]	50	[1,000, 2,000]	$N(0.3, 0.05^2)$
B	200	100	[3,6]	200	[200, 400]	100	[4,000, 6,000]	$N(0.5, 0.1^2)$
C	300	150	[4,7]	300	[300, 500]	150	[9,000, 12,000]	$N(0.7, 0.15^2)$
Real case	500	250		70,000	[1,000, 2,000]		[25,000, 30,000]	

where d is the number of decision variables. For RBF in AR-MOEA+SA, the size of hidden nodes is set to 6, and the scale of training data is set to $d_x * 12$, in which d_x indicates the data dimension. The results for each of MNSGA-II, SA-COSO, and AR-MOEA+SA are obtained after 20 independent runs of the corresponding algorithm. All the experiments are performed on a PC with a 3.40 GHz Intel Core i7 CPU 6700 and the Windows 7 SP1 64 bit operating system.

Table 6.7 presents the evacuation distances obtained by five compared algorithms on the four instances, averaging over 20 independent runs. For highlighting the performance of AR-MOEA+SA, three non-surrogate-assisted compared algorithms BD, AR-MOEA, and MNSGA-II are performed for 20,000 exact function evaluations, while two surrogate-assisted compared algorithms AR-MOEA+SA and SA-COSO are performed for 2000 exact function evaluations.

It can be observed from Table 6.7 that AR-MOEA+SA overall achieves the best performance in terms of the evacuation distances for both the undamage and damage scenarios on all test instances in comparison with four compared algorithms. Furthermore, AR-MOEA+SA also uses the smallest shelter construction cost on all instances. Considering that the above results of AR-MOEA+SA are obtained after only 2000 exact evaluations, AR-MOEA+SA in fact achieves the best performance on all instances by using significantly less computational cost in comparison with non-surrogate-assisted compared algorithms BD, MNSGA-II, and AR-MOEA.

For a deeper insight into the computational efficiency of AR-MOEA, Table 6.8 lists the runtime(s) of four compared algorithms and AR-MOEA+SA on four instances, where three non-surrogate-assisted compared algorithms still are performed for 20,000 exact evaluations, and two surrogate-assisted compared algorithms are performed for 2000 exact ones. It can be seen from the table that the two surrogate-assisted compared algorithms AR-MOEA+SA and SA-COSO achieve the highest computational efficiency on all instances in comparison with the other three non-surrogate-assisted ones. This is because of the fact that the

Table 6.7 Evacuation Distances Obtained by BD, AR-MOEA, MNSGA-II, SA-COSO, and AR-MOEA+SA on Instances A, B, C, and the Real Case, Where BD, AR-MOEA and MNSGA-II are Performed for 20,000 Exact Function Evaluations, and AR-MOEA+SA and SA-COSO Are Performed for 2000 Exact Function Evaluations. The Results of AR-MOEA, MNSGA-II, SA-COSO and AR-MOEA+SA Are Averaged over 20 Independent Runs.

	BD		AR-MOEA		MNSGA-II		SA-COSO		AR-MOEA+SA	
	TED1	TED2	TED1	TED2	TED1	TED2	TED1	TED2	TED1	TED2
Instance	Area		Area		Area		Area		Area	
A	1.32e+5	2.79e+5	1.33e+5	2.88e+5	1.31e+5	2.75e+5	1.35e+5	2.83e+5	**1.30e+5**	**2.75e+5**
	5.71e+4		5.80e+4		5.65e+4		5.87e+4		**5.63e+4**	
B	6.61e+5	9.01e+5	6.66e+5	9.06e+5	6.62e+5	8.98e+5	6.70e+5	9.21e+5	**6.59e+5**	**8.97e+5**
	4.45e+5		4.61e+5		4.44e+5		4.64e+5		**4.42e+5**	
C	1.52e+6	1.69e+6	1.54e+6	1.70e+6	1.53e+6	1.69e+6	1.67e+6	1.82e+6	**1.51e+6**	**1.67e+6**
	1.29e+6		1.34e+6		1.33e+6		1.38e+6		**1.29e+6**	
Real case	1.05e+9	1.19e+9	1.18e+9	1.38e+9	1.03e+9	1.18e+9	1.33e+9	1.42e+9	**1.02e+9**	**1.16e+9**
	1.33e+7		1.37e+7		1.34e+7		1.56e+7		**1.31+7**	

"TED1 (TED2)" denotes the total evacuation distance in the undamage (damage) scenario and "Area" denotes the total area of selected shelters.

Table 6.8 Runtime(s) of BD, AR-MOEA, MNSGA-II, SA-COSO, and AR-MOEA+SA on Instances A, B, C, and the Real Case for One Run, Where BD, AR-MOEA and MNSGA-II Are Performed for 20,000 Exact Function Evaluations, AR-MOEA+SA and SA-COSO Are Performed for 2000 Exact Function Evaluations. The Runtime(s) of AR-MOEA, MNSGA-II, SA-COSO, and AR-MOEA+SA Is Averaged over 20 Independent Runs.

Instance	BD	AR-MOEA	MNSGA-II	SA-COSO	AR-MOEA+SA
A	3.21e+4	3.53e+4	3.74e+4	3.33e+3	**3.28e+3**
B	1.31e+5	1.45e+5	1.55e+5	1.46e+4	**1.33e+4**
C	3.87e+5	4.10e+5	4.27e+5	4.19e+4	**3.91e+4**
Real case	1.23e+6	1.52e+6	1.70e+6	1.58e+5	**1.43e+5**

two surrogate-assisted algorithms use a considerable number of computationally efficient approximate evaluations to replace computationally expensive ones during their evolution, which thus leads to high computational efficiency of the two algorithms.

Furthermore, Fig. 6.5 shows the number of exact function evaluations performed by AR-MOEA, MNSGA-II, SA-COSO, and AR-MOEA+SA on four instances with the growth of running generations. It can be observed from Fig. 6.5 that the two non-surrogate-assisted algorithms AR-MOEA and MNSGA-II perform much more exact evaluations than the two surrogate-assisted algorithms SA-COSO and AR-MOEA+SA when they are in the same generations. This further validates that AR-MOEA+SA is efficient in addressing computationally expensive LSUSLPs.

It can also be seen from Fig. 6.5 that AR-MOEA+SA performs more exact evaluations than the other surrogate-assisted algorithm SA-COSO when they are at the same generations. Nevertheless, AR-MOEA+SA contrarily takes less runtime than SA-COSO on all instances, which is exhibited in Fig. 6.6. The reason lies in the fact that the surrogate model training of AR-MOEA+SA consumes much less time than that of SA-COSO during their evolution. The surrogate model training of SA-COSO assists in creating approximate evaluations that have much cheaper computational cost than exact ones, which is, however, relatively more expensive than the model training of AR-MOEA+SA. As a result, it is still necessary to develop surrogate-assisted EAs that simultaneously hold high model precision and cheap training cost.

In summary, it can be drawn from the comparison results that AR-MOEA+SA is an effective and efficient optimization algorithm for solving LSUSLPs in terms of both solution quality and computational efficiency.

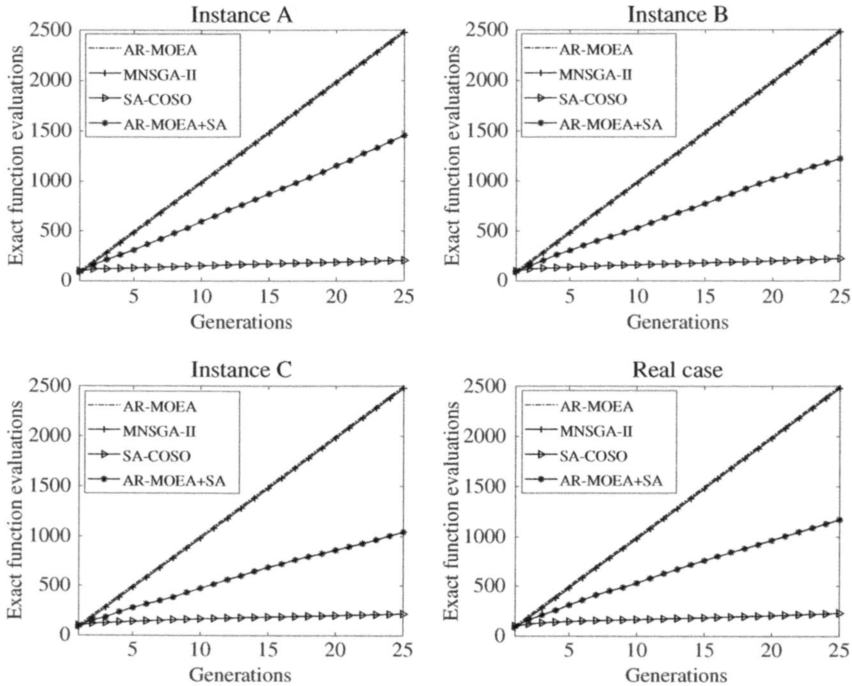

Figure 6.5 (a–d) Number of exact function evaluations of AR-MOEA, MNSGA-II, SA-COSO, and AR-MOEA+SA on instances A, B, C, and the real case under different generations.

6.4 Summary

This chapter presents a CVRP optimization algorithm and a USLP optimization algorithm for solving LSCVRPs and USLPs, respectively. Specifically, a divide-and-conquer-based heuristic algorithm, EMRG-HA is introduced for solving LSCVRPs, where an evolutionary multi-objective route grouping method called EMRG is developed to perform the grouping of a set of given routes. The competitiveness of EMRG-HA has also been empirically verified by comparing it with five existing heuristic algorithms for LSCVRPs in terms of both computational efficiency and solution quality.

A clustering-based surrogate-assisted multi-objective EA, termed AR-MOEA+SA, is detailed for addressing USLPs. To build an effective surrogate model for the LSUSLP, in the AR-MOEA+SA, a clustering-based strategy is developed to convert the surrogate modeling of high-dimensional problem into one of low-dimensional problem. Comparison with four state-of-the-art facility location optimization

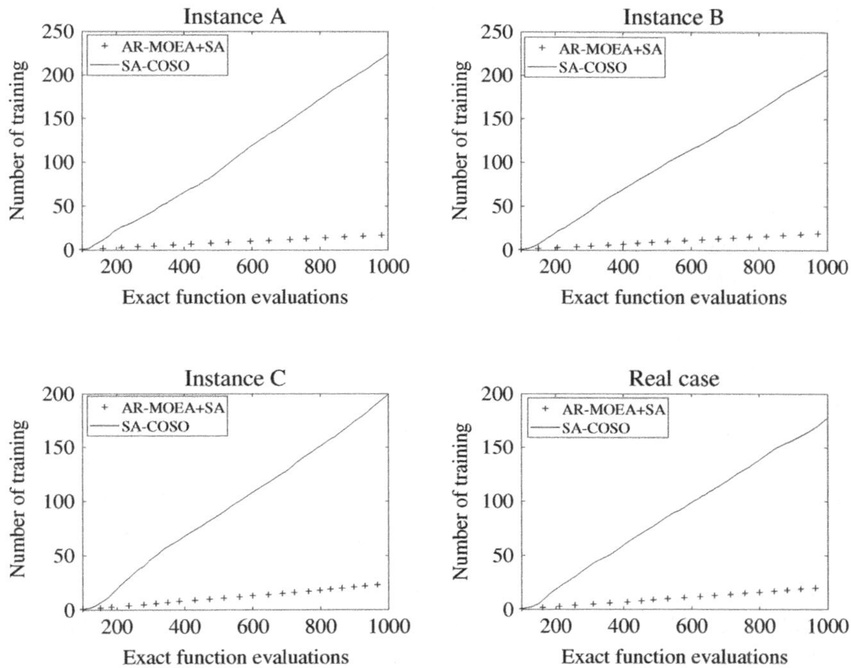

Figure 6.6 (a–d) Computational time (s) of training of SA-COSO and AR-MOEA+SA on instances A, B, C, and the real case.

algorithms demonstrates the superiority of AR-MOEA+SA in effectively and efficiently addressing USLPs.

References

1 K. Zheng, Z. Zhang, and B. Song, "E-commerce logistics distribution mode in big-data context: a case analysis of JD. COM," *Industrial Marketing Management*, vol. 86, no. 1, pp. 154–162, 2020.

2 R. Kain and A. Verma, "Logistics management in supply chain-an overview," *Materials Today: Proceedings*, vol. 5, no. 2, pp. 3811–3816, 2018.

3 R. Lahyani, L. C. Coelho, M. Khemakhem, G. Laporte, and F. Semet, "A multi-compartment vehicle routing problem arising in the collection of olive oil in Tunisia," *Omega*, vol. 51, pp. 1–10, 2015.

4 S. Wang and J. Watada, "A hybrid modified PSO approach to VaR-based facility location problems with variable capacity in fuzzy random uncertainty," *Information Sciences*, vol. 192, pp. 3–18, 2012.

5 Y. Li, M. Chen, and J. Huo, "A hybrid adaptive large neighborhood search algorithm for the large-scale heterogeneous container loading problem," *Expert Systems with Applications*, vol. 189, p. 115909, 2022.
6 C. Novoa and R. Storer, "An approximate dynamic programming approach for the vehicle routing problem with stochastic demands," *European Journal of Operational Research*, vol. 196, no. 2, pp. 509–515, 2009.
7 J. Chen, F. Ling, Y. Zhang, T. You, Y. Liu, and X. Du, "Coverage path planning of heterogeneous unmanned aerial vehicles based on ant colony system," *Swarm and Evolutionary Computation*, vol. 69, p. 101005, 2022.
8 H. Zhang, Q. Zhang, L. Ma, Z. Zhang, and Y. Liu, "A hybrid ant colony optimization algorithm for a multi-objective vehicle routing problem with flexible time windows," *Information Sciences*, vol. 490, pp. 166–190, 2019.
9 D. Cinalli, L. MartÃ-, N. Sanchez-Pi, and A. Garcia, "Extending collective intelligence evolutionary algorithms: a facility location problem application," in *Proceedings of the IEEE Congress on Evolutionary Computation*, 2020, pp. 1–8.
10 A. N. Letchford and J.-J. Salazar-González, "The capacitated vehicle routing problem: stronger bounds in pseudo-polynomial time," *European Journal of Operational Research*, vol. 272, no. 1, pp. 24–31, 2019.
11 X. Zhang, Y. Tian, R. Cheng, and Y. Jin, "A decision variable clustering-based evolutionary algorithm for large-scale many-objective optimization," *IEEE Transactions on Evolutionary Computation*, vol. 22, no. 1, pp. 97–112, 2016.
12 W. Hong, K. Tang, A. Zhou, H. Ishibuchi, and X. Yao, "A scalable indicator-based evolutionary algorithm for large-scale multiobjective optimization," *IEEE Transactions on Evolutionary Computation*, vol. 23, no. 3, pp. 525–537, 2018.
13 H. Zille, H. Ishibuchi, S. Mostaghim, and Y. Nojima, "A framework for large-scale multiobjective optimization based on problem transformation," *IEEE Transactions on Evolutionary Computation*, vol. 22, no. 2, pp. 260–275, 2017.
14 X. Zhang, K. Zhou, H. Pan, L. Zhang, X. Zeng, and Y. Jin, "A network reduction-based multiobjective evolutionary algorithm for community detection in large-scale complex networks," *IEEE Transactions on Cybernetics*, vol. 50, no. 2, pp. 703–716, 2018.
15 M. N. Omidvar, X. Li, Y. Mei, and X. Yao, "Cooperative co-evolution with differential grouping for large scale optimization," *IEEE Transactions on Evolutionary Computation*, vol. 18, no. 3, pp. 378–393, 2013.
16 É. Taillard, "Parallel iterative search methods for vehicle routing problems," *Networks*, vol. 23, no. 8, pp. 661–673, 1993.
17 R. He, W. Xu, J. Sun, and B. Zu, "Balanced K-means algorithm for partitioning areas in large-scale vehicle routing problem," in *Proceedings of the 3rd International Symposium on Intelligent Information Technology Application*, 2009, pp. 87–90.

18 J. Zhang, "An efficient density-based clustering algorithm for the capacitated vehicle routing problem," in *Proceedings of International Conference on Computer Network, Electronic and Automation*, 2017, pp. 465–469.

19 D. Mester, O. Bräysy, and W. Dullaert, "A multi-parametric evolution strategies algorithm for vehicle routing problems," *Expert Systems with Applications*, vol. 32, no. 2, pp. 508–517, 2007.

20 R. Bent and P. Van Hentenryck, "Randomized adaptive spatial decoupling for large-scale vehicle routing with time windows," in *Proceedings of AAAI Conference on Artificial Intelligence*, 2007, pp. 173–178.

21 V. Praveen, V. Hemalatha, and K. R. P. Gomathi, "A nearest centroid classifier-based clustering algorithm for solving vehicle routing problem," in *Proceedings of the Innovations in Electronics and Communication Engineering*, 2018, pp. 575–586.

22 J. Xiao, T. Zhang, J. Du, and X. Zhang, "An evolutionary multiobjective route grouping-based heuristic algorithm for large-scale capacitated vehicle routing problems," *IEEE Transactions on Cybernetics*, vol. 51, no. 8, pp. 4173–4186, 2019.

23 P. Toth and D. Vigo, "Models, relaxations and exact approaches for the capacitated vehicle routing problem," *Discrete Applied Mathematics*, vol. 123, no. 1–3, pp. 487–512, 2002.

24 K. Deb, A. Pratap, S. Agarwal, and T. Meyarivan, "A fast and elitist multiobjective genetic algorithm: NSGA-II," *IEEE Transactions on Evolutionary Computation*, vol. 6, no. 2, pp. 182–197, 2002.

25 B. L. Golden, E. A. Wasil, J. P. Kelly, and I. M. Chao, "The impact of metaheuristics on solving the vehicle routing problem: algorithms, problem sets, and computational results," in *Proceedings of the Fleet Management and Logistics*, 1998, pp. 33–56.

26 F. Li, B. Golden, and E. Wasil, "Very large-scale vehicle routing: new test problems, algorithms, and results," *Computer & Operations Research*, vol. 32, no. 5, pp. 1165–1179, 2005.

27 T. Vidal, T. G. Crainic, M. Gendreau, N. Lahrichi, and W. Rei, "A hybrid genetic algorithm for multidepot and periodic vehicle routing problems," *Operations Research*, vol. 60, no. 3, pp. 611–624, 2012.

28 J. Jin, T. G. Crainic, and A. Løkketangen, "A cooperative parallel metaheuristic for the capacitated vehicle routing problem," *Computer & Operations Research*, vol. 44, no. 2, pp. 33–41, 2014.

29 C. Groër, B. Golden, and E. Wasil, "A parallel algorithm for the vehicle routing problem," *INFORMS Journal of Computing*, vol. 23, no. 2, pp. 315–330, 2011.

30 A. Syrichas and A. Crispin, "Large-scale vehicle routing problems: quantum annealing, tunings and results," *Computer & Operations Research*, vol. 87, pp. 52–62, 2017.

31 Y. Xiao, Q. Zhao, I. Kaku, and N. Mladenovic, “Variable neighbourhood simulated annealing algorithm for capacitated vehicle routing problems,” *Engineering Optimization*, vol. 46, no. 4, pp. 562–579, 2014.

32 F. S. Salman and E. Yücel, “Emergency facility location under random network damage: insights from the Istanbul case,” *Computers & Operations Research*, vol. 62, pp. 266–281, 2015.

33 V. Bayram and H. Yaman, “Shelter location and evacuation route assignment under uncertainty: a benders decomposition approach,” *Transportation Science*, vol. 52, no. 2, pp. 416–436, 2018.

34 M. Saadatseresht, A. Mansourian, and M. Taleai, “Evacuation planning using multiobjective evolutionary optimization approach,” *European Journal of Operational Research*, vol. 198, no. 1, pp. 305–314, 2009.

35 Z. Chen, X. Chen, Q. Li, and J. Chen, “The temporal hierarchy of shelters: a hierarchical location model for earthquake-shelter planning,” *International Journal of Geographical Information Science*, vol. 27, no. 8, pp. 1612–1630, 2013.

36 A. Ghaderi, M. S. Jabalameli, F. Barzinpour, and R. Rahmaniani, “An efficient hybrid particle swarm optimization algorithm for solving the uncapacitated continuous location-allocation problem,” *Networks and Spatial Economics*, vol. 12, no. 3, pp. 421–439, 2012.

37 M. Karatas and E. Yakící, “An iterative solution approach to a multi-objective facility location problem,” *Applied Soft Computing*, vol. 62, pp. 272–287, 2018.

38 V. Arya, N. Garg, R. Khandekar, A. Meyerson, K. Munagala, and V. Pandit, “Local search heuristic for k-median and facility location problems,” in *Proceedings of the thirty-third annual ACM symposium on Theory of computing*, 2001, pp. 21–29.

39 X. Xiang, Y. Tian, J. Xiao, and X. Zhang, “A clustering-based surrogate-assisted multiobjective evolutionary algorithm for shelter location problem under uncertainty of road networks,” *IEEE Transactions on Industrial Informatics*, vol. 16, no. 12, pp. 7544–7555, 2019.

40 Y. Tian, R. Cheng, X. Zhang, F. Cheng, and Y. Jin, “An indicator based multi-objective evolutionary algorithm with reference point adaptation for better versatility,” *IEEE Transactions on Evolutionary Computation*, vol. 22, no. 4, pp. 609–622, 2017.

41 B. Ji, X. Yuan, and Y. Yuan, “Modified NSGA-II for solving continuous berth allocation problem: using multiobjective constraint handling strategy,” *IEEE Transactions on Cybernetics*, vol. 47, no. 9, pp. 2885–2895, 2017.

42 C. Sun, Y. Jin, R. Cheng, J. Ding, and J. Zeng, “Surrogate-assisted cooperative swarm optimization of high-dimensional expensive problems,” *IEEE Transactions on Evolutionary Computation*, vol. 21, no. 4, pp. 644–660, 2017.

7

Evolutionary Large-Scale Multi-Objective Optimization in Power Systems

7.1 Introduction

Power systems are becoming increasingly complex due to rapid advancements in technology and the growing need for efficient, reliable, and sustainable energy management. This complexity is further intensified by the integration of renewable energy sources, distributed generation, and smart grid technologies. Managing and optimizing power systems necessitate addressing multiple objectives simultaneously, as various aspects of system performance, such as power generation, transmission, distribution, and consumption, are often interconnected and interdependent.

In order to effectively tackle these challenges, advanced optimization techniques are required that can capture the intricate relationships between multiple objectives and provide high-quality solutions for the diverse array of problems encountered in power systems. Evolutionary algorithms are population-based meta-heuristics optimization algorithms, which are capable of obtaining a set of trade-off solutions in a single run and less vulnerable to local optima. In the past few decades, evolutionary algorithms have demonstrated remarkable success in solving multi-objective optimization problems across a wide range of domains. However, their performance usually deteriorates considerably when they are used to address large-scale multi-objective optimization problems (LSMOPs), which is attributed to the curse of dimensionality. In fact, the curse of dimensionality is also widely present in power systems. After all, the optimization problems in power systems usually have tens of thousands of or even more decision variables to be optimized.

This chapter presents an in-depth exploration of evolutionary algorithms applied to power systems, with a particular focus on a case study involving the ratio error estimation of voltage transformers (VTs). In the case study, first, we discuss the importance of ratio error estimation of VTs and the limitations of

Evolutionary Large-Scale Multi-Objective Optimization and Applications: Integrating Evolutionary Computation, Machine Learning, and Data Science, First Edition. Xingyi Zhang, Ran Cheng, Ye Tian, and Yaochu Jin.

existing methods. Then, we introduce the time-varying ratio error estimation (TREE) method as an innovative solution to address the shortcomings of existing methods. Next, we present a data-driven approach for formulating the TREE problem as a LSMOP, detailing the process of extracting statistical and physical principles from power distribution systems. Finally, a problem knowledge-driven coevolutionary algorithm is designed to solve the TREE problems, demonstrating the potential of evolutionary algorithms in addressing complex real-world problems in power systems.

7.2 Ratio Error Estimation of Voltage Transformers

7.2.1 Problem Introduction

VTs are fundamental devices widely employed in industrial power delivery systems, particularly in substations, for measuring voltage values.[1] They serve not only as measurement tools but also as transformation devices. By the end of 2011, over 120,000 VTs had been installed in the high-voltage grid of China's State Grid. According to statistics from Guangzhou, China, the VT failure rate has reached 0.8% per year over the past decade [1]. Ratio error drift fault constitutes one of the main reasons for VT failures, which can mislead downstream applications due to incorrect data in the grid, such as relaying, metering, and controlling tasks [2]. For instance, China's total electricity consumption in 2017 was 6.3077×10^{13} kilowatt-hour [3]. If the ratio errors of VTs were 0.001%, a metering error of 6×10^8 kilowatt-hour electricity would occur. Consequently, the estimation of the time-varying ratio error of VTs plays a crucial role in power delivery systems [4].

The ratio error (RE) of a VT can be mathematically expressed as [5]

$$\mathrm{RE} = \frac{V_m - V_t}{V_t}, \tag{7.1}$$

where V_m represents the primary voltage value measured by the VT, and V_t (always unknown) denotes the corresponding true primary voltage value. Figure 7.1 displays the diagram of a representative 110 kV substation with four sets of three-phase VTs in the power delivery system. Specifically, VT-1 and VT-2 measure the same three-phase primary voltage values A_0, B_0, C_0 in parallel, while VT-3 and VT-4 measure the same three-phase secondary voltage values a_0, b_0, c_0 in parallel. Consequently, the four VTs interact through the true voltage values. For example, VT-1 and VT-2 measure the same primary voltage value, while VT-3 and VT-4 measure the same secondary voltage value. Since the primary side

1 The proposed method in this work applies to general VTs, including the capacitor ones (i.e., CVTs).

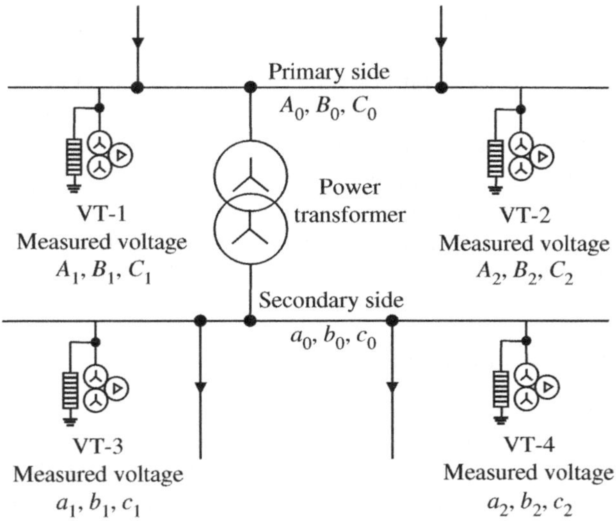

Figure 7.1 The diagram of the VTs in a representative power delivery system, where A_0, B_0, C_0 and a_0, b_0, c_0 are the true primary and secondary three-phase voltage values, respectively. Note that, the primary side and the secondary side are connected by the power transformer.

voltage value is proportional to the secondary side voltage value, VT-1 and VT-2 are related to VT-3 and VT-4. Meanwhile, we denote the primary and secondary voltage values obtained by the ith set of VTs as A_i, B_i, C_i and a_i, b_i, c_i, respectively. In this substation, the RE can be calculated based on one of the primary voltage values, such as $(A_i - A_0)/A_0$, $(B_i - B_0)/B_0$, or $(C_i - C_0)/C_0$.

To obtain the REs of VTs, there are primarily two types of methods: calibration methods and data-driven estimation methods. Calibration is conducted by de-energizing the power and manually comparing the uncalibrated VT with the standard one [6]. Typical methods include off-line and on-site calibrations [7–9]. In contrast, data-driven estimation methods construct specific models of the VTs and estimate the REs based on constraints derived from the electrical relationship and data collected from VTs, such as the adaptive detection method [10] and the line parameter estimation method [11]. However, both calibration and data-driven estimation methods suffer from practical deficiencies. While calibration methods provide accurate REs under standard working conditions, they are expensive in terms of time and cost and must be used periodically [6]. For instance, most substations conduct calibration every four years or even longer, resulting in economic losses (e.g., over 0.3 million CNY) and taking a long time (e.g., more than one day) [12]. As for data-driven estimation methods, despite their lower cost, their accuracy and versatility are unsatisfactory due to the limitations of the VT-specific models [13, 14].

To address the challenges associated with existing calibration and data-driven methods, Section 7.2.2 will introduce an innovative method that formulates the TREE problem as an LSMOP by taking into account the differences between the true voltage values and those measured by the VTs. Similar to calibration, the primary goal of the formulated problem is to estimate the true voltage values. The proposed method offers significant advantages over calibration and data-driven error estimation methods, including reduced costs and enhanced versatility.

7.2.2 Data-Driven Problem Formulation

7.2.2.1 Overview of the Problem Formulation

In this section, we present the data-driven problem formulation for six different TREE (TREE1 to TREE6) problems as LSMOPs with inequality constraints. These problems are categorized into three types based on the involved data: data with primary voltage values only (Type 1), data with both primary and secondary voltage values (Type 2), and data with both voltage and phase angle values (Type 3). Figure 7.2 illustrates the primary/secondary voltage values or the phase angle values measured by the uncalibrated VTs for each problem.

We use three symbols, K, P, and T, to formulate the TREE problems, with detailed settings provided in Table 7.1. K represents the total number of primary and/or secondary voltage phases, P denotes the number of VT sets involved in the problem, and T is the length of the measured sequential data.

In the formulation of the TREE problems, the decision variables are defined as follows:

$$\mathbf{x} = (x_{1,1}, \dots, x_{1,T}, \dots, x_{K,1}, \dots, x_{K,T}), \tag{7.2}$$

where $x_{i,j}$ represents the true ith phase voltage at time j (refer to V_t in Eq. (7.1)). The number of decision variables is given by $D = K \cdot T$. The data gathered from the ith set of VTs is expressed as

$$\mathbf{d}^i = (d^i_{1,1}, \dots, d^i_{1,T}, \dots, d^i_{K,1}, \dots, d^i_{K,T}), \tag{7.3}$$

with $i \in [1, 2, \dots, P]$ and $d_{p,q}$ denoting the pth measured data at time q. The REs of these VTs can be calculated based on Eq. (7.1) using Eq. (7.4)

$$\mathbf{e}^i = \left(\frac{d^i_{1,1} - x_{1,1}}{x_{1,1}}, \dots, \frac{d^i_{1,T} - x_{1,T}}{x_{1,T}}, \dots, \frac{d^i_{K,1} - x_{K,1}}{x_{K,1}}, \dots, \frac{d^i_{K,T} - x_{K,T}}{x_{K,T}} \right). \tag{7.4}$$

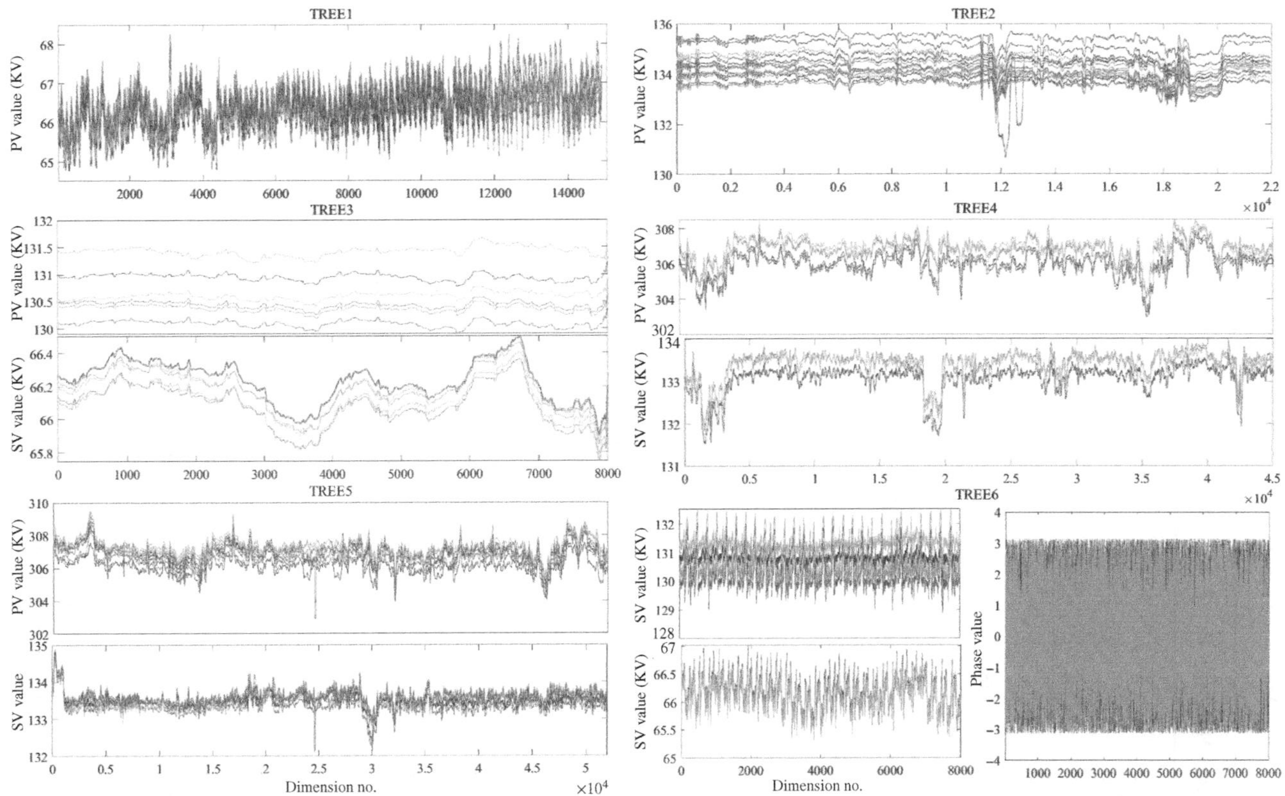

Figure 7.2 Primary/secondary voltage (PV and SV, respectively) values or phase angle values of TREE1 to TREE6 problems measured by real-world uncalibrated VTs. Dimension number denotes the sample index of the measurement.

Table 7.1 Settings of M, J, K, P, T, and the Type of Involved Data for Each TREE Problem, Where M and J Denote the Numbers of Objectives and Constraints, Respectively.

Problem	M	J	K	P	T	Data Type	Problem	M	J	K	P	T	Data Type
TREE1	2	3	3	3	14,918	Type 1	TREE4	2	4	6	2	45,156	Type 2
TREE2	2	3	3	12	21,950	Type 1	TREE5	2	4	6	3	52,029	Type 2
TREE3	2	3	6	2	8,000	Type 1	TREE6	3	6	12	2	8,000	Type 3

To simplify, the REs are denoted as

$$\mathbf{e}^i = \left(e^i_{1,1}, \ldots, e^i_{1,T}, \ldots, e^i_{K,1}, \ldots, e^i_{K,T}\right). \tag{7.5}$$

Additionally, the variations of REs over time are

$$\Delta\mathbf{e}^i = \Big(e^i_{1,2} - e^i_{1,1}, \ldots, e^i_{1,T} - e^i_{1,T-1}, \ldots, \\ e^i_{K,2} - e^i_{K,1}, \ldots, e^i_{K,T} - e^i_{K,T-1}\Big). \tag{7.6}$$

For simplification, the RE variation over time is represented as

$$\Delta\mathbf{e}^i = \left(\Delta\mathbf{e}^i_{1,1}, \ldots, \Delta\mathbf{e}^i_{1,T}, \ldots, \Delta\mathbf{e}^i_{K,2}, \ldots, \Delta\mathbf{e}^i_{K,T}\right). \tag{7.7}$$

In the case of TREE6, the phase angle information measured from the primary and secondary sides of the power transformer is denoted as

$$\mathbf{\Phi} = (\Delta\varphi^{12}, \Delta\varphi^{23}, \Delta\varphi^{13}), \tag{7.8}$$

with

$$\Delta\boldsymbol{\varphi}^{ij} = [(x_{i,1} - x_{i+3,1}) - (x_{j,1} - x_{j+3,1}), \ldots, \\ (x_{i,T} - x_{i+3,T}) - (x_{j,T} - x_{j+3,T})]. \tag{7.9}$$

Furthermore, the complex values of the measured voltage and phase angle can be calculated using:

$$c_{i,j} = x_{i,j} \cdot \exp(x_{i+\lfloor K/2 \rfloor,j} \cdot 1j), \tag{7.10}$$

and as a result,

$$\mathbf{c}^i = (c_{i,1}, c_{i,2}, \ldots, c_{i,T}). \tag{7.11}$$

In Sections 7.2.2.2 and 7.2.2.3, we present the details of the objective and constraint functions for the TREE problems.

7.2.2.2 Objective Functions

The TREE problem involves three objective functions, which are formulated based on the relationship between the true voltage values and the measurements obtained from the VTs. A smaller objective value implies a better estimation accuracy for each objective function.

(1) Total time-varying REs

The first objective is the total time-varying relative errors (REs) of all uncalibrated VTs, which indicates the degree of matching between the true voltage values and the measured ones. The goal is to minimize the time-varying REs over the VTs. The formulation for the first objective is:

$$f_1(\mathbf{x}, \mathbf{e}^1, \dots, \mathbf{e}^P) = \sum_{j=1}^{T} \sum_{k=1}^{K} \sum_{l=1}^{P} e^l_{k,j}, \tag{7.12}$$

where $\mathbf{e}^l$ is given in Eq. (7.5).

(2) Sum of RE variation over time

The second objective is the sum of the RE variation over time, which reflects the time-varying relationship between the true voltage values and the measured ones. The variance of the RE variations for different VTs should be minimized. The formulation for the second objective is:

$$f_2(\mathbf{x}, \mathbf{\Delta e}^1, \dots, \mathbf{\Delta e}^P) = \sum_{l=1}^{P} \sqrt{std(\Delta e^l_{1,1}, \cdots, \Delta e^l_{L,T})}, \tag{7.13}$$

where $\mathbf{\Delta e}^l$ is given in Eq. (7.7), and L is K for TREE1 to TREE5 and $\lfloor K/2 \rfloor$ for TREE6, respectively.

(3) Phase angle relationship

In contrast to the first two objectives, which only consider voltage values, the third objective focuses on the phase angle relationship among the true voltage values, the measured phase angle values, and the measured voltage values. The variance of the phase angle RE variations for different VTs should be minimized. The formulation for the third objective is:

$$f_3(\mathbf{x}, \mathbf{\Delta e}^1, \dots, \mathbf{\Delta e}^P) = \sum_{l=1}^{P} \sqrt{std((\Delta e^l_{\lfloor K/2 \rfloor,1}, \dots, \Delta e^l_{K,T}))}. \tag{7.14}$$

Although the formulations for the second and third objectives are similar, they are not merged due to the presence of different types of data.

7.2.2.3 Constraint Functions

In a power delivery system, VTs in the same substation are connected through conductors, and the measurement of voltage and phase angle values is a long-term process. Therefore, three main constraints can be formulated to restrict the feasible area of the problem: topology constraint, time series constraint, and phase constraint.

(1) Topology constraints

Topology constraints are derived from the three-phase balance principle, stating that the true three-phase primary voltages should be kept in balance if the grid is under control [15]. This balance is described by the voltage unbalance factor (VUF), which indicates the balance between three-phase impedance. Since the three-phase impedance is balanced and varies slowly (especially in the high-voltage grid) [16], the VUF value and its fluctuation can be limited within a range as

$$\sum_{i=1}^{K}(\tau_i - C_1) < 0. \tag{7.15}$$

Moreover, because the primary side voltage value is proportional to the secondary side voltage value, all proportions for three-phase voltages should be the same. We can derive the second constraint from its fluctuation by

$$\sum_{i=1}^{K-1}[std(\tau_2 - \tau_1, \cdots, \tau_K - \tau_{K-1}) - C_2] < 0. \tag{7.16}$$

The detailed formulation of VUF is given as

$$\boldsymbol{\tau} = (\tau_1, \tau_2, \cdots, \tau_K), \tag{7.17}$$

where

$$\tau_i = \max_{j \in [1,2,\ldots,T]} |x_{i,j} - \mu_j| / \mu_j,$$
$$\mu_j = \frac{1}{3}\sum_{i=1}^{3} x_{i,j}. \tag{7.18}$$

(2) Series constraints

The series constraints mainly describe the time-varying relationships of the same decision variable. The basic assumption is that the decision variable changes slowly and continuously, rather than abruptly. The variation of the decision vector $\mathbf{x}$ at time j is defined as

$$\boldsymbol{\Delta}\mathbf{x}^j = \left(\frac{x_{1,j+1} - x_{1,j}}{x_{1,j}}, \frac{x_{2,j+1} - x_{2,j}}{x_{2,j}}, \ldots, \frac{x_{K,j+1} - x_{i,j}}{x_{K,j}}\right), \tag{7.19}$$

where $i \in [1, 2, 3]$ and $j \in [1, 2, \ldots, T]$. The corresponding constraint can be derived as

$$\sum_{j=1}^{T-1}\left[std(\boldsymbol{\Delta}\mathbf{x}^j) - C_3\right] < 0. \tag{7.20}$$

Additionally, another constraint for the primary and secondary voltages (i.e., TREE3 to TREE6) is derived. This constraint reveals the imbalance between the primary and secondary sides, and its formulation is:

$$\sum_{j=1}^{T-1} \left[std(\mathbf{t}^j) - C_4 \right] < 0, \tag{7.21}$$

where

$$\mathbf{t}^j = \left(\frac{x_{1,j} \cdot x_{4,j}}{x_{3,j} \cdot x_{2,j}}, \frac{x_{2,j} \cdot x_{6,j}}{x_{4,j} \cdot x_{3,j}}, \frac{x_{3,j} \cdot x_{4,j}}{x_{6,j} \cdot x_{1,j}} \right). \tag{7.22}$$

(3) Phase constraint

The VT is an inductive sensor in the power delivery system, and the primary voltage phasors will be lagged after being transformed to the secondary side. Besides, the phase angle difference between every two phasors mainly depends on the equivalent impedance of the power transformer and the load current through it. The basic assumption of the phase constraint is that the difference between three-phase angles in a power transformer system is held constant. Therefore, the first phase constraint indicates the tolerance of phase angle imbalance, and its formulation is:

$$std(\mathbf{\Phi}) - C_5 < 0, \tag{7.23}$$

where $\mathbf{\Phi}$ is given in Eq. (7.8).

Furthermore, we can obtain the following matrix according to the phase and voltage balance in the power delivery system:

$$\begin{bmatrix} IA & Ia \\ IB & Ib \\ IC & Ic \end{bmatrix} = \frac{1}{3} \times \begin{bmatrix} 1 & \exp\left(\frac{2\pi}{3}j\right) & \exp\left(\frac{4\pi}{3}j\right) \\ 1 & \exp\left(\frac{4\pi}{3}j\right) & \exp\left(\frac{2\pi}{3}j\right) \\ 1 & 1 & 1 \end{bmatrix} \begin{bmatrix} \mathbf{c}^1 & \mathbf{c}^4 \\ \mathbf{c}^2 & \mathbf{c}^5 \\ \mathbf{c}^3 & \mathbf{c}^6 \end{bmatrix}, \tag{7.24}$$

where $\mathbf{c}^i$ is given in Eq. (7.11). Then, we can obtain

$$\begin{aligned} \boldsymbol{\psi}^1 &= IB/IA = (\psi_{1,1}, \psi_{1,2}, \cdots, \psi_{1,T}), \\ \boldsymbol{\psi}^2 &= Ib/Ia = (\psi_{2,1}, \psi_{2,2}, \cdots, \psi_{2,T}). \end{aligned} \tag{7.25}$$

The variation of $\boldsymbol{\psi}$ over time can be derived as

$$\begin{aligned} \mathbf{\Delta}\boldsymbol{\psi}^1 &= (\psi_{1,2} - \psi_{1,1}, \cdots, \psi_{1,T} - \psi_{1,T-1}), \\ \mathbf{\Delta}\boldsymbol{\psi}^2 &= (\psi_{2,2} - \psi_{2,1}, \cdots, \psi_{2,T} - \psi_{2,T-1}). \end{aligned} \tag{7.26}$$

Thus, the final constraint, which reveals the tolerance of phase angle variation, can be derived as

$$\max std(\mathbf{\Delta}\boldsymbol{\psi}^1), std(\mathbf{\Delta}\boldsymbol{\psi}^2) - C_6 < 0. \tag{7.27}$$

With the objective functions and constraints established, the TREE problems can be solved using evolutionary algorithms. The goal is to obtain the best estimation of the true voltage values and phase angles, while satisfying the constraints derived from the power delivery system's characteristics. This will lead to a more accurate and reliable operation of the power grid, ensuring that the VTs perform their crucial role in monitoring and controlling the system effectively.

7.2.3 Problem Analysis

From the elaborations on the proposed TREE problems, it can be found that these problems are scalable to have a very vast search space, causing the curse of dimensionality. Besides, the TREE problems have multiple complicated constraints, which reduce the size of feasible search space considerably. When the above characteristics of TREE problems are known, some other problem characteristics deserve to be explored, including the relationships among decision variables, objectives, and constraints. After all, an LSMOP can be easily solved by a divide-and-conquer method if its decision variables are fully separable. By contrast, when the decision variables have intricate interactions, the difficulty of the problem usually increases considerably.

In this section, we explore the relationships among decision variables, objectives, and constraints using three different methods, including the differential grouping method from [17], and the variable interaction analysis methods in multi-objective evolutionary algorithm with decision variable analysis (MOEA/DVA) [18] and large-scale multi-objective evolutionary algorithm (LMEA) [19]. For these analyses, T is set to 200 for each problem, and the number of decision variables is 600, 600, 1200, 1200, 1200, and 2400 for TREE1 through TREE6, respectively.

Table 7.2 presents the numbers of interacting and independent decision variables associated with each objective/constraint for each TREE problem. It can be seen that the first objectives of TREE1 through TREE5 are fully separable, while the second ones can be partially divided into three equal-sized groups without any separable variables. Additionally, all constraints of TREE1 through TREE5, except for TREE6, are fully nonseparable. In summary, the proposed test suite encompasses both separable and nonseparable components.

Table 7.3 presents the numbers of different types of decision variables obtained by MOEA/DVA and LMEA. The analysis results indicate that almost all decision variables of TREE1 through TREE5 are position variables, and approximately half of the decision variables of TREE6 are distance variables concerning relationships between decision variables and objectives. However, significant differences arise when applying MOEA/DVA and LMEA methods to analyze relationships between decision variables and constraints. This discrepancy can be attributed to the fact that both methods were designed for analyzing relationships between decision variables and objectives and might not perform effectively when handling constraints.

Table 7.2 The Decision Variable Interaction in Association with Different Objective and Constraint Functions.

Problem	Part	f_1	f_2	f_3	g_1	g_2	g_3	g_4	g_5	g_6
TREE1	Part 1	0	200:200:200	—	600	600	600	—	—	—
	Part 2	600	0	—	0	0	0	—	—	—
TREE2	Part 1	0	400:400:400	—	1200	1200	1200	—	—	—
	Part 2	1200	0	—	0	0	0	—	—	—
TREE3	Part 1	0	200:200:200	—	600	600	600	—	—	—
	Part 2	600	0	—	0	0	0	—	—	—
TREE4	Part 1	0	400:400:400	—	1200	1200	1200	1200	—	—
	Part 2	1200	0	—	0	0	0	0	—	—
TREE5	Part 1	0	400:400:400	—	1200	1200	1200	1200	—	—
	Part 2	1200	0	—	0	0	0	0	—	—
TREE6	Part 1	0	300:150:150	300:150:150	1200	1200	1200	1200	600	600:600
	Part 2	1200	600	600	0	0	0	0	600	0

Table 7.3 The Numbers of Position Variables and Distance Variables of Each TREE Problem Obtained by MOEA/DVA and LMEA.

	MOEA/DVA		LMEA	
Problem	**Objectives**	**Constraints**	**Objectives**	**Constraints**
TREE1	600:0	598:2	600:0	337:263
TREE2	1199:1	1200:0	1200:0	341:859
TREE3	597:3	600:0	600:0	14:586
TREE4	1197:3	1200:0	1200:0	126:1074
TREE5	599:1	1200:0	1200:0	199:1001
TREE6	1193:1207	2380:20	1200:1200	2346:54

7.3 Problem Knowledge-Driven Coevolutionary Algorithm for Time-Varying Ratio Error Estimation

Based on the problem analysis as above, it can be noted that the TREE problems pose stiff challenges to the existing MOEAs tailored for general LSMOPs. However, this does not mean that evolutionary algorithms cannot effectively address the TREE problems. For this specific problem, as indicated in previous work, it may be very useful to utilize the information of the problem in designing effective MOEAs. For this reason, the following observation can be found. Despite that the TREE problems are characterized by the vast search space, complicated constraints, and intricate decision variable interactions, another important characteristic of TREE problems is that most VTs work under normal operating conditions, which means that there only exist a small proportion of VTs having relatively large errors. Therefore, it can be inferred that most decision variables in each TREE problem have similar values, which inspires us to exploit this problem knowledge to design a novel search strategy, thus improving the effectiveness and efficiency for solving the challenging TREE problems.

7.3.1 Problem Knowledge Utilization

When we exploit the problem knowledge (i.e., most decision variables in each TREE problem have similar values) to improve the search efficiency of evolutionary algorithms in solving TREE problems, the idea is to coevolve two populations, where one population consists of non-sparse solutions and the other population is composed of sparse solutions. As shown in Fig. 7.3, when a sparse solution and a non-sparse solution are selected as parent solutions for offspring generation, most decision variables in the generated offspring solution have similar values, which enables the Pareto optimal solution sets of TREE problems to be handled effectively. Thus, in the case that many algorithms can be used to evolve the population that contains non-sparse solutions, the main task of solving TREE problems changes to design a novel search strategy to evolve the population that contains sparse solutions.

As introduced in Chapter 4, in recent years, a lot of works have been reported for sparse large-scale multi-objective optimization. Considering that the search space of TREE problems is more complicated, in this section, we propose a new operator to improve the efficiency in generating sparse solutions. In the operator, decision variables are first divided into multiple groups. Then, a group of decision variables are selected randomly to perform grouped and linked polynomial mutation [20], which has been verified to be effective to generate sparse solutions in SparseEA2 [21]. However, in the grouped and linked polynomial mutation, the direction of mutation is determined randomly, which is prone to mutate a decision variable from zero to nonzero, thus affecting the efficiency of generating sparse solutions. To address this issue, in the proposed operator, a unidirectional mutation operator is incorporated, which always mutates decision variables along the direction of coming close to zero. In this way, the effectiveness and efficiency of generating sparse solutions can be improved greatly. In Fig. 7.4, two illustrative examples are presented, which show that promising sparse solutions can be obtained by performing the proposed unidirectional mutation operator.

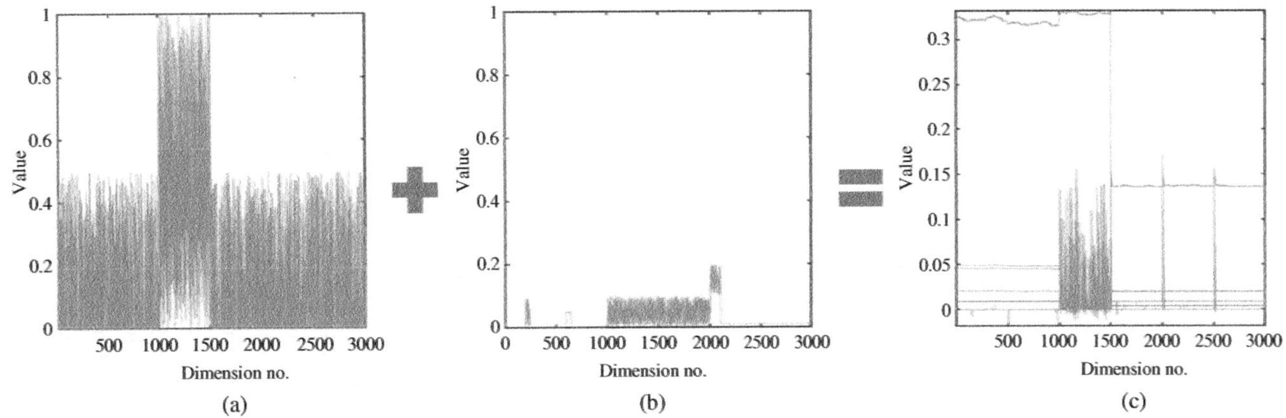

Figure 7.3 The offspring solution generated by a sparse parent solution and a non-sparse parent solution. (a) Non-sparse parent solution. (b) Sparse parent solution. (c) Offspring solution.

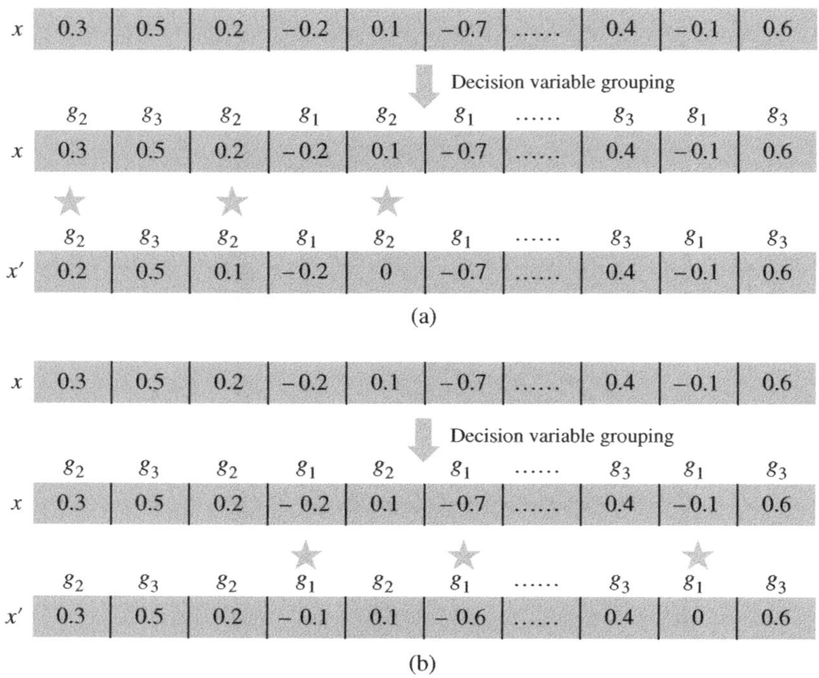

Figure 7.4 Examples of the proposed unidirectional mutation operator. (a) Mutating selected decision variables along the direction of decreasing their values to zero. (b) Mutating selected decision variables along the direction of increasing their values to zero.

To describe the proposed operator in more detail, its pseudocode is presented in Algorithm 7.1. To begin with, the mutation operator divides decision variables into k groups by the grouping method GM. Then, a group of decision variables are selected randomly to perform mutation operator. Specifically, a rand number r_1 is first generated and compared with 0.5. If r_1 is smaller than 0.5, the unidirectional mutation is performed; otherwise, the bidirectional mutation (i.e., grouped and linked polynomial mutation) is performed. In the unidirectional mutation, the direction of mutation is determined in the following way. Decision variables in the selected decision variable group gj are first sorted. Then, the decision variable with the median value is selected and termed as x_{flag}. If x_{flag} is larger than zero, which means that at least a half of decision variables in gj have values larger than zero, all decision variables in gj are mutated along the direction of decreasing their values to zero. Otherwise, there exist at least a half decision variables in gj that their values are smaller than zero; thus, all decision variables in gj are mutated along the direction of increasing their values to zero.

Algorithm 7.1: Improved Grouped and Linked Polynomial Mutation

---

Input: $\mathbf{x}$ (solution to be mutated), *disM* (distribution index of mutation), *GM*(decision variable grouping method)

Output: $\mathbf{x}'$ (mutated solution)

$[g1, \ldots, gk] \leftarrow$ Divide decision variables into k groups by *GM*;

$gj \leftarrow$ Randomly select a group of decision variables from $[g1, \ldots, gk]$ for mutation;

$r1 \leftarrow rand(0, 1)$;

if $r1 < 0.5$ **then**

  //Unidirectional Mutation

  $x_{flag} \leftarrow$ Sort the decision variables in gj and select a decision variable with the median value;

  if $x_{flag} > 0$ **then**

    for all decision variables x_i *in* gj **do**

      $\delta = \frac{x_i - l_i}{u_i - l_i}$;

      $\delta_q = (2\mu + (1 - 2\mu)(1 - \delta)^{disM+1})^{\frac{1}{disM+1}} - 1$;

  else

    for all decision variables x_i *in* gj **do**

      $\delta = \frac{u_i - x_i}{u_i - l_i}$;

      $\delta_q = 1 - (2(1 - \mu) + 2(\mu - 0.5)(1 - \delta)^{disM+1})^{\frac{1}{disM+1}}$;

else

  //Bidirectional Mutation

  $r2 \leftarrow rand(0, 1)$;

  if $r2 < 0.5$ **then**

    for all decision variables x_i *in* gj **do**

      $\delta = \frac{x_i - l_i}{u_i - l_i}$;

      $\delta_q = (2\mu + (1 - 2\mu)(1 - \delta)^{disM+1})^{\frac{1}{disM+1}} - 1$;

  else

    for all decision variables x_i *in* gj **do**

      $\delta = \frac{u_i - x_i}{u_i - l_i}$;

      $\delta_q = 1 - (2(1 - \mu) + 2(\mu - 0.5)(1 - \delta)^{disM+1})^{\frac{1}{disM+1}}$;

$\mathbf{x}_\mathbf{i}' = \mathbf{x}_\mathbf{i} + \delta_q(u_i - l_i)$;

$\mathbf{x}_\mathbf{i}' = \min(\max(\mathbf{x}_\mathbf{i}', l_i), u_i)$;

for all decision variables x_i *not in* gj **do**

  $\mathbf{x}_\mathbf{i}' = x_i$;

return mutated solution $\mathbf{x}'$;

7.3.2 Framework of PKCEA

This section gives the framework of the proposed problem knowledge-driven coevolutionary algorithm, namely, PKCEA. As shown in Fig. 7.5, the proposed

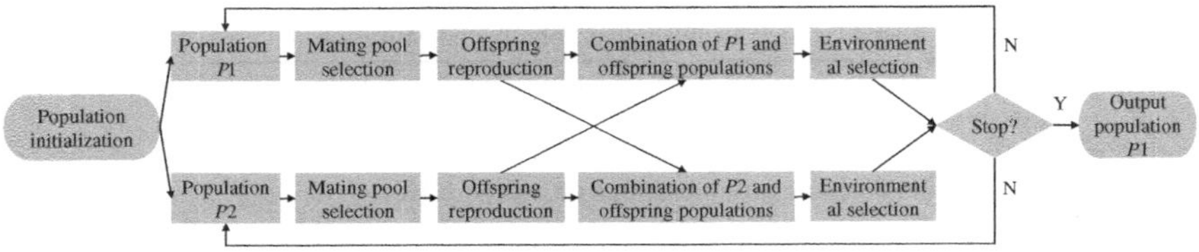

Figure 7.5 Framework of the proposed problem knowledge-driven coevolutionary algorithm PKCEA.

algorithm contains two optimization tasks, namely, the main optimization task and the auxiliary optimization task. The main optimization task is addressed by population $P1$, which uses the conventional genetic operators (i.e., simulated binary crossover [22] and polynomial mutation [23]) to generate offspring solutions. The auxiliary optimization task is addressed by population $P2$, which uses the improved grouped and linked polynomial mutation operator to generate offspring solutions. In fact, a lot of similar frameworks [24–27] have been reported in recent years, which demonstrate their effectiveness to solve multi-objective optimization problems with constraints by introducing auxiliary optimization. However, the majority of existing algorithms devote their efforts to solving medium and small-scale multi-objective optimization problems, and the auxiliary optimization is designed by easing the difficulty brought by constraints [28], thus improving the effectiveness and efficiency of environmental selection. By contrast, the algorithm PKCEA is proposed to solve LSMOPs and the auxiliary optimization is designed by focusing on generating sparse solutions, which enables the generated solutions similar to the Pareto optimal solution sets as much as possible, thus improving the effectiveness and efficiency in solving TREE problems.

Algorithm 7.2 gives the detailed procedure of PKCEA. To begin with, two initial populations $P1$ and $P2$ are generated, where population $P1$ is generated by the random method and population $P2$ is generated by Algorithm 4.1. When solutions in the two populations are evaluated, the algorithm enters the loop. In the loop, first, two mating parent pools, namely, $S1$ and $S2$, are obtained by the binary tournament selection method, where the non-dominated front number and crowding distance of solutions are used as the selection criteria. Then, two offspring populations are generated by genetic operators. Specially, the offspring in $O1$ are generated by the simulated binary crossover and polynomial mutation operators, the offspring in $O2$ are generated by the improved grouped and linked polynomial mutation operator. Next, the two offspring populations are combined with $P1$ and $P2$ to obtain populations $H1$ and $H2$. Finally, when the solutions in $H1$ and $H2$ are evaluated, the environmental selection strategy of

Algorithm 7.2: Procedure of the Proposed PKCEA

Input: *MaxFEs* (the maximum number of function evaluations), *N* (population size)
Output: population *P*1

*P*1 ← Generate *N* solutions randomly;
*P*2 ← Generate *N* sparse solutions by Algorithm 4.1;
Evaluate population *P*1 and population *P*2;
$FE = 2N$;
while $FE < MaxFEs$ **do**
    S1← Select *N* parents from *P1* by the binary tournament selection method;
    S2← Select *N* parents from *P2* by the binary tournament selection method;
    O1← Generate *N/2* offspring based on *S1* by the simulated binary crossover and polynomial mutation operators;
    O2← Generate *N/2* offspring based on *S2* by the improved grouped and linked polynomial mutation operator;
    $H1 \leftarrow P1 \cup O1 \cup O2$;
    $H2 \leftarrow P2 \cup O1 \cup O2$;
    Evaluate population *H*1 and population *H*2;
    $FE = FE + N$;
    P1← Select *N* solutions from *H1* by the environmental selection strategy of NSGA-II;
    P2← Select *N* solutions from *H2* by the environmental selection strategy of NSGA-II;
return population *P*1;

non-dominated sorting based genetic algorithm-II (NSGA-II) [29] is performed to update populations *P*1 and *P*2. The loop iterates until the number of function evaluations *FE* is not smaller than the maximum number of function evaluations *MaxFEs*, and the algorithm terminates and outputs population *P*1.

7.3.3 Experimental Comparisons

In this section, experimental comparisons between the proposed algorithm and several representative MOEAs are performed on the formulated TREE test problems, namely, TREE1–TREE5. For each test problem, the number of decision variables is set to 3000, 6000, 9000, 12,000, and 15,000; thus, there are totally 25 test instances used for performance assessment. The proposed algorithm is compared with five MOEAs, including NSGA-II [29], large-scale multi-objective

Table 7.4 The HV Results Obtained by NSGA-II, LSMOF, LMOCSO, CLS-RVEA, FLEA, and PKCEA on 25 TREE Test Instances. The Best Result in Each Row is Highlighted.

Problem	*D*	NSGA-II	LSMOF	LMOCSO	CLS-RVEA	FLEA	PKCEA
TREE1	3,000	5.5609e−1 (7.08e−3)	8.3744e−1 (6.35e−5)	7.2430e−1 (1.15e−2)	7.4358e−1 (1.24e−2)	8.3587e−1 (2.82e−3)	8.4733e−1 (7.10e−4)
	6,000	5.5082e−1 (2.88e−3)	8.3684e−1 (5.43e−5)	7.3589e−1 (1.36e−2)	7.6230e−1 (8.63e−3)	8.3633e−1 (5.24e−4)	8.4796e−1 (4.60e−4)
	9,000	0.00E+00(0.00E+00)	8.3725e−1 (4.32e−5)	7.3928e−1 (1.22e−2)	7.6258e−1 (1.06e−2)	8.3693e−1 (6.71e−4)	8.4780e−1 (6.30e−4)
	12,000	0.00E+00(0.00E+00)	8.3252e−1 (2.26e−5)	7.2139e−1 (2.28e−2)	7.5578e−1 (9.81e−3)	8.3045e−1 (6.09e−3)	8.4261e−1 (3.26e−4)
	15,000	0.00E+00(0.00E+00)	8.3498e−1 (2.89e−5)	7.3851e−1 (2.35e−2)	7.6048e−1 (1.12e−2)	8.3495e−1 (2.15e−4)	8.4477e−1 (4.24e−4)
TREE2	3,000	6.2957e−1 (5.60e−3)	8.4982e−1 (1.76e−5)	7.4414e−1 (1.51e−2)	7.6648e−1 (9.90e−3)	8.5135e−1 (1.48e−3)	8.5598e−1 (5.56e−5)
	6,000	6.2247e−1 (3.40e−3)	8.4986e−1 (1.03e−5)	7.4794e−1 (1.17e−2)	7.7872e−1 (7.01e−3)	8.5106e−1 (1.56e−3)	8.5593e−1 (4.37e−5)
	9,000	6.1745e−1 (3.18e−3)	8.5007e−1 (5.61e−6)	7.4418e−1 (2.11e−2)	7.8524e−1 (4.68e−3)	8.5163e−1 (6.63e−4)	8.5617e−1 (5.23e−5)
	12,000	6.1612e−1 (6.81e−4)	8.4947e−1 (6.93e−6)	7.5074e−1 (1.58e−2)	7.8256e−1 (5.74e−3)	8.5149e−1 (6.36e−4)	8.5603e−1 (9.15e−5)
	15,000	0.00E+00(0.00E+00)	8.4941e−1 (5.99e−6)	7.5056e−1 (1.76e−2)	7.8304e−1 (5.54e−3)	8.5154e−1 (6.16e−4)	8.5610e−1 (5.47e−5)
TREE3	3,000	0.00E+00(0.00E+00)	8.8733e−1 (9.69e−7)	0.00E+00(0.00E+00)	0.00E+00(0.00E+00)	8.8675e−1 (1.32e−3)	8.8749e−1 (3.24e−6)
	6,000	0.00E+00(0.00E+00)	8.8743e−1 (1.05e−6)	0.00E+00(0.00E+00)	0.00E+00(0.00E+00)	8.8742e−1 (1.42e−5)	8.8760e−1 (1.93e−5)
	9,000	0.00E+00(0.00E+00)	8.8797e−1 (3.96e−7)	0.00E+00(0.00E+00)	0.00E+00(0.00E+00)	8.8796e−1 (3.62e−7)	8.8813e−1 (8.36e−6)
	12,000	0.00E+00(0.00E+00)	8.8948e−1 (1.51e−6)	0.00E+00(0.00E+00)	0.00E+00(0.00E+00)	8.8948e−1 (4.63e−7)	8.8963e−1 (6.65e−6)
	15,000	0.00E+00(0.00E+00)	8.8988e−1 (6.86e−7)	0.00E+00(0.00E+00)	0.00E+00(0.00E+00)	8.8988e−1 (5.26e−8)	8.9004e−1 (6.94e−6)
TREE4	3,000	0.00E+00(0.00E+00)	9.6117e−1 (5.67e−6)	0.0000e+0 (0.00e+0)	4.8583e−2 (7.47e−2)	9.5705e−1 (9.18e−3)	9.6109e−1 (4.07e−5)
	6,000	0.00E+00(0.00E+00)	9.6439e−1 (2.53e−6)	0.0000e+0 (0.00e+0)	2.9504e−2 (5.49e−2)	9.6433e−1 (1.12e−4)	9.6438e−1 (3.23e−5)
	9,000	0.00E+00(0.00E+00)	9.6574e−1 (2.39e−6)	0.00E+00(0.00E+00)	1.3989e−2 (4.12e−2)	9.6050e−1 (1.62e−2)	9.6571e−1 (1.85e−5)
	12,000	0.00E+00(0.00E+00)	9.6651e−1 (1.57e−6)	0.00E+00(0.00E+00)	2.6069e−2 (4.33e−2)	9.6648e−1 (3.24e−5)	9.6648e−1 (1.79e−5)
	15,000	0.00E+00(0.00E+00)	9.6696e−1 (8.68e−7)	0.00E+00(0.00E+00)	1.1774e−2 (3.30e−2)	8.8464e−1 (2.01e−1)	9.6696e−1 (9.73e−6)
TREE5	3,000	0.00E+00(0.00E+00)	9.2734e−1 (1.68e−5)	6.1514e−1 (0.00e+0)	6.7691e−1 (1.20e−2)	9.2253e−1 (1.05e−2)	9.3518e−1 (5.84e−4)
	6,000	0.00E+00(0.00E+00)	9.2742e−1 (1.53e−5)	6.3378e−1 (6.46e−3)	6.8636e−1 (7.21e−3)	9.2721e−1 (4.82e−4)	9.3549e−1 (4.10e−4)
	9,000	0.00E+00(0.00E+00)	9.2735e−1 (1.67e−5)	6.4223e−1 (0.00e+0)	6.9261e−1 (9.40e−3)	9.2663e−1 (2.40e−3)	9.3568e−1 (3.29e−4)
	12,000	0.00E+00(0.00E+00)	9.2756e−1 (9.06e−6)	6.4643e−1 (0.00e+0)	7.0572e−1 (1.73e−2)	9.2755e−1 (4.00e−5)	9.3608e−1 (1.76e−4)
	15,000	0.00E+00(0.00E+00)	9.2735e−1 (9.48e−6)	6.4696e−1 (0.00e+0)	6.9637e−1 (1.32e−2)	9.2607e−1 (4.48e−3)	9.3625e−1 (0.00e+0)

optimization framework (LSMOF) [30], large-scale multi-objective competitive swarm optimizer (LMOCSO) [31], constrained large-scale reference vector guided evolutionary algorithm (CLS-RVEA) [32], and FLEA [33].

In the experiments, the maximum number of function evaluations is adopted as the termination criterion, which is set to $20 \times D$ for each algorithm on each test problem with D denoting the number of decision variables. For all algorithms, the population size is set to 100. For the proposed PKCEA, the ordered grouping method is used, and the number of decision variable groups is set to four. For the compared algorithms, their algorithm-specific parameters are set to the values in their original papers. Considering that the Pareto fronts of the TREE problems are unknown, the hypervolume (HV) [34] indicator is used for performance assessment. It is worth noting that the experiments as mentioned above are performed on PlatEMO [35].

Table 7.4 shows the HV results obtained by the six algorithms on the 25 test instances. From the table, it can be seen that the proposed algorithm PKCEA achieves the best overall performance, having obtained the best result on most test instances. LSMOF and FLEA show inferior overall performance than the proposed PKCEA, but they can always obtain a nonzero HV value on each test instance. By contrast, the HV values obtained by the other three algorithms on some test instances are zero, due to the fact that they fail to obtain any feasible solution. To be specific, NSGA-II fails to obtain any feasible solution on the TREE3, TREE4, TREE5, high-dimensional TREE1, and high-dimensional TREE2 problems, LMOCSO cannot obtain any feasible solution on the TREE3 and TREE4 problems, and CLS-RVEA does not obtain any feasible solution on TREE3.

Figure 7.6 plots the populations and convergence curves with median HV values obtained by the six algorithms on the TREE1, TREE3, and TREE5 problems with 3000 decision variables. It is worth noting that the populations and convergence curves of NSGA-II, LMOCSO, and CLS-RVEA on some problems are not provided since they fail to converge to feasible regions. From the figure, it can be seen that the proposed algorithm exhibits better performance.

7.4 Summary

In this chapter, a data-driven problem formulation method is proposed for ratio error estimation of VTs, which offers several advantages, including lower cost compared to on-site/off-line calibration, real-time estimation for early warnings of potential faulty VTs, and versatility in handling different substation types and VT models. These advantages make the method practical and valuable for real-world applications. Based on the data-driven problem formulation method, a number of test problems are proposed. In the proposed TREE problems, a

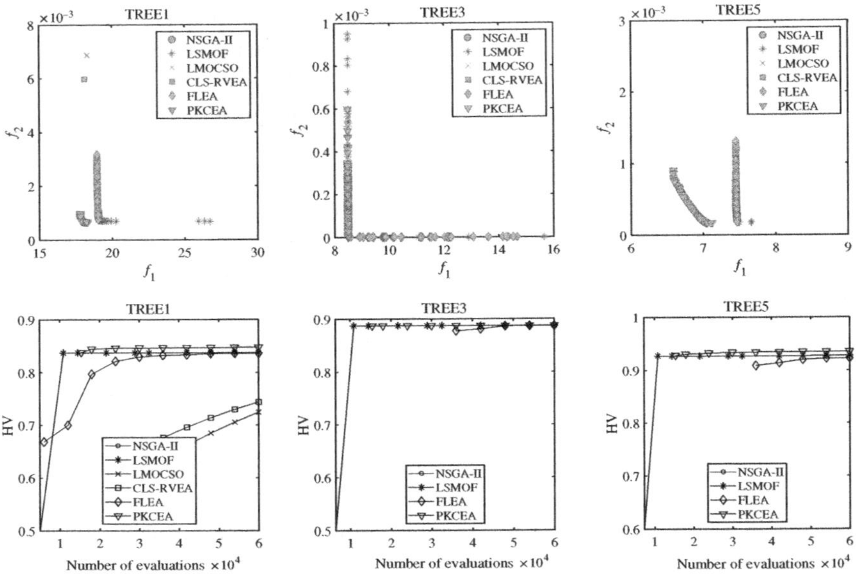

Figure 7.6 Populations and convergence curves with median HV values obtained by the six algorithms on the TREE1, TREE3, and TREE5 problems.

lot of challenges are associated, including the vast search space, the complex constraints, and the intricate interactions between decision variables.

To solve the challenging TREE problems, a problem knowledge-driven coevolutionary algorithm is reported, which introduces a sparse large-scale multi-objective optimization as the auxiliary optimization. In the auxiliary optimization, a novel genetic operator is proposed to improve the efficiency of generating sparse solutions, thus enabling the Pareto optimal solution sets of TREE problems to be solved effectively. Experimental studies show that the proposed algorithm is superior over the compared MOEAs.

References

1 L. Guangmao, "Operation analysis of CVT in Guangzhou Grid from 2005 to 2015," *Power Capacitor & Reactive Power Compensation*, vol. 38, no. 3, pp. 110–115, 2017.

2 H. Zhang, H. Shao, J. Wang, W. Wei, F. Lin, T. Sun, Z. Wei, C. Li, and Y. Wu, "Voltage ratio traceability of 10 kV low-voltage excited two-stage voltage transformer," *IEEE Transactions on Instrumentation & Measurement*, vol. 66, no. 6, pp. 1405–1410, 2017.

3 C. E. Council, "2017–2018 national power supply and demand situation analysis and forecast report," *Electro Industries*, vol. 207, no. 2, pp. 11–15, 2018.

4 B. Chen, L. Du, K. Liu, X. Chen, F. Zhang, and F. Yang, "Measurement error estimation for capacitive voltage transformer by insulation parameters," *Energies*, vol. 10, no. 3, p. 357, 2017.

5 IEEE Team, "IEEE standard requirements for instrument transformers," *IEEE Std C*, vol. 57, pp. 1–96, 2016.

6 JJG 314-2010, *Verification Regulation of Instrument Voltage Transformers*, Beijing, China, 2010.

7 M. Zhang, K. Li, J. Wang, and S. He, "An on-site calibration system for electronic instrument transformers based on LabVIEW," *Metrology and Measurement Systems*, vol. 21, no. 2, pp. 257–270, 2014.

8 C. Hu, M. Chen, H. Li, Z. Zhang, Y. Jiao, and H. Shao, "An accurate on-site calibration system for electronic voltage transformers using a standard capacitor," *Measurement Science and Technology*, vol. 29, no. 5, p. 055901, 2018.

9 A. Pal, P. Chatterjee, J. S. Thorp, and V. A. Centeno, "Online calibration of voltage transformers using synchrophasor measurements," *IEEE Transactions on Power Delivery*, vol. 31, no. 1, pp. 370–380, 2015.

10 D. Shi, D. J. Tylavsky, and N. Logic, "An adaptive method for detection and correction of errors in PMU measurements," *IEEE Transactions on Smart Grid*, vol. 3, no. 4, pp. 1575–1583, 2012.

11 K. V. Khandeparkar, S. A. Soman, and G. Gajjar, "Detection and correction of systematic errors in instrument transformers along with line parameter estimation using PMU data," *IEEE Transactions on Power Systems*, vol. 32, no. 4, pp. 3089–3098, 2016.

12 C. Zhang, H. Li, and Q. Chen, "Detection of the ratio error drift in CVT considering AVC," *Measurement*, vol. 138, pp. 425–432, 2019.

13 P. Chatterjee, A. Pal, J. S. Thorp, J. De La Ree Lopez, and V. A. Centeno, "Error reduction of phasor measurement unit data considering practical constraints," *IET Generation, Transmission & Distribution*, vol. 12, no. 10, pp. 2332–2339, 2018.

14 C. Wang, V. A. Centeno, K. D. Jones, and D. Yang, "Transmission lines positive sequence parameters estimation and instrument transformers calibration based on PMU measurement error model," *IEEE Access*, vol. 7, pp. 145 104–145 117, 2019.

15 Y.-J. Kim, "Development and analysis of a sensitivity matrix of a three-phase voltage unbalance factor," *IEEE Transactions on Power Systems*, vol. 33, no. 3, pp. 3192–3195, 2018.

16 Y.-J. Wang and L. Pierrat, "A method integrating deterministic and stochastic approaches for the simulation of voltage unbalance in electric power distribution systems," *IEEE Transactions on Power Systems*, vol. 16, no. 2, pp. 241–246, 2001.

17 M. N. Omidvar, M. Yang, Y. Mei, X. Li, and X. Yao, "DG2: A faster and more accurate differential grouping for large-scale black-box optimization," *IEEE Transactions on Evolutionary Computation*, vol. 21, no. 6, pp. 929–942, 2017.

18 X. Ma, F. Liu, Y. Qi, X. Wang, L. Li, L. Jiao, M. Yin, and M. Gong, "A multiobjective evolutionary algorithm based on decision variable analyses for multiobjective optimization problems with large-scale variables," *IEEE Transactions on Evolutionary Computation*, vol. 20, no. 2, pp. 275–298, 2015.

19 X. Zhang, Y. Tian, R. Cheng, and Y. Jin, "A decision variable clustering-based evolutionary algorithm for large-scale many-objective optimization," *IEEE Transactions on Evolutionary Computation*, vol. 22, no. 1, pp. 97–112, 2016.

20 H. Zille, H. Ishibuchi, S. Mostaghim, and Y. Nojima, "Mutation operators based on variable grouping for multi-objective large-scale optimization," in *Proceedings of 2016 IEEE Symposium Series on Computational Intelligence*, 2016, pp. 1–8.

21 Y. Zhang, Y. Tian, and X. Zhang, "Improved SparseEA for sparse large-scale multi-objective optimization problems," *Complex & Intelligent Systems*, vol. 9, pp. 1127–1142, 2021.

22 K. Deb and R. B. Agrawal, "Simulated binary crossover for continuous search space," *Complex Systems*, vol. 9, no. 2, pp. 115–148, 1995.

23 K. Deb and M. Goyal, "A combined genetic adaptive search (GeneAS) for engineering design," *Computer Science and Informatics*, vol. 26, pp. 30–45, 1996.

24 Y. Tian, T. Zhang, J. Xiao, X. Zhang, and Y. Jin, "A coevolutionary framework for constrained multiobjective optimization problems," *IEEE Transactions on Evolutionary Computation*, vol. 25, no. 1, pp. 102–116, 2020.

25 K. Qiao, K. Yu, B. Qu, J. Liang, H. Song, and C. Yue, "An evolutionary multitasking optimization framework for constrained multiobjective optimization problems," *IEEE Transactions on Evolutionary Computation*, vol. 26, no. 2, pp. 263–277, 2022.

26 M. Ming, A. Trivedi, R. Wang, D. Srinivasan, and T. Zhang, "A dual-population-based evolutionary algorithm for constrained multiobjective optimization," *IEEE Transactions on Evolutionary Computation*, vol. 25, no. 4, pp. 739–753, 2021.

27 Y. Tian, Z. Shi, Y. Zhang, L. Zhang, H. Zhang, and X. Zhang, "Solving optimal power flow problems via a constrained many-objective co-evolutionary algorithm," *Frontiers in Energy Research*, vol. 11, p. 1293193, 2023.

28 Y. Zhang, Y. Tian, H. Jiang, X. Zhang, and Y. Jin, "Design and analysis of helper-problem-assisted evolutionary algorithm for constrained multiobjective optimization," *Information Sciences*, vol. 648, p. 119547, 2023.

29 K. Deb, A. Pratap, S. Agarwal, and T. Meyarivan, “A fast and elitist multiobjective genetic algorithm: NSGA-II,” *IEEE Transactions on Evolutionary Computation*, vol. 6, no. 2, pp. 182–197, 2002.

30 C. He, L. Li, Y. Tian, X. Zhang, R. Cheng, Y. Jin, and X. Yao, “Accelerating large-scale multiobjective optimization via problem reformulation,” *IEEE Transactions on Evolutionary Computation*, vol. 23, no. 6, pp. 949–961, 2019.

31 Y. Tian, X. Zheng, X. Zhang, and Y. Jin, “Efficient large-scale multiobjective optimization based on a competitive swarm optimizer,” *IEEE Transactions on Cybernetics*, vol. 50, no. 8, pp. 3696–3708, 2019.

32 C. Fan, J. Wang, L. T. Yang, L. Xiao, and Z. Ai, “Efficient constrained large-scale multi-objective optimization based on reference vector-guided evolutionary algorithm,” *Applied Intelligence*, vol. 53, pp. 21027–21049, 2023.

33 L. Li, C. He, R. Cheng, H. Li, L. Pan, and Y. Jin, “A fast sampling based evolutionary algorithm for million-dimensional multiobjective optimization,” *Swarm and Evolutionary Computation*, vol. 75, p. 101181, 2022.

34 L. While, P. Hingston, L. Barone, and S. Huband, “A faster algorithm for calculating hypervolume,” *IEEE Transactions on Evolutionary Computation*, vol. 10, no. 1, pp. 29–38, 2006.

35 Y. Tian, R. Cheng, X. Zhang, and Y. Jin, “PlatEMO: A MATLAB platform for evolutionary multi-objective optimization [educational forum],” *IEEE Computational Intelligence Magazine*, vol. 12, no. 4, pp. 73–87, 2017.

8

Evolutionary Large-Scale Multi-Objective Optimization in Radiotherapy Planning

8.1 Introduction

Since X-ray was found by Röntgen in the 1890s, ionizing radiation has received increasing attention in the therapy of malignant tumors. With the invention of linear accelerators in the 1950s, computed tomography and magnetic resonance imaging in the 1970s, and multileaf collimators (MLCs) in the 1980s, radiotherapy has become an important modality used in cancer treatment. Together with surgery and chemotherapy, nearly 70% cancer patients have received radiotherapy during their course of illness [1]. To provide conformal dose distribution for different clinical cases, intensity-modulated radiotherapy (IMRT) has served as an effective technique for radiotherapy, able to generate individual-specific therapeutic fields by adjusting the angles, intensities, and shapes of radiation [2].

8.1.1 Challenges of IMRT Planning

As illustrated in Fig. 8.1, the therapeutic fields are formed by the photons or protons transmitted through a MLC, where the photons or protons are emitted by radiation beams produced by a linear accelerator [3]. The therapeutic fields form heterogeneous dose distribution, so as to meet the clinical requirements of a given cancer case, i.e., maximizing the tumor control probability while minimizing the normal tissue complication probability [4]. Therefore, the core issue of IMRT planning is essentially an optimization problem, where the decision variables include the angles, intensities, and shapes of radiation, and the objectives include the satisfaction of the dose distribution in the planning tumor volume (PTV consisting of tumor cells) and the minimization of the dose distribution in the organ at risk (OAR consisting of healthy cells).

IMRT planning introduces challenging optimization problems characterized by discrete variables, high-dimensional decision spaces, many objectives, and

Evolutionary Large-Scale Multi-Objective Optimization and Applications: Integrating Evolutionary Computation, Machine Learning, and Data Science,
First Edition. Xingyi Zhang, Ran Cheng, Ye Tian, and Yaochu Jin.

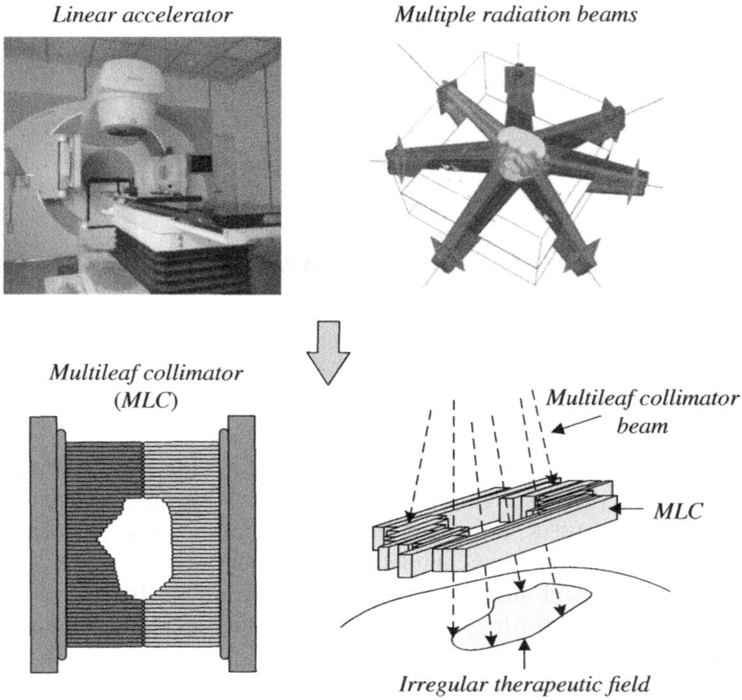

Figure 8.1 The therapeutic fields generated in intensity-modulated radiotherapy (IMRT).

computationally expensive functions. First, the therapeutic fields are partitioned into many small grids to ease the calculation of dose distribution; hence, their shapes are determined by integer variables rather than continuous variables. Second, since all the small grids can only have the same intensity at a time, many apertures with different shapes and intensities are required to form a therapeutic field with heterogeneous dose distribution, which leads to thousands of shape-related variables. Third, a clinical case usually has multiple PTV and OAR regions and each region corresponds to multiple clinical requirements, which leads to dozens of conflicting objectives to be optimized simultaneously. Lastly, the calculation of dose distribution involves the multiplication of large matrices, where too many function evaluations are unaffordable in clinical practice. Due to these difficulties of IMRT planning, a variety of methods have been developed in the last three decades, and some of which are introduced in the following.

8.1.2 IMRT Planning Methods

Existing IMRT planning methods can be roughly grouped into two categories, including two-step optimization methods and direct aperture optimization methods. The two-step optimization methods aim to obtain the optimal dose distribution and then obtain the corresponding apertures, where the first step is known as fluence map optimization and the second step is known as MLC sequencing. The task of fluence map optimization is to obtain the optimal intensity for each small grid in the therapeutic fields, and thus the therapeutic fields can form heterogeneous dose distribution in the PTV and OAR regions. Fluence map optimization is a continuous optimization problem that has been handled by various advanced techniques, including analytical algorithms (e.g., Fourier transform [5] and Taylor series expansion [6]), mathematical programming (e.g., mixed integer programming [7] and quadratic programming [8]), machine learning (e.g., regression forests [9] and deep neural networks [10]), and metaheuristics (e.g., simulated annealing [11] and genetic algorithms [12]). While the small grids with different intensities cannot be implemented directly, the task of MLC sequencing is to obtain the optimal apertures, where the small grids in each aperture share the same intensity. Thus, MLC sequencing aims to minimize the difference between the fluence map and the superposition of all apertures, which is usually handled by column generation methods [13, 14].

The direct aperture optimization methods aim to obtain the optimal apertures directly, so as to eliminate the distortion of intensities caused by MLC sequencing. Obviously, direct aperture optimization is more challenging, since it is a combinatorial optimization problem. Some methods have been suggested for direct aperture optimization, such as enumeration [15], gradient descent [16], random search [17], local search [18], and genetic algorithms [19]. Nevertheless, direct aperture optimization is so complex that it cannot be solved in a totally black-box manner, where problem-dependent search strategies are required to avoid invalid solutions brought by the physical limitations of MLC.

The difficulty encountered by the above methods mainly lies in the balance between many conflicting objectives (i.e., clinical requirements). Basically, larger intensities are more effective for killing tumor cells but also damage more healthy cells [20]. Therefore, MOEAs have been suggested for IMRT planning in recent research [21, 22], which can provide many candidate solutions with diverse trade-offs between objectives, thus highly reducing the human expertise required by medical physicists. Nevertheless, IMRT planning involves a large number of discrete variables, where MOEAs should be equipped with customized search strategies for high efficiency [23, 24].

8.2 Problem Formulation

IMRT planning is essentially a combinatorial LSMOP, which aims to find the optimal angles, intensities, and shapes of multiple apertures for meeting many clinical requirements of a given cancer case. The clinical requirements are related to the dose distribution determined by apertures, where the apertures are two-dimensional regions partitioned into many small grids and the dose distribution corresponds to a three-dimensional region partitioned into many voxels (i.e., small cubes). Here, the angles of apertures are fixed by experience while the intensities and shapes of apertures are optimized by optimizers, and thus the dose deposition coefficient of the vth voxel in terms of the gth small grid D_{gv} can be calculated by Monte Carlo methods [25] in advance. Then, the dose of the vth voxel $dose_v$ can be calculated by

$$dose_v = \sum_{i=1}^{N_B} \sum_{g \in G_i} D_{gv} \sum_{j=1}^{N_A} w_{ij} s_{ijg}, \tag{8.1}$$

where N_B is the number of radiation beams, G_i is the set of small grids constituting the therapeutic field of the ith radiation beam, N_A is the number of apertures of a therapeutic field, w_{ij} is the intensity of the jth aperture of the ith therapeutic field, and $s_{ijg} \in \{0, 1\}$ indicates whether the gth small grid in the jth aperture of the ith therapeutic field is shielded by MLC leaves or not. The decision vector of a solution $\mathbf{x}$ consists of intensity-related variables and shape-related variables, which determine the values of w_{ij} and s_{ijg}, respectively. An illustrative example of the calculation of dose distribution based on multiple apertures of multiple radiation beams is given in Fig. 8.2, where a darker grid indicates a larger intensity and a white grid indicates that it is shielded by MLC leaves.

After the calculation of dose distribution, the clinical requirements of a region of PTV or OAR can be mathematically represented by the following four types of objective functions [26–28]. The first objective defines the minimum dose $dose_{min}$ of all voxels in a region V, which is defined as

$$MinDose(\mathbf{x}, V, dose_{min}) = \frac{1}{|V|} \sum_{v \in V} (\max\{0, dose_{min} - dose_v\})^2. \tag{8.2}$$

The second objective defines the maximum dose $dose_{max}$ of all voxels in a region V, which is defined as

$$MaxDose(\mathbf{x}, V, dose_{max}) = \frac{1}{|V|} \sum_{v \in V} (\max\{0, dose_v - dose_{max}\})^2. \tag{8.3}$$

The third objective defines the prescribed dose $dose_{pre}$ of all voxels in a region V, which is defined as

$$PreDose(\mathbf{x}, V, dose_{pre}) = \frac{1}{|V|} \sum_{v \in V} (dose_v - dose_{pre})^2, \tag{8.4}$$

The last objective ensures at most r voxels whose doses are larger than $dose_{dvc}$ in a region V, which is defined as

$$\begin{aligned} DVC(\mathbf{x}, V, dose_{dvc}, r) = & \frac{1}{|V|} \sum_{v \in V} (\max\{0, dose_v - dose_{dvc}\})^2 \\ & \cdot (\max\{0, dose_r - dose_v\}), \end{aligned} \tag{8.5}$$

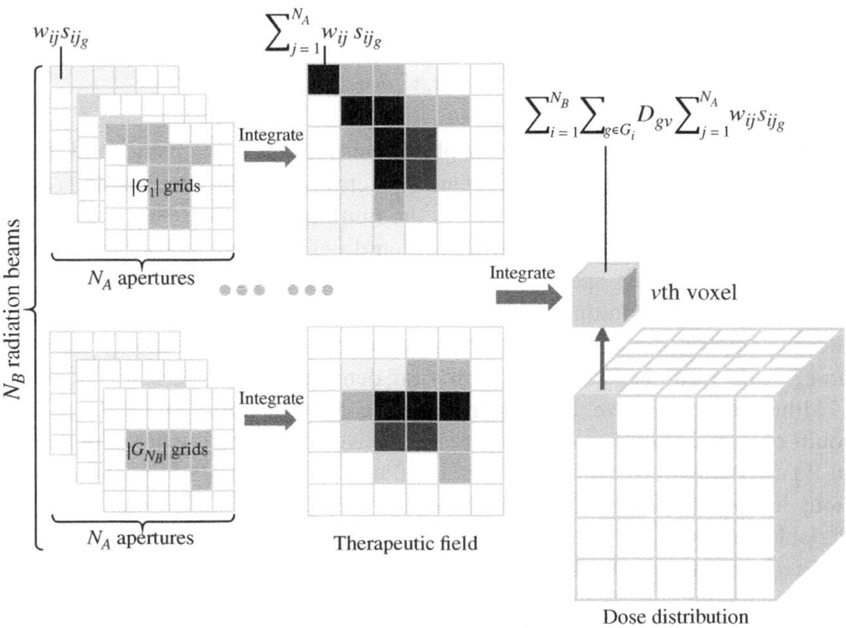

Figure 8.2 Illustration of the calculation of dose distribution based on multiple apertures of multiple radiation beams.

where $dose_r$ denotes the rth largest dose in the region. Clinically, an IMRT planning task contains dozens of the above objectives for multiple regions of PTV and OAR, which constitute a many-objective optimization problem that is difficult to distinguish between solutions by algorithms and decision-makers. Considering that some objectives are not conflicting with each other, all the objectives are aggregated into five new objectives to be optimized simultaneously as suggested in [24]. As a consequence, IMRT planning is formulated as the following LSMOP:

$$
\begin{aligned}
\text{Minimize}\, \mathbf{f}(\mathbf{x}) &= (f_1(\mathbf{x}), \ldots, f_5(\mathbf{x})),\\
f_1(\mathbf{x}) &= \sum_{V\in PTV} MinDose\left(\mathbf{x}, V, dose^V_{min}\right),\\
f_2(\mathbf{x}) &= \sum_{V\in PTV} MaxDose(\mathbf{x}, V, dose^V_{max})\\
&\quad + \sum_{V\in PTV} DVC\left(\mathbf{x}, V, dose^V_{dvc}, r^V\right),\\
f_3(\mathbf{x}) &= \sum_{V\in OAR} MaxDose\left(\mathbf{x}, V, dose^V_{max}\right),
\end{aligned}
$$

$$f_4(\mathbf{x}) = \sum_{V \in OAR} DVC\left(\mathbf{x}, V, dose^V_{dvc}, r^V\right),$$
$$f_5(\mathbf{x}) = \sum_{V \in PTV} PreDose\left(\mathbf{x}, V, dose^V_{pre}\right), \tag{8.6}$$

where $\mathbf{x}$ is a solution (i.e., a treatment plan) determining the intensities and shapes of all apertures, whose encoding scheme is determined by the adopted IMRT planning method. The objective f_1 specifies the minimum doses of PTV regions, the objective f_2 specifies the maximum doses and dose–volume constraints of PTV regions, the objective f_3 specifies the maximum doses of OAR regions, the objective f_4 specifies the dose–volume constraints of OAR regions, and the objective f_5 specifies the prescribed doses of PTV regions. Besides, the parameters $dose^V_{min}$, $dose^V_{max}$, $dose^V_{pre}$, $dose^V_{dvc}$, and r^V for each region V are determined by experience.

Although there have been many generic MOEAs tailored for large-scale multi-objective optimization, they are ineffective for solving the problem defined in Eq. (8.6) due to two main reasons. First, most generic MOEAs evolve populations in continuous decision spaces, where the search strategies (e.g., decision variable grouping and decision space reduction) cannot handle the discrete variables related to aperture shapes. Second, many solutions generated in a stochastic search paradigm are likely to be invalid due to the physical limitations of MLC; moreover, the repair of invalid solutions is highly disruptive to the evolution of populations. To address these issues, several MOEAs have been suggested for IMRT planning with customized search strategies in recent years [23, 24], where the first MOEA tailored for IMRT planning is introduced in Section 8.3.

8.3 Bi-Encoding Coevolutionary Algorithm for IMRT Planning

The multi-objective evolutionary algorithm for intensity-modulated radiotherapy planning (MOEA/IMRT) suggests a bi-encoding coevolutionary framework for solving the problem defined in Eq. (8.6) [24]. More specifically, one population represented in fine encoding is evolved for the local exploitation of complex and irregular aperture shapes, and another population represented in rough encoding is evolved for the global exploration of simple and regular aperture shapes. The population represented in fine encoding serves as the final output, which provides diverse candidate treatment plans for medical physicists and doctors. On the other hand, the population represented in rough encoding forms a low-dimensional decision space for the acceleration of the convergence of the population represented in fine encoding, where the transfer of useful genetic material has been verified to be effective in handling complex constraints [29].

Furthermore, MOEA/IMRT customizes two local search strategies for repairing invalid solutions and accelerating the convergence, respectively.

8.3.1 Bi-Encoding Scheme of MOEA/IMRT

While the intensities of apertures can be directly represented by real variables, the first issue that should be addressed is to represent the shapes of apertures by proper variables. Intuitively, as shown in Fig. 8.2, a binary matrix can be associated with an aperture to indicate whether each small grid is shielded or not. However, most solutions represented by this encoding scheme will violate the physical limitations of MLC shown in Fig. 8.1, as the unshielded small grids must be successive in each row. Therefore, the fine encoding scheme of MOEA/IMRT uses many integer variables to represent the positions of MLC leaves for each aperture, and the rough encoding scheme of MOEA/IMRT uses four integer variables to determine a rectangular region of unshielded small grids.

Figure 8.3 illustrates the bi-encoding scheme used in MOEA/IMRT. For an IMRT task with N_B radiation beams and N_A apertures for each beam, a solution

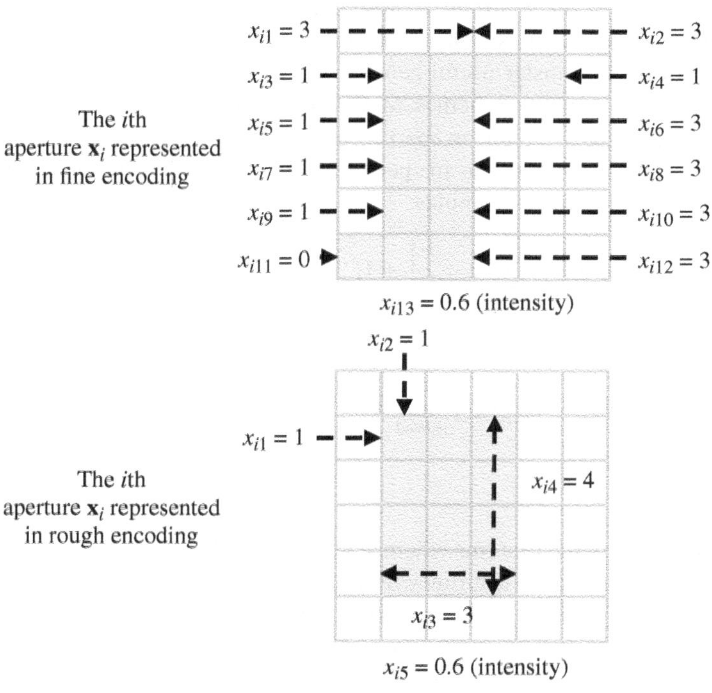

Figure 8.3 Illustration of the bi-encoding scheme used in MOEA/IMRT.

$\mathbf{x} = (\mathbf{x}_1, \dots, \mathbf{x}_{N_B \times N_A})$ is the concatenation of $N_B \times N_A$ vectors, where vector $\mathbf{x}_i$ encodes the intensity and shape of the ith aperture. To be specific, for an aperture $\mathbf{x}_i = (x_{i1}, \dots, x_{i(2K+1)})$ represented in fine encoding, it consists of $2K + 1$ decision variables with K denoting the number of rows of the aperture, where the integer $x_{i(2j-1)}$ denotes the position of the jth left leaf of the MLC, the integer $x_{i(2j)}$ denotes the position of the jth right leaf of the MLC, and $1 \leq j \leq K$. Besides, the real number $x_{i(2K+1)}$ denotes the intensity of the aperture. For an aperture $\mathbf{x}_i = (x_{i1}, \dots, x_{i5})$ represented in rough encoding, it consists of five decision variables, where the real number x_{i5} denotes the intensity of the aperture and the integers x_{i1}, x_{i2}, x_{i3}, and x_{i4} denote the horizontal position of the upper left corner, the vertical position of the upper left corner, the width, and the height of the rectangular region of unshielded small grids, respectively. In short, a solution represented in fine encoding contains $N_B \times N_A \times (2K + 1)$ variables forming a high-dimensional decision space, and a solution represented in rough encoding contains $N_B \times N_A \times 5$ variables forming a low-dimensional decision space.

As can be observed from Fig. 8.3, the solutions represented in fine encoding can search for irregular apertures to be used in clinical practice, while the solutions represented in rough encoding can only search for rectangular apertures to assist the optimization of the solutions represented in fine encoding. Thus, the exploitation and exploration can be balanced by the cooperation of the two encoding schemes. In order to transfer useful genetic material between the solutions represented in different encoding schemes, MOEA/IMRT suggests the conversion strategies presented in Fig. 8.4. First, an aperture $\mathbf{x}_i = (x_{i1}, \dots, x_{i(2K+1)})$ represented in fine encoding can be converted into an aperture $\mathbf{y}_i = (y_{i1}, \dots, y_{i5})$ represented in rough encoding by the following formulas:

$$\begin{aligned}
y_{i1} &= \left\lceil \frac{1}{rear - head + 1} \sum_{j=head}^{rear} x_{i(2j-1)} \right\rceil + 1, \\
y_{i2} &= head, \\
y_{i3} &= c - y_{i1} - \left\lceil \frac{1}{rear - head + 1} \sum_{j=head}^{rear} x_{i(2j)} \right\rceil + 1, \\
y_{i4} &= rear - head + 1, \\
y_{i5} &= x_{i(2K+1)},
\end{aligned} \tag{8.7}$$

where *head* is the index of the first row where at least one grid is unshielded in the aperture, *rear* is the index of the last row where at least one grid is unshielded in the aperture, and c is the number of columns (i.e., maximum value of MLC positions) of the aperture. Second, an aperture $\mathbf{y}_i = (y_{i1}, \dots, y_{i5})$ represented in rough encoding can be converted into an aperture $\mathbf{x}_i = (x_{i1}, \dots, x_{i(2K+1)})$ represented in

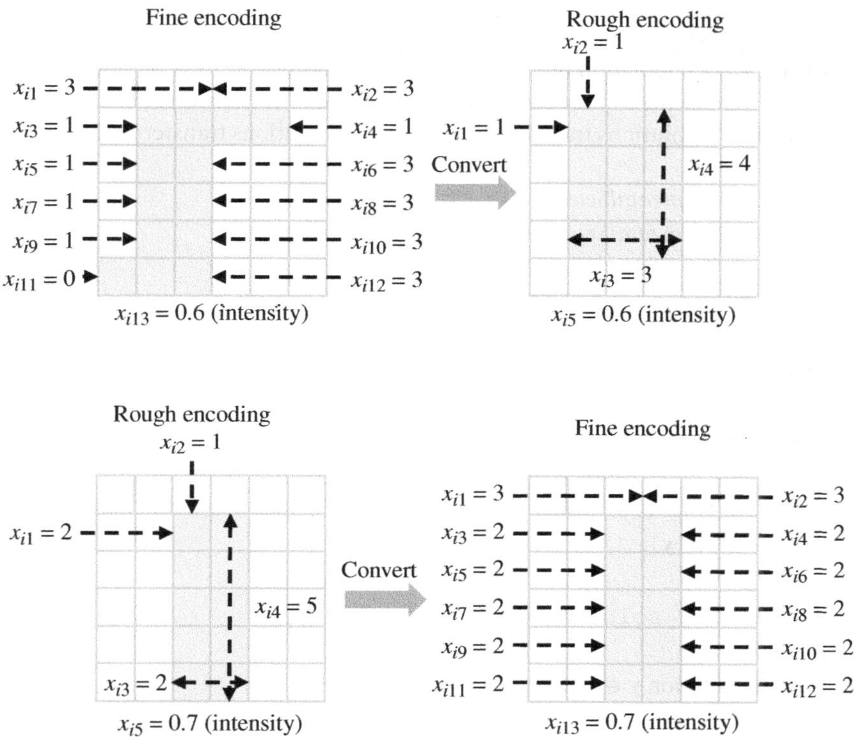

Figure 8.4 Illustration of the conversion between find encoding and rough encoding used in MOEA/IMRT.

fine encoding by the following formulas:

$$\begin{aligned} x_{i(2j-1)} &= \left\lceil \frac{c}{2} \right\rceil, \quad j < y_{i2} \text{ or } j \geq y_{i2} + y_{i4}, \\ x_{i(2j-1)} &= y_{i1} - 1, \quad y_{i2} \leq j < y_{i2} + y_{i4}, \\ x_{i(2j)} &= \left\lceil \frac{c}{2} \right\rceil, \quad j < y_{i2} \text{ or } j \geq y_{i2} + y_{i4}, \\ x_{i(2j)} &= c - y_{i1} - y_{i3} + 1, \ y_{i2} \leq j < y_{i2} + y_{i4} \\ x_{i(2K+1)} &= y_{i5}. \end{aligned} \tag{8.8}$$

Considering a population P_1 represented in fine encoding and a population P_2 represented in rough encoding, Algorithm 8.1 details the procedure of the conversion between the two populations. To begin with, L solutions are selected from each of P_1 and P_2 by using the environmental selection strategy of MOEA/IMRT (Lines 1 and 2). The algorithm then exchanges the encoding schemes of the $2L$

Algorithm 8.1: *Conversion*(P_1, P_2, L)

Input: P_1 (a population represented in fine encoding), P_2 (a population represented in rough encoding), L (number of transferred solutions)

Output: P'_1 (solutions transferred from P_2), P'_2 (solutions transferred from P_1)

1 $P_1^L \leftarrow EnvironmentalSelection(P_1, L)$; `//Algorithm 8.6`

2 $P_2^L \leftarrow EnvironmentalSelection(P_2, L)$; `//Algorithm 8.6`

`//Transfer solutions in` P_1^L `to` P'_2

3 $P'_2 \leftarrow \emptyset$;

4 **foreach** solution $\mathbf{x} \in P_1^L$ **do**

5     $\mathbf{y} \leftarrow \emptyset$;

6   **for** $i = 1, \dots, N_B \times N_A$ **do**

7   Convert $\mathbf{x}_i$ into $\mathbf{y}_i$ by Eq. (8.7);

8         $\mathbf{y} \leftarrow \mathbf{y} \cup \{\mathbf{y}_i\}$;

9     $P'_2 \leftarrow P'_2 \cup \{\mathbf{y}\}$;

`//Transfer solutions in` P_2^L `to` P'_1

10 $P'_1 \leftarrow \emptyset$;

11 **foreach** solution $\mathbf{y} \in P_2^L$ **do**

12     $\mathbf{x} \leftarrow \emptyset$;

13   **for** $i = 1, \dots, N_B \times N_A$ **do**

14   Convert $\mathbf{y}_i$ into $\mathbf{x}_i$ by Eq. (8.8);

15         $\mathbf{x} \leftarrow \mathbf{x} \cup \{\mathbf{x}_i\}$;

16     $P'_1 \leftarrow P'_1 \cup \{\mathbf{x}\}$;

17 **return** P'_1 and P'_2;

solutions rather than all the solutions, since the transferred solutions need to be reevaluated and only the best solutions (i.e., useful genetic material) are expected to be transferred for the assistance of convergence. For each solution represented in fine encoding (Line 4), each of its apertures is converted into a new aperture by Eq. (8.7) (Lines 6–8), and all the new apertures constitute a new solution represented in rough encoding to be stored in P'_2 (Line 9). For each solution represented in rough encoding (Line 11), each of its apertures is converted into a new aperture by Eq. (8.8) (Lines 13–15), and all the new apertures constitute a new solution represented in fine encoding to be stored in P'_1 (Line 16). Lastly, all the transferred solutions are returned (Line 17), where the encoding scheme of P'_1 is transferred from rough encoding to fine encoding, and the encoding scheme of P'_2 is transferred from fine encoding to rough encoding.

8.3.2 Local Search Strategies of MOEA/IMRT

The two encoding schemes consider the physical nature of MLC leaves, but they are still unable to prevent invalid apertures such as those with very small shapes or disconnected shapes. Besides, the co-evolution of two populations with different encoding schemes is not efficient enough to converge in the high-dimensional decision space. To handle these problems, MOEA/IMRT develops a local search strategy for repairing invalid solutions and a local search strategy for finding better solutions.

As illustrated in Fig. 8.5, an aperture represented in fine encoding may be too small (i.e., the number of unshielded grids is less than a threshold) or disconnected (i.e., some of the unshielded grids are not connected), and an aperture represented in rough encoding may also be too small. To repair such apertures, a local search strategy is developed in MOEA/IMRT for ensuring clinical practicality. Specifically, given an aperture $\mathbf{x}_i = (x_{i1}, \dots, x_{i(2K+1)})$ represented in fine encoding, it is repaired by the following formulas:

$$
\begin{aligned}
x_{i(2j)} &= \min\{x_{i(2j)}, c - x_{i(2j-1)}\}, \quad 1 \le j \le K, \\
x_{i(2j-1)} &= x_{i(2j-1)} - \max\left\{0, \left\lfloor \frac{S_{min} - S}{2(rear - head + 1)} \right\rfloor\right\}, \\
&\quad head \le j \le rear, \\
x_{i(2j)} &= x_{i(2j)} - \max\left\{0, \left\lfloor \frac{S_{min} - S}{2(rear - head + 1)} \right\rfloor\right\}, \\
&\quad head \le j \le rear, \\
x_{i(2j-1)} &= \min\{x_{i(2j-1)}, c - x_{i(2j-2)} - 1\}, \ head < j \le rear, \\
x_{i(2j)} &= \min\{x_{i(2j)}, c - x_{i(2j-3)} - 1\}, \ head < j \le rear,
\end{aligned}
\tag{8.9}
$$

where *head* is the index of the first row where at least one grid is unshielded in the aperture, *rear* is the index of the last row where at least one grid is unshielded in the aperture, c is the number of columns of the aperture, $S = \sum_{j=1}^{K} c - x_{i(2j-1)} - x_{i(2j)}$ counts the number of unshielded grids in the aperture, and S_{min} is the minimum number of unshielded grids. That is, the first formula ensures that the jth left leaf does not intersect the jth right leaf, the second and third formulas ensure that the aperture has at least S_{min} unshielded grids, and the fourth formula ensures that all the unshielded grids are connected. Given an aperture $\mathbf{y}_i = (y_{i1}, \dots, y_{i5})$

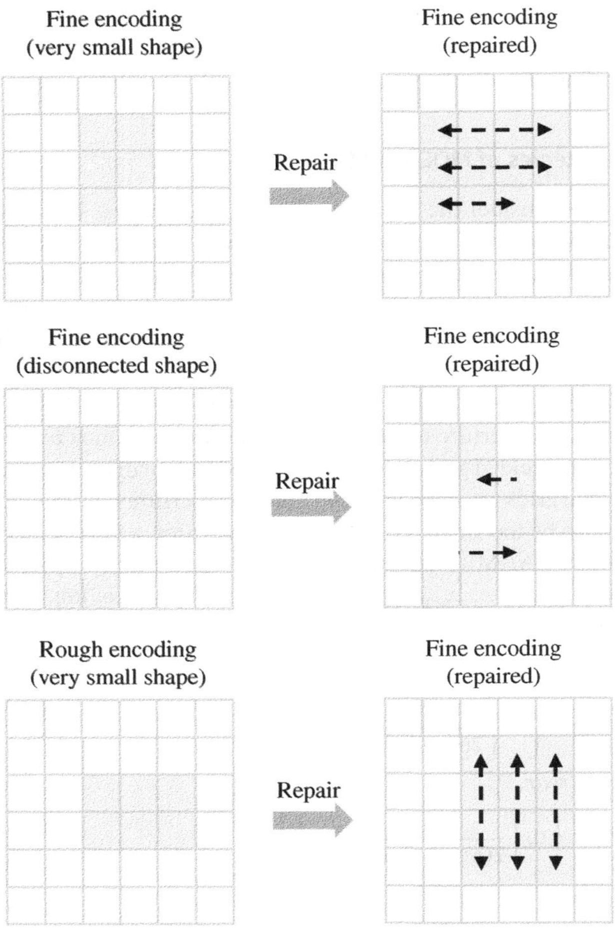

Figure 8.5 Illustration of the local search strategy for repairing invalid solutions in MOEA/IMRT.

represented in rough encoding, it is repaired by the following formulas:

$$\begin{aligned} y_{i1} &= y_1 - \max\left\{0, \left\lceil \frac{S_{min} - S}{2y_{i4}} \right\rceil\right\}, \quad rand \le 0.5, \\ y_{i3} &= y_3 + \max\left\{0, \left\lceil \frac{S_{min} - S}{y_{i4}} \right\rceil\right\}, \quad rand \le 0.5, \\ y_{i2} &= y_2 - \max\left\{0, \left\lceil \frac{S_{min} - S}{2y_{i3}} \right\rceil\right\}, \quad rand > 0.5, \\ y_{i4} &= y_4 + \max\left\{0, \left\lceil \frac{S_{min} - S}{y_{i3}} \right\rceil\right\}, \quad rand > 0.5, \end{aligned} \tag{8.10}$$

where $S = y_{i3} \times y_{i4}$ counts the number of unshielded grids in the aperture, and *rand* is a uniformly distributed random number within $[0, 1]$. These formulas ensure that the aperture has at least S_{min} unshielded grids, by means of expanding the aperture shape vertically or horizontally at random. Considering a population P_1 represented in fine encoding and a population P_2 represented in rough encoding, Algorithm 8.2 elaborates on the procedure of repairing invalid solutions in the two populations. For each solution represented in fine encoding (Line 1), each of its apertures is repaired by Eq. (8.9) (Lines 2 and 3). While for each solution represented in rough encoding (Line 4), each of its apertures is repaired by Eq. (8.10) (Lines 5 and 6). Lastly, the repaired populations are directly returned (Line 7).

The aforementioned bi-encoding scheme and repair strategy of MOEA/IMRT focus on the optimization of aperture shapes. To further improve the convergence speed, the second local search strategy of MOEA/IMRT aims to enhance the optimization of aperture intensities, whose procedure is described in Algorithm 8.3. Given a population P_1 represented in fine encoding, the non-dominated solutions in P_1 are firstly copied to a new set P_N (Line 1), whose shape-related variables are retained while intensity-related variables are to be optimized. Then, a new population Q with N_Q solutions having only aperture intensities is randomly initialized (Line 2), i.e., each solution has only one real variable. To evaluate Q on the objectives of IMRT planning, each solution in Q is associated with a

Algorithm 8.2: *LocalSearchA*(P_1, P_2)

Input: P_1 (a population represented in fine encoding), P_2 (a population represented in rough encoding)

Output: P_1 (repaired population represented in fine encoding), P_2 (repaired population represented in rough encoding)

```
  //Repair apertures in fine encoding
1 foreach solution x ∈ P_1 do
2     for i = 1, ..., N_B × N_A do
3         Repair x_i by Eq. (8.9);

  //Repair apertures in rough encoding
4 foreach solution y ∈ P_2 do
5     for i = 1, ..., N_B × N_A do
6         Repair y_i by Eq. (8.10);

7 return P_1, P_2;
```

Algorithm 8.3: *LocalSearchB*(P_1, N_Q)

Input: P_1 (a population represented in fine coding), N_Q (population size for intensity optimization)
Output: Q (enhanced solutions represented in fine coding)

1 $P_N \leftarrow$ Non-dominated solutions in P_1;
2 $Q \leftarrow$ Randomly generate N_Q solutions with only aperture intensities;
3 Randomly select a solution from P_N for each solution in Q, then evaluate Q;
4 $Fitness \leftarrow CalFitness(Q)$; `//Algorithm 8.5`
5 **while** termination criterion not fulfilled **do**
6     $Q' \leftarrow TournamentSelection(Q, N_Q, Fitness)$; `//Algorithm 1.4`
7     $O \leftarrow Variation(Q')$; `//Algorithm 1.5`
8   Randomly select a solution from P_N for each solution in O, then evaluate O;
9     $[Q, Fitness] \leftarrow EnvironmentalSelection(Q \cup O, N_Q)$; `//Algorithm 8.6`
10 **return** Q;

solution randomly selected from P_N (Line 3). In other words, the decision vector to be evaluated consists of the shape-related variables of a solution from P_N and the intensity-related variable of each solution in Q, and thus the aperture shapes are retained in P_N and the aperture intensities are optimized by evolving Q. The evolution of Q is conducted for a predefined number of generations, by means of the mating selection strategy (Line 6), variation operators (Line 7), and environmental selection strategy (Line 9) of MOEA/IMRT, whose procedures are elaborated in Section 8.3.3. It should be noted that the optimization of aperture intensities consumes additional function evaluations, hence it is just conducted with a population size $N_Q = 20$ and 100 generations after every 100 main generations of MOEA/IMRT. This way, the local search does not deteriorate the efficiency of the whole algorithm obviously.

8.3.3 Procedure of MOEA/IMRT

With the bi-encoding scheme and two local search strategies, the procedure of MOEA/IMRT is described in Algorithm 8.4. First, a population P_1 with N solutions represented in fine encoding is randomly initialized (Line 1), and a population P_2 with N solutions represented in rough encoding is randomly initialized (Line 2). Then, the solutions in the two populations are repaired by the repair strategy of MOEA/IMRT (Line 3), and the fitness values of all the solutions are calculated to be used in selection strategies (Lines 4 and 5). MOEA/IMRT adopts the fitness

Algorithm 8.4: Procedure of MOEA/IMRT

Input: N (population size), L (number of transferred solutions)
Output: P_1 (final population)

$P_1 \leftarrow$ Randomly generate N solutions represented in fine encoding;
$P_2 \leftarrow$ Randomly generate N solutions represented in rough encoding;
$[P_1, P_2] \leftarrow \mathit{LocalSearchA}(P_1, P_2)$; `//Algorithm 8.2`
$\mathit{Fitness}_1 \leftarrow \mathit{CalFitness}(P_1)$; `//Algorithm 8.5`
$\mathit{Fitness}_2 \leftarrow \mathit{CalFitness}(P_2)$; `//Algorithm 8.5`
while termination criterion not fulfilled **do**
  $P'_1 \leftarrow \mathit{TournamentSelection}(P_1, N, \mathit{Fitness}_1, \emptyset)$; `//Algorithm 1.4`
  $O_1 \leftarrow \mathit{Variation}(P'_1)$; `//Algorithm 1.5`
  foreach solution $\mathbf{x} \in O_1$ **do**
    for $i = 1, \dots, N_B \times N_A$ **do**
      for $j = 1, \dots, 2K$ **do**
        $x_{ij} \leftarrow \lceil x_{ij} \rceil$;
  $P'_2 \leftarrow \mathit{TournamentSelection}(P_2, N, \mathit{Fitness}_2, \emptyset)$; `//Algorithm 1.4`
  $O_2 \leftarrow \mathit{Variation}(P'_2)$; `//Algorithm 1.5`
  foreach solution $\mathbf{y} \in O_2$ **do**
    for $i = 1, \dots, N_B \times N_A$ **do**
      for $j = 1, \dots, 4$ **do**
        $y_{ij} \leftarrow \lceil y_{ij} \rceil$;
  $[O_1, O_2] \leftarrow \mathit{LocalSearchA}(O_1, O_2)$; `//Algorithm 8.2`
  $[O'_1, O'_2] \leftarrow \mathit{Conversion}(O_1, O_2, L)$; `//Algorithm 8.1`
  $[P_1, \mathit{Fitness}_1] \leftarrow \mathit{EnvironmentalSelection}(P_1 \cup O_1 \cup O'_1, N)$; `//Algorithm 8.6`
  $[P_2, \mathit{Fitness}_2] \leftarrow \mathit{EnvironmentalSelection}(P_2 \cup O_2 \cup O'_2, N)$; `//Algorithm 8.6`
  if condition fulfilled **then**
    $Q \leftarrow \mathit{LocalSearchB}(P_1, 20)$; `//Algorithm 8.3`
    $[P_1, \mathit{Fitness}_1] \leftarrow \mathit{EnvironmentalSelection}(P_1 \cup Q, N)$; `//Algorithm 8.6`
return P_1;

assignment strategy of SPEA2 with shift-based density estimation [30] as described in Algorithm 8.5, which is the same as that of the original SPEA2 [31] except for a new method to measure the distances between solutions. When calculating the Euclidean distances to a solution $\mathbf{x}$ in the objective space, all the objective values of the other solutions are shifted to be not less than the objective values of $\mathbf{x}$ (Line 7 of Algorithm 8.5). Such a shift-based density estimation can consider both the convergence and diversity of solutions in the Euclidean distances, which is particularly effective for many-objective optimization problems including IMRT planning where general selection strategies suffer from the Pareto dominance resistance phenomenon [32].

Algorithm 8.5: *CalFitness(P)*

Input: P (population)
Output: *Fitness* (fitness of each solution)

1 **for** $i = 1, \dots, |P|$ **do**
2     $S_i \leftarrow$ Number of solutions dominated by P_i in P;
3 **for** $i = 1, \dots, |P|$ **do**
4     $R_i \leftarrow \sum_{P_j \in P, P_j \prec P_i} S_j$;
5 **foreach** $\mathbf{x} \in P$ **do**
6   **foreach** $\mathbf{y} \in P$ **do**
7         $Dis_{ij} \leftarrow \sqrt{\sum_{i=1}^{M} \max\{0, (f_i(\mathbf{y}) - f_i(\mathbf{x}))^2\}}$;
8 **for** $i = 1, \dots, |P|$ **do**
9     $j \leftarrow$ Index of the $\lfloor \sqrt{|P|} \rfloor$th closest solution to the ith solution according to *Dis*;
10     $D_i \leftarrow \frac{1}{Dis_{ij}+2}$;
11 $Fitness \leftarrow R + D$;
12 **return** *Fitness*;

At each generation, the mating pool P'_1 is constructed by selecting N parents from the population P_1 via binary tournament selection according to the fitness of solutions (Line 7). Then, N offspring solutions are generated by simulated binary crossover [33] and polynomial mutation [34] (Line 8), and the first $2K$ decision variables of each offspring solution are rounded since these shape-related variables can only be integers (Lines 9–12). Similarly, the mating pool P'_2 is constructed by selecting N parents from the population P_2 via binary tournament selection according to the fitness of solutions (Line 13). Then, N offspring solutions are generated (Line 14) and the first four decision variables of each offspring solution are rounded since these shape-related variables can only be integers (Lines 15–18). Afterward, the offspring solutions represented in two encoding schemes are repaired by the repair strategy of MOEA/IMRT (Line 19) and transferred by the conversion strategy of MOEA/IMRT (Line 20). After combing each of P_1 and P_2 with the offspring solutions represented in the same encoding scheme, P_1 and P_2 are truncated by the environmental selection strategy of SPEA2 with shift-based density estimation (Lines 21 and 22). As presented in Algorithm 8.6, the procedure of environmental selection is the same as that of the original SPEA2 except for the shift-based density estimation-based Euclidean distance calculation. Lastly, the local search strategy for intensity optimization is conducted periodically to further enhance the convergence of P_1 (Lines 23–25), and P_1 is returned as the output of the algorithm (Line 26).

Algorithm 8.6: *EnvironmentalSelection*(P, N)

Input: P (combined population), N (population size)

Output: P (population for next generation), *Fitness* (fitness of each solution)

Fitness ← *CalFitness*(P); `//Algorithm 8.5`

if $|\{i|Fitness_i < 1\}| < N$ **then**

  P ← Select N solutions with the smallest *Fitness* from P;

else

  P ← Delete the solutions with $Fitness_i \geq 1$ from P;

  while $|P| > N$ **do**

    $\mathbf{p} \leftarrow \text{argmin}_{\mathbf{x} \in P} \min_{\mathbf{y} \in P \setminus \{\mathbf{x}\}} \|\mathbf{f}(\mathbf{x}) - \mathbf{f}(\mathbf{y})\|$;

    $P \leftarrow P \setminus \{\mathbf{p}\}$;

Fitness ← Retain the elements in *Fitness* whose corresponding solutions are in P;

return P and *Fitness*;

At each generation of MOEA/IMRT, the time complexity is $O(N)$ for mating pool selection, $O(ND)$ for offspring generation, $O(ND)$ for offspring repair, $O(MN|V||G|N_BN_A)$ for offspring evaluation, $O(MN^2 + ND)$ for offspring conversion, and $O(MN^2)$ for environmental selection, where D denotes the number of decision variables, M denotes the number of objectives, $|V|$ denotes the number of voxels in a region, $|G|$ denotes the number of small grids in a therapeutic field, N_B denotes the number of radiation beams, and N_A denotes the number of apertures. Besides, the time complexity of the local search strategy for intensity optimization is $O(G_{sub}MN_Q|V||G|N_BN_A)$ with G_{sub} being the number of generations for intensity optimization, but it is not included in the total time complexity since it is rarely conducted. As a result, the time complexity of one generation of

MOEA/IMRT is $O(ND + MN|V||G|N_B N_A)$, which is mainly determined by the scales of dose distribution regions and apertures of IMRT planning tasks.

8.4 Experimental Studies

To study the performance of MOEA/IMRT, it is compared with several representative IMRT methods and MOEAs on a number of clinical cases. The compared methods include Shepard's method [17], NSGA-II [35], CCGDE3 [36], LSMOF-NSGA-II [37], LMOCSO [38], and MOEA/IMRT, where Shepard's method is the first direct aperture optimization method based on simulated annealing and mathematical programming, NSGA-II is a classical MOEA, CCGDE3 is a decision variable grouping-based MOEA, LSMOF-NSGA-II is a decision space reduction-based MOEA, and LMOCSO is a novel search strategy-based MOEA. The population size is set to 100, the number of function evaluations is set to 100,000, and the number of transferred solutions L in MOEA/IMRT is set to 10. For Shepard's method generating a single solution in one run, it is executed for 100 independent runs with 100 uniformly distributed weight vectors for aggregating all the objectives, and the number of function evaluations is also equally divided. For NSGA-II, CCGDE3, LSMOF-NSGA-II, and LMOCSO, they represent solutions by using the fine encoding scheme given in Fig. 8.3 and repair invalid solutions by using the local search strategy given in Algorithm 8.2. Five clinical cases are taken from the real world, whose statistics are listed in Table 8.1. It can be found that the number of objectives varies from 9 to 25 and the number of decision variables varies from 1950 to 2715, which results in typical combinatorial LSMOPs. The populations obtained by the methods are assessed by HV [39] since the true Pareto fronts are unknown.

Table 8.2 lists the experimental results averaged over 30 runs, where the best results are shown in bold. It can be found from the table that Shepard's method obtains the worst performance since it generates only a single solution in one run and cannot maintain a good diversity for the solutions obtained in different runs. NSGA-II obtains the second worst performance among the six compared methods, since it does not contain any strategy for searching in high-dimensional decision spaces. By contrast, the three large-scale MOEAs exhibit significantly better performance, and the LMOCSO based on a novel competitive swarm optimizer seems very suitable for IMRT planning tasks. Nevertheless, they are underperformed by the customized method MOEA/IMRT, which suggests a bi-encoding scheme and two local search strategies leading to the best performance on all the cases.

For a visual comparison, Fig. 8.6 plots the dose–volume histograms of the treatment plans obtained by six methods on the Sneezer case, which is a popular visualization method for physicists. A dose–volume histogram contains several curves

Table 8.1 Statistics of Five Clinical Cases.

Statistics		RightLung	Kidney	Throat	Sneezer	Esophagus
PTV	# *MinDose*	2	2	2	1	1
	# *MaxDose*	2	2	2	3	1
	# *PreDose*	2	2	2	1	1
	# *DVC*	2	1	2	5	1
OAR	# *MaxDose*	1	3	11	15	1
	# *DVC*	6	2	0	0	4
	# Beams (N_B)	5	5	5	7	5
# Apertures per beam (N_A)		9, 10, 11, 9, 10	10, 10, 10, 10, 10	10, 9, 11, 10, 9	6, 7, 7, 8, 7, 5, 9	10, 11, 10, 10, 9
# Rows per aperture (K)		27, 28, 27, 27, 27	19, 19, 19, 19, 19	20, 22, 21, 21, 20	24, 24, 24, 24, 24, 23, 23	21, 21, 21, 21, 21
# Columns per aperture (c)		25, 29, 27, 26, 23	44, 37, 41, 46, 34	41, 38, 40, 41, 41	49, 35, 51, 53, 33, 47, 52	37, 33, 34, 37, 42
# Original objectives (M)		15	12	19	25	9
# Decision variables (D)		2715	1950	2087	2373	2150

Table 8.2 HV Values Obtained by Six Methods on Six Clinical Cases of IMRT Planning.

Case	Shepard	NSGA-II	CCGDE3	LSMOF-NSGA-II	LMOCSO	MOEA/IMRT
RightLung	2.3654e−1−	5.7293e−1−	7.3805e−1−	8.9422e−1−	9.3368e−1 ≈	**9.5267e−1**
Kidney	7.0790e−1−	5.7918e−1−	6.5530e−1−	7.6012e−1 ≈	7.3094e−1−	**7.8767e−1**
Throat	5.4048e−1−	6.2943e−1−	3.7606e−1−	6.8341e−1−	3.6544e−1−	**7.7271e−1**
Sneezer	6.4944e−1−	5.9476e−1−	4.2480e−1−	7.7933e−1 ≈	7.8312e−1 ≈	**8.0482e−1**
Esophagus	5.8674e−1−	5.7212e−1−	7.4805e−1−	9.1631e−1 ≈	9.1109e−1 ≈	**9.4172e−1**
+/ − / ≈	0/5/0	0/5/0	0/5/0	0/2/3	0/2/3	

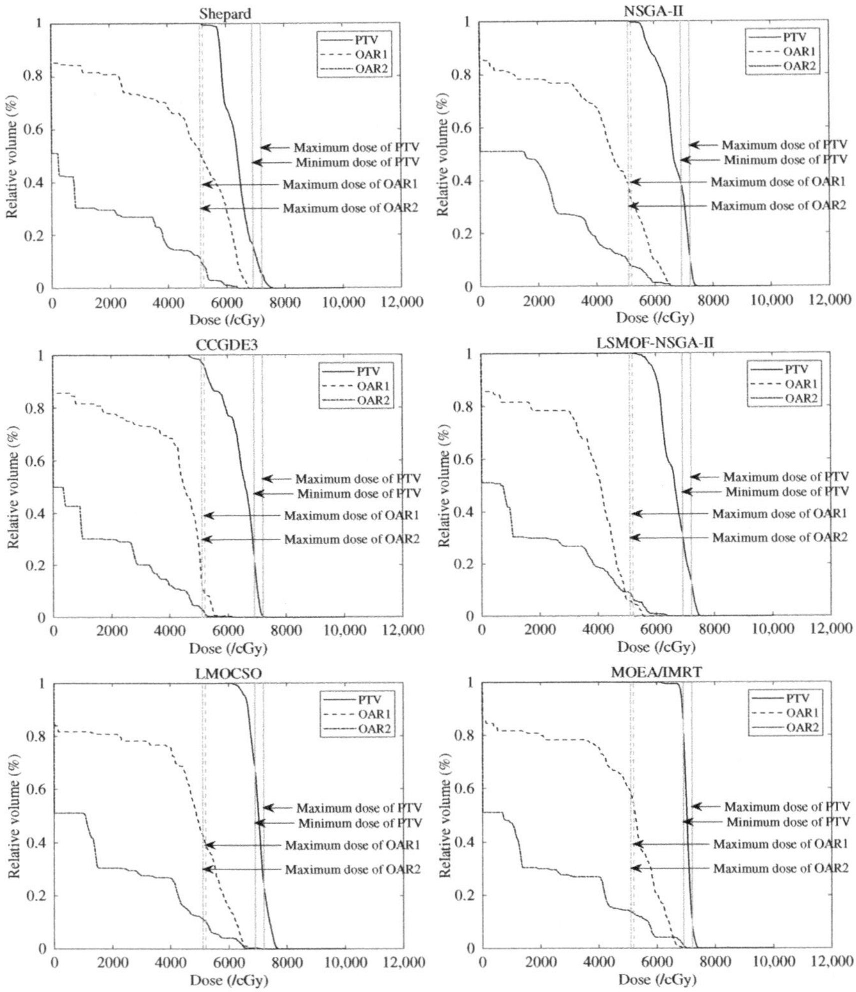

Figure 8.6 Dose–volume histograms of the treatment plans obtained by six methods on the Sneezer case.

representing the cumulative doses in each region, where a point (x, y) on a curve indicates that the proportion of voxels with doses larger than x is y. Figure 8.6 plots the curves of three regions (i.e., PTV, OAR1, and OAR2) of the treatment plan having the minimum sum of all objectives in the population with the median HV value obtained by each method and also plots the vertical lines of maximum dose of PTV, minimum dose of PTV, maximum dose of OAR1, and maximum dose of OAR2. Ideally, the curves should decrease sharply, and the vertical parts of the curves should be to the left of the lines of maximum doses and to the right of the lines of minimum doses. As can be observed from the figure, the vertical parts of the curves of PTV obtained by Shepard's method, NSGA-II, CCGDE3, and LSMOF-NSGA-II are mostly to the left of the line of minimum dose of PTV, which means that the obtained treatment plans are ineffective to kill tumor cells. By contrast, the vertical parts of the curves of PTV obtained by LMOCSO and MOEA/IMRT are mostly in between the lines of minimum dose and maximum dose of PTV, which means that the obtained treatment plans are obviously more effective. In particular, the curve of PTV obtained by MOEA/IMRT decreases more sharply than that obtained by LMOCSO, which further evidences the superiority of customized search strategies over generic large-scale MOEAs for IMRT planning.

8.5 Summary

This chapter introduces an emerging application of evolutionary large-scale multi-objective optimization, namely, IMRT planning. The challenges of this task lie in a large number of discrete variables and dozens of objectives, and the strict limitations of MLC also lead to many invalid solutions. Mathematical programming methods [7, 16], heuristics [6, 15], and deep learning methods [10, 40] are incapable of handling the conflicting objectives for providing diverse treatment plans, while general evolutionary algorithms [12, 22] are ineffective to optimize the large number of variables within a limited number of function evaluations. Therefore, several MOEAs have been developed for IMRT planning with customized search strategies [23, 24], such as the bi-encoding scheme and local search strategies of MOEA/IMRT. We can conclude from the experimental results that large-scale MOEAs are indeed more effective than conventional optimizers including general MOEAs for IMRT planning. However, the research of evolutionary multi-objective optimization mostly focuses on continuous optimization, while MOEA/IMRT can further improve the performance on IMRT planning by developing customized search strategies for combinatorial optimization.

References

1 R. Baskar, K. A. Lee, R. Yeo, and K.-W. Yeoh, "Cancer and radiation therapy: current advances and future directions," *International Journal of Medical Sciences*, vol. 9, no. 3, p. 193, 2012.

2 B. W. Fischer-Valuck, Y. J. Rao, and J. M. Michalski, "Intensity-modulated radiotherapy for prostate cancer," *Translational Andrology and Urology*, vol. 7, no. 3, pp. 297–307, 2018.

3 R. Bokrantz, "Multicriteria optimization for managing tradeoffs in radiation therapy treatment planning," Ph.D. dissertation, KTH Royal Institute of Technology, 2013.
4 A. N. Hasto and W. Rena, "Optimization of time-dose fractionation radiotherapy scheme by simulated annealing with consideration of biological factor," *Journal of Physics: Conference Series*, vol. 1090, no. 1, p. 012040, 2018.
5 A. Brahme, "Optimization of stationary and moving beam radiation therapy techniques," *Radiotherapy and Oncology*, vol. 12, no. 2, pp. 129–140, 1988.
6 B. K. Lind, "Properties of an algorithm for solving the inverse problem in radiation therapy," *Inverse Problems*, vol. 6, no. 3, pp. 415–426, 1990.
7 G. J. Lim, J. Choi, and R. Mohan, "Iterative solution methods for beam angle and fluence map optimization in intensity modulated radiation therapy planning," *OR Spectrum*, vol. 30, no. 2, pp. 289–309, 2008.
8 A. T. Redpath, B. L. Vickery, and D. H. Wright, "A new technique for radiotherapy planning using quadratic programming," *Physics in Medicine & Biology*, vol. 21, no. 5, p. 781, 1976.
9 C. Mclntosh and T. G. Purdie, "Voxel-based dose prediction with multi-patient atlas selection for automated radiotherapy treatment planning," *Physics in Medicine & Biology*, vol. 62, no. 2, p. 415, 2016.
10 J. Ma, D. Nguyen, T. Bai, M. Folkerts, X. Jia, W. Lu, L. Zhou, and S. Jiang, "A feasibility study on deep learning-based individualized 3D dose distribution prediction," *Medical Physics*, vol. 48, no. 8, pp. 4438–4447, 2021.
11 S. Webb, "Optimization by simulated annealing of three-dimensional conformal treatment planning for radiation fields defined by a multileaf collimator," *Physics in Medicine & Biology*, vol. 36, no. 9, pp. 1201–1226, 1991.
12 G. A. Ezzell, "Genetic and geometric optimization of three-dimensional radiation therapy treatment planning," *Medical Physics*, vol. 23, no. 3, pp. 293–305, 1996.
13 P. Xia and L. J. Verhey, "Multileaf collimator leaf sequencing algorithm for intensity modulated beams with multiple static segments," *Medical Physics*, vol. 25, no. 8, pp. 1424–1434, 1998.
14 M. Ehrgott, Ç. Güler, H. W. Hamacher, and L. Shao, "Mathematical optimization in intensity modulated radiation therapy," *Annals of Operations Research*, vol. 175, pp. 309–365, 2010.
15 J. Wang, X. Pei, R. Cao, L. Hu, and Y. Wu, "A fast direct aperture optimization method for intensity modulated radiation therapy," *Chinese Journal of Medical Physics*, vol. 32, no. 1, pp. 4–7, 2015.
16 J. Yang, P. Zhang, L. Zhang, and Z. Gui, "A gradient-based direct aperture optimization," *Journal of Biomedical Engineering*, vol. 35, no. 3, pp. 358–367, 2018.

17 D. M. Shepard, M. A. Earl, X. A. Li, S. Naqvi, and C. Yu, "Direct aperture optimization: a turnkey solution for step-and-shoot IMRT," *Medical Physics*, vol. 29, no. 6, pp. 1007–1018, 2002.

18 M. Moyano and G. Cabrera-Guerrero, "Local search for the direct aperture optimisation in IMRT," in *Proceedings of the International Conference of the Chilean Computer Science Society*, 2020, pp. 1–6.

19 R. Cao, X. Pei, H. Zheng, L. Hu, and Y. Wu, "Direct aperture optimization based on genetic algorithm and conjugate gradient in intensity modulated radiation therapy," *Chinese Medical Journal*, vol. 127, no. 23, pp. 4152–4153, 2014.

20 S. Breedveld, D. Craft, R. V. Haveren, and B. Heijmen, "Multi-criteria optimisation and decision-making in radiotherapy," *European Journal of Operational Research*, vol. 277, no. 1, pp. 1–19, 2019.

21 G. Li and D. Cao, "A multi-objective particle swarm algorithm for the optimization of IMRT inverse planning," in *Proceedings of the 3rd International Conference on Biomedical Engineering and Informatics*, 2010, pp. 1327–1330.

22 R.-F. Cao, Y.-C. Wu, X. Pei, J. Jing, G.-L. Li, M.-Y. Cheng, G. Li, and L.-Q. Hu, "Multi-objective optimization of inverse planning for accurate radiotherapy," *Chinese Physics C*, vol. 35, no. 3, p. 313, 2011.

23 R. Cao, L. Si, X. Li, Y. Guang, C. Wang, Y. Tian, X. Pei, and X. Zhang, "A conjugate gradient-assisted multi-objective evolutionary algorithm for fluence map optimization in radiotherapy treatment," *Complex & Intelligent Systems*, vol. 8, no. 5, pp. 4051–4077, 2022.

24 Y. Tian, Y. Feng, C. Wang, R. Cao, X. Zhang, X. Pei, K. C. Tan, and Y. Jin, "A large-scale combinatorial many-objective evolutionary algorithm for intensity-modulated radiotherapy planning," *IEEE Transactions on Evolutionary Computation*, vol. 26, no. 6, pp. 1511–1525, 2022.

25 A. M. Bergman, K. Bush, M.-P. Milette, I. A. Popescu, K. Otto, and C. Duzenli, "Direct aperture optimization for IMRT using Monte Carlo generated beamlets," *Medical Physics*, vol. 33, no. 10, pp. 3666–3679, 2006.

26 S. V. Spirou and C. S. Chui, "A gradient inverse planning algorithm with dose-volume constraints," *Medical Physics*, vol. 25, no. 3, pp. 321–333, 1998.

27 Y. Chen, D. Michalski, C. Houser, and J. M. Galvin, "A deterministic iterative least-squares algorithm for beam weight optimization in conformal radiotherapy," *Physics in Medicine & Biology*, vol. 47, no. 10, pp. 1647–1658, 2002.

28 J. Yang, P. Zhang, L. Zhang, H. Shu, B. Li, and Z. Gui, "Particle swarm optimizer for weighting factor selection in intensity-modulated radiation therapy optimization algorithms," *Physica Medica*, vol. 33, pp. 136–145, 2017.

29 Y. Tian, T. Zhang, J. Xiao, X. Zhang, and Y. Jin, "A coevolutionary framework for constrained multiobjective optimization problems," *IEEE Transactions on Evolutionary Computation*, vol. 25, no. 1, pp. 102–116, 2020.

30 M. Li, S. Yang, and X. Liu, "Shift-based density estimation for Pareto-based algorithms in many-objective optimization," *IEEE Transactions on Evolutionary Computation*, vol. 18, no. 3, pp. 348–365, 2013.

31 E. Zitzler, M. Laumanns, and L. Thiele, "SPEA2: Improving the strength pareto evolutionary algorithm," *TIK Report*, vol. 103, 2001.

32 Y. Tian, H. Wang, X. Zhang, and Y. Jin, "Effectiveness and efficiency of non-dominated sorting for evolutionary multi- and many-objective optimization," *Complex & Intelligent Systems*, vol. 3, pp. 247–263, 2017.

33 K. Deb and R. B. Agrawal, "Simulated binary crossover for continuous search space," *Complex Systems*, vol. 9, no. 2, pp. 115–148, 1995.

34 K. Deb and M. Goyal, "A combined genetic adaptive search (GeneAS) for engineering design," *Computer Science and Informatics*, vol. 26, pp. 30–45, 1996.

35 K. Deb, A. Pratap, S. Agarwal, and T. Meyarivan, "A fast and elitist multiobjective genetic algorithm: NSGA-II," *IEEE Transactions on Evolutionary Computation*, vol. 6, no. 2, pp. 182–197, 2002.

36 L. M. Antonio and C. A. Coello Coello, "Use of cooperative coevolution for solving large scale multiobjective optimization problems," in *Proceedings of the 2013 IEEE Congress on Evolutionary Computation*, 2013, pp. 2758–2765.

37 C. He, L. Li, Y. Tian, X. Zhang, R. Cheng, Y. Jin, and X. Yao, "Accelerating large-scale multiobjective optimization via problem reformulation," *IEEE Transactions on Evolutionary Computation*, vol. 23, no. 6, pp. 949–961, 2019.

38 Y. Tian, X. Zheng, X. Zhang, and Y. Jin, "Efficient large-scale multiobjective optimization based on a competitive swarm optimizer," *IEEE Transactions on Cybernetics*, vol. 50, no. 8, pp. 3696–3708, 2019.

39 L. While, P. Hingston, L. Barone, and S. Huband, "A faster algorithm for calculating hypervolume," *IEEE Transactions on Evolutionary Computation*, vol. 10, no. 1, pp. 29–38, 2006.

40 J. Fan, J. Wang, Z. Chen, C. Hu, Z. Zhang, and W. Hu, "Automatic treatment planning based on three-dimensional dose distribution predicted from deep learning technique," *Medical Physics*, vol. 46, no. 1, pp. 370–381, 2019.

9

Evolutionary Large-Scale Multi-Objective Optimization in Deep Learning

9.1 Introduction

Deep learning has made significant success in tackling various machine learning tasks, including image processing and computer vision [1], natural language processing [2], and object detection [3]. To achieve desirable performance, it is widely recognized that the weights and architecture of deep neural networks (DNNs) are very crucial. Hence, optimizing the weights and designing the architectures of DNNs remain a hot research field in recent years.

Regarding optimizing weights of DNNs, a number of learning algorithms have been developed, including the stochastic gradient descent (SGD) [4], RMSProp [5], and Adam [6], among several others, most of which are gradient-based methods. The success of DNNs has proven that gradient-based methods are efficient in training DNNs due to the rapid convergence speed provided by the gradient information. But, they also suffer from several limitations [7–11]. First, they may easily get trapped into local optimums or saddle points due to their greedy nature [7, 8]. Second, a regularization term usually needs to be included to alleviate overfitting [12], which may not always lead to the desired training performance, since it is nontrivial to specify the hyperparameter for regularization. Finally, a number of other hyperparameters, such as the learning rate, momentum, and drop rate, must be carefully predefined [11].

Regarding designing architectures of DNNs, early efforts are devoted to manually designing various neural architectures. Examples of this category include the ResNet [13], DenseNet [14], and GoogLeNet [15, 16]. Due to the requirements of human domain knowledge and expertise in handcrafted neural architectures, a hot research direction called neural architecture search (NAS) has arisen since it first appeared in [17]. To solve the NAS task, a variety of optimization techniques have been adopted for NAS to yield high-performance architectures, including reinforcement learning (RL) [18, 19], gradient optimization [20, 21], Bayesian optimization [22], and others [23].

Evolutionary Large-Scale Multi-Objective Optimization and Applications: Integrating Evolutionary Computation, Machine Learning, and Data Science, First Edition. Xingyi Zhang, Ran Cheng, Ye Tian, and Yaochu Jin.

As a class of population-based meta-heuristics, evolutionary algorithms (EAs) have been shown effective in solving the weight optimization of neural networks (NNs) (i.e., the training of NNs) and the architecture search of DNNs (i.e., NAS). Compared to gradient-based approaches to training NNs and other optimization approaches to NAS, EAs are more insensitive to local optimums due to the attractive exploratory ability [11] and perform better in handling multiple objectives without extra hyperparameters [24], which can provide a set of potential solutions for decision. Nevertheless, it is still difficult for EAs to solve large-scale optimization problems for DNNs (e.g., training DNNs) [11], since EAs suffer from poor scalability to the search dimension; besides, it is also challenging for EAs to solve the NAS task, because equipping EAs with an effective encoding scheme and genetic operators for NAS are not easily attainable, where the population characteristics of EAs further makes NAS need more computational resources. This motivates many researchers to improve EAs for large-scale multi-objective optimization problems in deep learning and this chapter presents two state-of-the-art EAs for two typical multi-objective optimization problems for deep learning: a gradient-guided multi-objective EA for training DNNs (termed GEMONN) and an action command-based surrogate-assisted multi-objective EA for searching neural architectures of DNNs (termed ACEncoding).

9.2 Gradient-Guided Multi-Objective Evolutionary Algorithm for Training Deep Neural Networks

To obtain better weights for NNs, many EAs have been suggested for training NNs [10, 11, 25–31]. For instance, Whitley *et al.* [27] adopted a genetic algorithm to optimize the weights of NNs, where experimental results demonstrated its competitiveness on training NNs. Montana and Davis [26] trained a feedforward NNs with a total of 126 weights via a tailored genetic algorithm, which was shown to outperform gradient-based approaches. Despite their superiority in training NNs, EAs are less efficient in solving the training of DNNs due to the curse of dimensionality [32]. To address this issue, some attempts have been made to enhance the scalability of EAs in training large-scale NNs, especially DNNs. For example, Gong *et al.* [31] suggested a bi-objective EA to optimize the weights of sparse DNNs by only considering the biases instead of all weights, mainly because the biases are the main factor for controlling the sparsity of hidden layers; Liu *et al.* [10] suggested an evolutionary layer-wise structure learning approach for optimizing network weights, which reduces the large search space by limiting the weight value to 0 or 1 and making multiple weights in the same region share the same value; Sun *et al.* [11] developed a single-objective EA to optimize the weights of DNNs, where the weight optimization was converted into the optimization of orthogonal bias vector in weight space; instead of simplifying the optimization problem, Such *et al.* [33] used the parallel computing devices to enhance the search ability, aiming to improve the scalability of EAs in optimizing the weights of DNNs, by which millions of weights can be effectively optimized on a single modern desktop; Cui *et al.* [8] proposed an evolutionary approach to optimize the weights by combining the SGD and an EA, where the EA enhances the stability of the SGD and guarantees steady reduction of loss function. Most of these evolutionary approaches for training large-scale NNs aim to address the curse of dimensionality by reducing the search space (i.e., number of weights to be optimized), which may miss the optimal search region and increase the probability of getting trapped in local optima [34].

To remedy this limitation, this section introduces a gradient-guided evolutionary approach to training DNNs (called GEMONN), which can take advantages of the strength of both gradient descent (GD) and EAs. The presented crossover is a gradient-guided simulated binary crossover operator [35], termed gSBX, for optimizing the weights of DNNs. The search direction of gSBX is defined to be the same as the gradient of the parent to strike a balance between exploitation and exploration in optimizing the weights of DNNs. gSBX is then embedded in an evolutionary multi-objective algorithm to train DNNs, which minimizes the training loss and the network sparsity at the same time. To validate the performance of the proposed approach, GEMONN is adopted to train the weights of four types of NNs, namely, auto-encoder (AE), stacked auto-encoder (SAE), long short-term memory (LSTM), and convolutional neural networks (CNNs). Experimental results indicate that GEMONN is able to enhance the scalability of EAs in training large-scale neural networks having up to millions of weights.

In the following, we first present the formal definition of neural network training, then the GEMONN for training the DNNs is elaborated. Finally, experimental comparisons between the GEMONN and state-of-the-art approaches for network training are showcased and discussed.

9.2.1 Problem Formulation of Multi-Objective Neural Network Training

Given a training set D and a network *net*, EAs usually search for the optimal values of the weights and/or the optimal architecture [36–38] by minimizing the following bi-objective optimization problem [9]:

$$\min_{\mathbf{x}} \mathbf{F}(\mathbf{x}) = \begin{cases} f_1(\mathbf{x}) = f_{complexity}(net, \mathbf{x}), \\ f_2(\mathbf{x}) = f_{loss}(net, D, \mathbf{x}), \end{cases} \tag{9.1}$$

where $\mathbf{x}$ is the set of all the weights to be optimized, $f_1(\mathbf{x})$ denotes the complexity of *net* with the weights being $\mathbf{x}$, and $f_2(\mathbf{x})$ denotes the training loss of *net* averaged over all the samples in D. In practice, the network complexity can be measured by the number of nonzero weights [39] or the sum of absolute weights [32], while the training loss can be classification error as [9] or reconstruction error [10].

Since each solution contains two objective values, the quality of solutions is usually determined by the Pareto dominance relation. Specifically, a solution $\mathbf{x}_1$ is said to Pareto dominate solution $\mathbf{x}_2$, denoted as $\mathbf{x}_1 \prec \mathbf{x}_2$, if and only if the following conditions hold:

$$\begin{cases} \forall i = 1, 2, \dots, m, & f_i(\mathbf{x}_1) \leq f_i(\mathbf{x}_2), \\ \exists j = 1, 2, \dots, m, & f_j(\mathbf{x}_1) < f_j(\mathbf{x}_2), \end{cases} \tag{9.2}$$

where m is the number of objectives and equals to two in this section. The goal of multi-objective neural network training is to find a set of solutions approximating the Pareto front, presenting trade-offs between the neural network complexity and training loss.

9.2.2 Gradient-Guided Evolutionary Multi-Objective Training of Deep Neural Networks

9.2.2.1 Motivations

In spite of the promising performance of existing evolutionary approaches to training NNs, most of them encounter difficulties in optimizing a large number of weights in DNNs [11]. The limitation of these approaches can mainly be attributed to the dimension reduction strategies they employ, which typically can address the curse of dimensionality to some extent but is easily vulnerable to get trapped in local optima [34]. Hence, these approaches have to reduce the number of optimized weights due to the limitation of the search ability of EAs for large-scale optimization [32]. Besides, there are some approaches [8, 31] directly combining EAs and the GD to conduct search in the original search space, where the weights are successively updated by a traditional genetic operator and GD. However, the poor scalability of EAs to large-scale optimization problems is also not overcame. In other words, the EA in these approaches can only guarantee that the best fitness in the population does not degrade, while the GD is mainly responsible for finding better solutions in the original search space.

To better understanding the limitation of traditional genetic operators, Fig. 9.1 illustrates the difference between the search behaviors of genetic operators and GD in optimizing a multimodal function $f(x)$. More specifically, the popular genetic operator simulated binary crossover (SBX) [35] is taken as an example in Fig. 9.1(a), where the generated offspring solutions distribute on both sides of the parent $\mathbf{x}$, since SBX determines the search direction randomly and enables more unknown areas to be explored. In other words, SBX makes many unnecessary search efforts, which becomes even more serious in optimizing the weights of DNNs in a high-dimensional search space. The GD method, by contrast, has

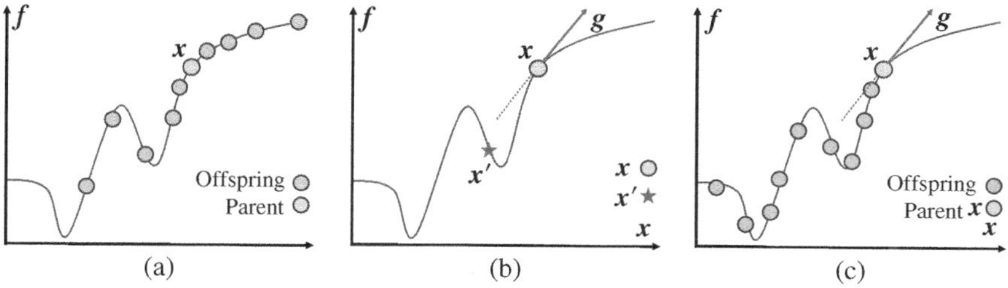

Figure 9.1 Search behaviors of the simulated binary crossover (SBX) and the gradient descent on a minimization problem $f(\mathbf{x})$. (a) The offspring solutions generated by the genetic operator (SBX). (b) The offspring solution generated by the gradient descent. (c) The offspring solutions generated by combining SBX with the gradient descent.

shown better exploitation ability since the search direction is always opposite to the gradient of $\mathbf{x}$, as shown in Fig. 9.1(b). However, an unsuitable step size (determined by the learning rate) is likely to drive $\mathbf{x}$ to a local optimum.

As illustrated in Fig. 9.1(c), the weakness of SBX, i.e., it generates offspring solutions only in the expected direction, can be remedied by the GD. Vice versa, the vulnerability of the gradient approach to local optimums can be alleviated by the genetic operator. To be more specific, the introduction of gradient information into the genetic operator can guide the search behavior toward promising directions, thereby significantly enhancing the search efficiency. Meanwhile, the generation of multiple offspring solutions can considerably reduce the probability of getting trapped into local optima and improve the exploration ability of the GD. In addition, this combination also eliminates the use of learning rate, which is difficult to be predefined or adapted.

Following the above basic idea, a gradient-guided evolutionary search approach to training DNNs will be presented below. In contrast to existing approaches that are usually tailored for a specific type of DNN, the proposed generic approach is applicable to different types of DNNs.

9.2.2.2 Simulated Binary Crossover

In recent years, many crossover operators have been proposed for real-coded EAs, such as SBX, UNDX [40], and DE [41]. Among these crossovers, SBX exhibits better performance on most test problems as demonstrated in [42]. Therefore, we adopt SBX as the basic genetic operator, which is widely used for solving continuous optimization problems. Formally, given two parents $\mathbf{x}_1$ and $\mathbf{x}_2$ with n decision variables, SBX generates two offspring solutions $\mathbf{c}_1$ and $\mathbf{c}_2$ by

$$\begin{cases} \mathbf{c}_1^i = [(1+\beta)\mathbf{x}_1^i + (1-\beta)\mathbf{x}_2^i]/2 \\ \mathbf{c}_2^i = [(1-\beta)\mathbf{x}_1^i + (1+\beta)\mathbf{x}_2^i]/2 \end{cases}, \quad 1 \le i \le n, \tag{9.3}$$

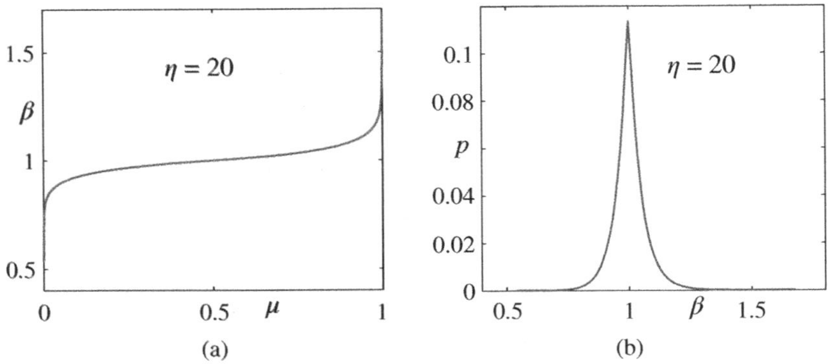

Figure 9.2 The relationship between β and μ in SBX. (a) Relationship between β and μ. (b) Probability distribution of β.

where $\mathbf{x}_1^i$ denotes the ith variable of decision vector $\mathbf{x}_1$ and β is a random value that needs to be regenerated for each dimension by

$$\beta = \begin{cases} (2\mu)^{\frac{1}{\eta+1}}, & \mu \leq 0.5, \\ (2(1-\mu))^{-\frac{1}{\eta+1}}, & \mu > 0.5, \end{cases} \tag{9.4}$$

where μ is a uniformly distributed random value sampled in $[0, 1]$ and η is a parameter controlling the distribution of the offspring solution. According to [35], the value of η is suggested to be 20. Figure 9.2(a) plots the relationship between β and μ, and Fig. 9.2(b) plots the probability distribution of β. Apparently, β is very likely to be around 1, where $\beta = 1$ will make $\mathbf{c}_1 = \mathbf{x}_1$ and $\mathbf{c}_2 = \mathbf{x}_2$ according to (9.3).

For a visual observation of the search behavior of SBX, Fig. 9.3(a) depicts the probability distribution of the offspring solutions generated by SBX, based on two one-dimensional parents $x_1 = 0.2$ and $x_2 = 0.7$. It can be seen that the offspring solutions locate on the left and right sides of each parent with the same probability. By taking into account of the gradient information, as shown in Figure 9.3b, the offspring solutions should only locate on the left side of the first parent x_1 and the right side of the second parent x_2, where the gradient g_1 of the first parent is assumed to be positive and the gradient g_2 of the second parent is assumed to be negative. Next, we elaborate how the gradient information can be incorporated into SBX.

9.2.2.3 Core Idea of Gradient-Based SBX (gSBX)

By rewriting Eq. (9.3), we have

$$\begin{cases} \mathbf{c}_1^i = \mathbf{x}_1^i + \alpha_1(\mathbf{x}_1^i - \mathbf{x}_2^i)/2 \\ \mathbf{c}_2^i = \mathbf{x}_2^i - \alpha_2(\mathbf{x}_1^i - \mathbf{x}_2^i)/2 \end{cases}, \quad 1 \leq i \leq n \tag{9.5}$$

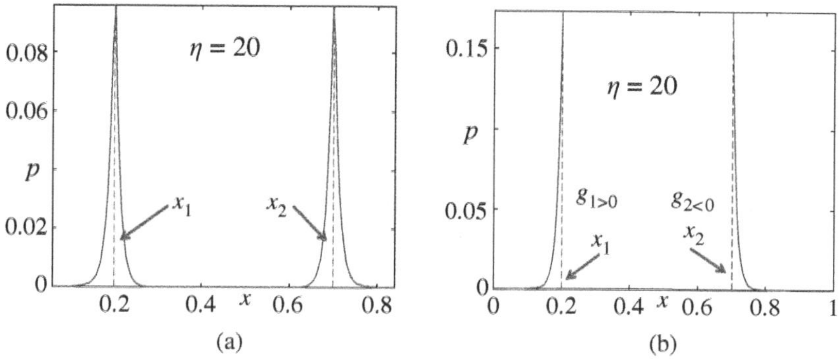

Figure 9.3 Probability distribution of the offspring solutions generated based on two parents $x_1 = 0.2$ and $x_2 = 0.7$ by SBX and a modified SBX. (a) Offspring solutions generated by SBX. (b) Offspring solutions generated by a modified SBX.

with

$$\begin{aligned} \alpha_1 &= \beta_1 - 1, \\ \alpha_2 &= \beta_2 - 1, \\ \beta_1 &= \beta_2 = \beta. \end{aligned} \tag{9.6}$$

To incorporate the gradient $\mathbf{g}_1$ and $\mathbf{g}_2$ into the genetic operator, we can get the following:

$$\begin{cases} \mathbf{c}^i_{1,2} < \mathbf{x}^i_{1,2}, \mathbf{g}^i_{1,2} > 0 \\ \mathbf{c}^i_{1,2} \geq \mathbf{x}^i_{1,2}, \mathbf{g}^i_{1,2} \leq 0 \end{cases}, \quad 1 \leq i \leq n, \tag{9.7}$$

where $\mathbf{c}_{1,2}$ stands for $\mathbf{c}_1$ or $\mathbf{c}_2$, $\mathbf{x}_{1,2}$ denotes $\mathbf{x}_1$ or $\mathbf{x}_2$, and $\mathbf{g}_{1,2}$ represents $\mathbf{g}_1$ or $\mathbf{g}_2$.

According to Eq. (9.5), $\boldsymbol{\alpha}_1(\mathbf{x}^i_1 - \mathbf{x}^i_2)$ should have a different sign with $\mathbf{g}^i_1$, whereas $\boldsymbol{\alpha}_2(\mathbf{x}^i_1 - \mathbf{x}^i_2)$ should have the same sign as $\mathbf{g}^i_2$. Since the value of $(\mathbf{x}^i_1 - \mathbf{x}^i_2)$ is given, the sign of $\boldsymbol{\alpha}^i_1$ and $\boldsymbol{\alpha}^i_2$ can be changed as desired by tuning μ according to (9.4) and (9.6). Consequently, the gradient information can be included in the genetic operator in gSBX by setting μ in the following way:

$$\mu_1 = \begin{cases} \text{sample in } [0, 0.5], & \mathbf{g}^i_1\left(\mathbf{x}^i_1 - \mathbf{x}^i_2\right) > 0 \\ \text{sample in } (0.5, 1], & \mathbf{g}^i_1\left(\mathbf{x}^i_1 - \mathbf{x}^i_2\right) < 0 \\ \text{sample in } [0, 1], & \text{otherwise} \end{cases} \tag{9.8}$$

and

$$\mu_2 = \begin{cases} \text{sample in } [0, 0.5], & \mathbf{g}^i_2\left(\mathbf{x}^i_1 - \mathbf{x}^i_2\right) < 0, \\ \text{sample in } (0.5, 1], & \mathbf{g}^i_2\left(\mathbf{x}^i_1 - \mathbf{x}^i_2\right) > 0, \\ \text{sample in } [0, 1], & \text{otherwise,} \end{cases} \tag{9.9}$$

where μ_1 and μ_2 are used for generating α_1 and α_2 by (9.4) and (9.6), respectively. Note that they need to be regenerated for each dimension. Besides, μ_1 and μ_2 are sampled in the original interval [0, 1] if the gradient is equal to zero or unavailable, which is the same as done in the original SBX.

9.2.2.4 Uniform Crossover in gSBX

The idea of uniform crossover is also adopted in gSBX, which has been used in many existing genetic operators for improving the diversity [43]. Specifically, the uniform crossover exchanges the decision variables in each dimension of the two offspring solutions $\mathbf{c}_1$ and $\mathbf{c}_2$ at a probability of 0.5. According to (9.5), we have

$$\begin{aligned} \mathbf{c}_1^i &= \mathbf{x}_i^1 + (\beta_1 - 1)\left(\mathbf{x}_1^i - \mathbf{x}_2^i\right)/2 \\ &= (1-\beta_1)\mathbf{x}_2^i/2 + (1+\beta_1)\mathbf{x}_1^i/2 \\ &= \mathbf{x}_2^i - (-\beta_1 - 1)\left(\mathbf{x}_1^i - \mathbf{x}_2^i\right)/2. \end{aligned} \tag{9.10}$$

Hence, $\mathbf{c}_2^i$ can be changed to $\mathbf{c}_1^i$ by setting $\alpha_2 = -\beta_1 - 1$. Similarly, $\mathbf{c}_1^i$ can be changed to $\mathbf{c}_2^i$ by setting $\alpha_1 = -\beta_2 - 1$.

Additional attention should be paid when exchanging $\mathbf{c}_1^i$ and $\mathbf{c}_2^i$. For example, if gradient $\mathbf{g}_1^i > 0$, it is unreasonable to change $\mathbf{c}_1^i$ to $\mathbf{c}_2^i$ when $\mathbf{x}_1^i < \mathbf{x}_2^i$, since $\mathbf{x}_1^i$ should be moved along the negative direction of $\mathbf{g}_1^i$. Therefore, some restrictions should be applied to limit the uniform crossover, where the uniform crossover can be performed only when $\mathbf{g}_1^i(\mathbf{x}_1^i - \mathbf{x}_2^i) > 0$ and $\mathbf{g}_2^i(\mathbf{x}_1^i - \mathbf{x}_2^i) < 0$. To sum up, the proposed gSBX sets α_1 and α_2 as

$$\alpha_1 = \begin{cases} -\beta_2 - 1, \lambda \le 0.5 & \text{and} \quad \mathbf{g}_1^i(\mathbf{x}_1^i - \mathbf{x}_2^i) > 0 \\ & \text{and} \quad \mathbf{g}_2^i(\mathbf{x}_1^i - \mathbf{x}_2^i) < 0 \\ \beta_1 - 1, & \text{otherwise} \end{cases} \tag{9.11}$$

and

$$\alpha_2 = \begin{cases} -\beta_1 - 1, \lambda \le 0.5 & \text{and} \quad \mathbf{g}_1^i(\mathbf{x}_1^i - \mathbf{x}_2^i) > 0, \\ & \text{and} \quad \mathbf{g}_2^i(\mathbf{x}_1^i - \mathbf{x}_2^i) < 0, \\ \beta_2 - 1, & \text{otherwise,} \end{cases} \tag{9.12}$$

where λ is sampled from a uniform distribution $U[0, 1]$ for each dimension.

9.2.2.5 Sparsity in gSBX

To maintain the sparsity of the offspring solutions in gSBX, it is intuitive to make the offspring solutions have the same or a higher degree of sparsity than their parents. For this purpose, gSBX sets each decision variable of an offspring solution to zero if the variable of the corresponding parent is zero. Thus, the number of zero weights in the offspring solution is always equal to or larger than the number of zero weights in the parent. According to (9.5), $\mathbf{c}_1^i = \mathbf{x}_1^i = 0$ holds if we set $\alpha_1 = 0$. As a result, the proposed gSBX further repairs α_1 and α_2 by

$$\alpha_1 = \begin{cases} 0, & \mathbf{x}_1^i = 0 \\ \alpha_1, & \text{otherwise} \end{cases} \tag{9.13}$$

and

$$\alpha_2 = \begin{cases} 0, & \mathbf{x}_2^i = 0, \\ \alpha_2, & \text{otherwise}. \end{cases} \tag{9.14}$$

9.2.2.6 Procedure of gSBX

The main steps of gSBX are listed in Algorithm 9.1. It can be seen from the algorithm that two parents, $\mathbf{x}_1$ and $\mathbf{x}_2$, are randomly selected from population P, which are used to generate two offspring solutions $\mathbf{c}_1$ and $\mathbf{c}_2$. For each dimension, μ_1, μ_2, β_1, β_2, α_1, and α_2 are determined in sequence, and the decision variables $\mathbf{c}_1^i$ and $\mathbf{c}_2^i$ are calculated accordingly. All solutions in P go through this procedure, and all newly generated solutions form the offspring population.

A preliminary comparative study is performed to verify the advantage of the proposed gSBX over original SBX. In the comparison, both gSBX and SBX are implemented in a simple genetic algorithm to minimize the following function:

$$\min_{\mathbf{x}} \mathbf{f}(\mathbf{x}) = \sum_{i=1}^{40} x_i^2 - \frac{1}{2}\sum_{i=41}^{60} x_i \sin\left(\sqrt{|x_i|}\right), \tag{9.15}$$

where $\mathbf{x} = (x_1, x_2, \ldots, x_{60}) \in [-500, 500]^{60}$. The population size is set to 100 and the number of generations is 100. Besides, the GD with a learning rate of 0.1 is also adopted to minimize the function for 10,000 epochs. The minimum objective values found by the steepest descent method, genetic algorithm with SBX, and genetic algorithm with the proposed gSBX are plotted in Fig. 9.4. It can be found

Algorithm 9.1: *gSBX(P)*

SetKwCommentComment//**Input**: P (Parent)
Output: O (Offspring population)
$O \leftarrow \emptyset$;
while $P \neq \emptyset$ **do**
  Randomly select two solutions $\mathbf{x}_1$ and $\mathbf{x}_2$ from P;
  $P \leftarrow P \setminus \{\mathbf{x}_1, \mathbf{x}_2\}$;
  for $i = 1$ *to* n **do**
    Sample μ_1 and μ_2 by (9.8) and (9.9);
    Calculate β_1 and β_2 by (9.4);
    Calculate α_1 and α_2 by (9.11) and (9.12);
    Repair α_1 and α_2 by (9.13) and (9.14);
    Calculate $\mathbf{c}_1^i$ and $\mathbf{c}_2^i$ by (9.5);
  $O \leftarrow O \cup \{\mathbf{c}_1, \mathbf{c}_2\}$;
return O;

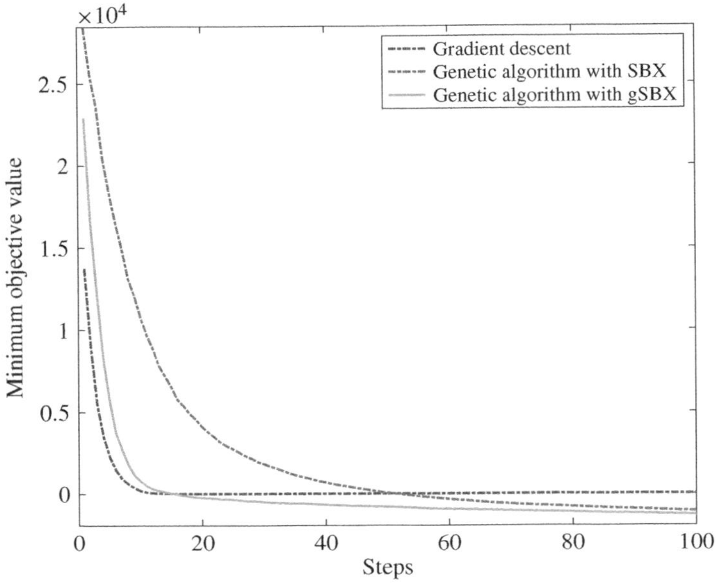

Figure 9.4 Profiles of the minimum objective value found by the gradient method, a genetic algorithm with SBX, and a genetic algorithm with the proposed gSBX on an objective function with mixed landscape.

that the steepest descent method quickly gets stuck a local optimum, while SBX is able to find an acceptable solution, although it converges slower than the steepest descent method. The proposed gSBX has a similar convergence speed to the gradient method, and finds a similar objective value to SBX. From these results, we can conclude that gSBX can strike a balance between exploitation and exploration. Therefore, it is expected to be efficient in training DNNs with a large number of weights.

Based on the proposed gSBX, a gradient-guided evolutionary approach is proposed to train DNNs, the details of which are presented in the next subsection.

9.2.3 Framework of GEMONN

Algorithm 9.2 presents the procedure of the proposed GEMONN. To begin with, a population with predefined size is initialized, where the decision variables (i.e., weights) of each solution are set to a random value within $[-1, 1]$ or 0 with

Algorithm 9.2: Procedure of the Proposed GEMONN

Input: *Gen* (Maximum number of generations), *Pop* (Population size), *Net* (A DNN to be trained)
Output: $\mathbf{x}$ (Weights of *Net*)

//Weights Optimization
$P \leftarrow$ Randomly initialize *Pop* solutions;
Calculate the objective values and gradient of each solution in P;
for $g = 1$ *to Gen* **do**
  Parent $\leftarrow$ Select *Pop* solutions from P by the mating selection of NSGA-II;
  $O \leftarrow gSBX(Parent)$;
  Calculate the objective values and gradient of each solution in O;
  $P \leftarrow$ Select *Pop* solutions from $P \cup O$ by the environmental selection of NSGA-II;

//Pareto Optimal Solutions Update
$P \leftarrow$ Delete the dominated solutions from P;
for *each* $\mathbf{x} \in P$ **do**
  $\mathbf{x} \leftarrow$ Fine-tune $\mathbf{x}$ by sparse stochastic gradient descent;

//Final Solution Selection
$\mathbf{p} \leftarrow$ Select one solution from P in the knee area;
return p;

the same probability. Then, the objective values of each solution are evaluated and the gradient in terms of the training loss is calculated. In each generation of GEMONN, a number of solutions is selected from the population by the mating selection strategy of NSGA-II [44] and used to generate the same number of offspring solutions by the proposed gSBX. Then, the offspring population is combined with the parent population, and half the solutions in the combined population are retained for the next generation by the environmental selection strategy of NSGA-II. After the evolution terminates, a single solution should be picked up from the final population as the output. For this purpose, strategies for updating Pareto optimal solutions and for selecting a final solution are developed.

A flowchart of the proposed GEMONN is plotted in Fig. 9.5. In the following, the details of the main components of GEMONN, including solution evaluation, Pareto optimal solution update, and final solution selection, are described separately.

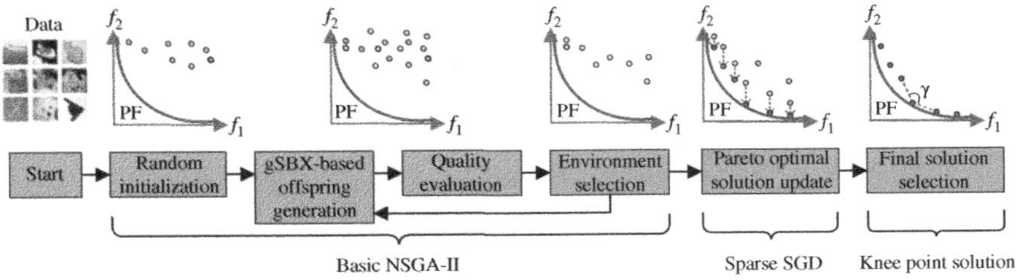

Figure 9.5 Overall flowchart of the proposed GEMONN.

9.2.3.1 Quality Evaluation

Given a training set $D = \{(\mathbf{s}_i, \mathbf{y}_i)\}_{i=1}^N$ and a DNN *net*, the proposed GEMONN calculates the following two objective values for each solution $\mathbf{x}$:

$$\min_{\mathbf{x}} \mathbf{F}(\mathbf{x}) = \begin{cases} f_1(\mathbf{x}) = ||\mathbf{x}||_0, \\ f_2(\mathbf{x}) = \dfrac{1}{N}\displaystyle\sum_{i=1}^{N} L(\hat{\mathbf{y}}_i), \end{cases} \tag{9.16}$$

where the decision vector $\mathbf{x}$ denotes the set of all the weights in *net*, and $\hat{\mathbf{y}}$ is the output of *net* with the weights being $\mathbf{x}$. The first objective f_1 indicates the complexity of the DNN, where a smaller f_1 means to a sparser network. Here, the l_0-norm is adopted to measure the complexity, which is believed to be effective since it directly counts the number of nonzero weights [9, 43]. The second objective f_2 indicates the training loss averaged over all the training samples, where the cross entropy loss (CE) [45] is adopted for supervised learning tasks and the mean squared error (MSE) [9] is employed for unsupervised learning tasks. Specifically, f_2 is defined as

$$f_2(\mathbf{x}) = \begin{cases} L_{CE}(\hat{\mathbf{y}}_i) = -\displaystyle\sum_{j}^{l} \mathbf{y}_i^j \log\left(\hat{\mathbf{y}}_i^j\right), \\ L_{MSE}(\hat{\mathbf{y}}_i) = \dfrac{1}{m}\displaystyle\sum_{j}^{m}\left(\mathbf{s}_i^j - \hat{\mathbf{y}}_i^j\right)^2, \end{cases} \tag{9.17}$$

where $\mathbf{y}_i^j$ denotes the jth label of the ith sample, $\mathbf{s}_i^j$ denotes the jth feature of the ith sample, l is the total number of labels, and m is the total number of features.

Since the network may overfit the training data if only the training loss is minimized, we consider the network complexity to alleviate overfitting. As shown in [9], the two objectives f_1 and f_2 are conflicting with each other since a network with lower training loss usually corresponds to a higher complexity. Owing to the delicate selection strategy provided by evolutionary multi-objective optimization, however, a set of solutions striking a good trade-off between the two objectives can be found, and a network with acceptable training loss without overfitting the training set is expected to be obtained.

9.2.3.2 Update of Pareto Optimal Solutions

Once the evolutionary optimization process is completed, a strategy for updating Pareto optimal solutions is designed for the final exploitation. To be specific, the dominated solutions in the population are removed, and then each non-dominated solution is fine-tuned by SGD [46] with a learning rate of 0.1 for one epoch. It is worth noting that only the nonzero weights are updated here, while the zero weights are kept unchanged. Such the proposed trick can decrease the training loss while maintaining the network sparsity, and we call it sparse stochastic gradient descent (SSGD) hereafter.

9.2.3.3 Final Solution Selection

Last but not the least, a single solution should be selected from the fine-tuned population as the output. Intuitively, it is unreasonable to select the solution with the lowest training loss or complexity, as the former tends to overfit the training set and the latter is usually not well trained. Since the solutions in the knee area of the population have a better trade-off between the two objectives [9], they are usually preferred in many scenarios [10, 31]. Based on this idea, the proposed GEMONN selects a solution $\mathbf{p}$ according to the following criteria:

$$\begin{aligned} X_1 &= \{\mathbf{x} | \text{int}(\log_{10} f_2(\mathbf{x})) - \text{int}(\log_{10} f_2(\mathbf{x}^+)) < 1\}, \\ X_2 &= \{\mathbf{x} | f_2(\mathbf{x}) < 2 * f_2(\mathbf{x}^+), \mathbf{x} \in X_1\}, \\ \mathbf{p} &= \arg\min_{\mathbf{x}} \gamma(\mathbf{x}), \mathbf{x} \in X_2. \end{aligned} \tag{9.18}$$

The first criterion aims to find the solutions whose training loss has the same order of magnitude as $\mathbf{x}^+$, where $\mathbf{x}^+$ is the solution having the lowest training loss in the population, i.e., $\mathbf{x}^+ = \arg\ \min_{\mathbf{x}\in P} f_2(\mathbf{x})$. The second criterion also aims to guarantee the training performance of the selected solution. The third criterion is used to find the final solution according to $\gamma(\mathbf{x})$, which calculates the angle between the left and right neighbors of the solution [47].

9.2.3.4 Time Complexity Analysis

The time complexity of proposed GEMONN is mainly determined by two components, i.e., the calculation of gradients and the optimization process of NSGA-II. According to [48], the time complexity for calculating the gradients is $O(N \times m)$, where N and m are the size of a mini-batch and feature dimensionality, respectively. The time complexity of one generation of NSGA-II is $O(Pop^2)$ [44], where Pop is the population size. Therefore, the overall time complexity of GEMONN is $O(N \times m \times Pop \times Gen) + O(Pop^2 \times Gen)$. Since $N \times m \gg Pop$ in general, the time complexity of GEMONN can be regarded as $O(N \times m \times Pop \times Gen)$, which is same as that of SGD [48].

9.2.4 Empirical Studies

9.2.4.1 Neural Networks to be Trained

The performance of the proposed GEMONN is verified on four types of NNs, including AE, SAE [49], LSTM [50], and CNNs [18]. A brief introduction to the training of the four types of NNs is given in the following.

1. Auto-encoder (AE). As a fully connected (FC) feedforward NN with a single hidden layer, AE is often used for dimensionality reduction, where the output is expected to be the same as the input and the hidden layer can be regarded as a meaningful and reduced representation of the training sample. As a result, AE can be trained in an unsupervised manner, and many existing EAs have been adopted to train it [10, 11, 31]. To train AE by the proposed GEMONN, the MSE is adopted as the second objective. Besides, the sigmoid activation function is applied in the hidden layer.
2. Stacked auto-encoder (SAE). By using AE as the building blocks, SAE can reduce the training sample layer by layer and perform classification at last. With the greedy layer-wise training process, the dependency of weights in different layers can be properly handled, and the weights in each layer can be optimized separately. The training of SAE consists of two phases [11, 51], i.e., pre-training and fine-tuning. The pre-training phase trains a set of AEs with different numbers of hidden nodes by minimizing the reconstruction error. Then, these AEs are stacked together and tailed by an FC layer, and the fine-tuning phase optimizes all the weights for minimizing the classification error. To train SAE by the proposed GEMONN, the MSE is adopted as the second objective in the pre-training phase and the CE is adopted in the fine-tuning phase.
3. Long short-term memory (LSTM). LSTM is currently one of the most popular recurrent NNs for many deep-learning tasks, in which the connections between layers are much more complex than those in feedforward NNs. Since all the weights in LSTM should be optimized together, it is much more difficult than training SAE whose weights are optimized layer by layer, and none of the existing evolutionary approaches have been applied to it. To demonstrate the effectiveness of the proposed genetic operator in training DNNs without dimensionality reduction, we also use the proposed GEMONN to train LSTM with its structure shown in Fig. 9.6(a), where the CE is adopted as the second objective.

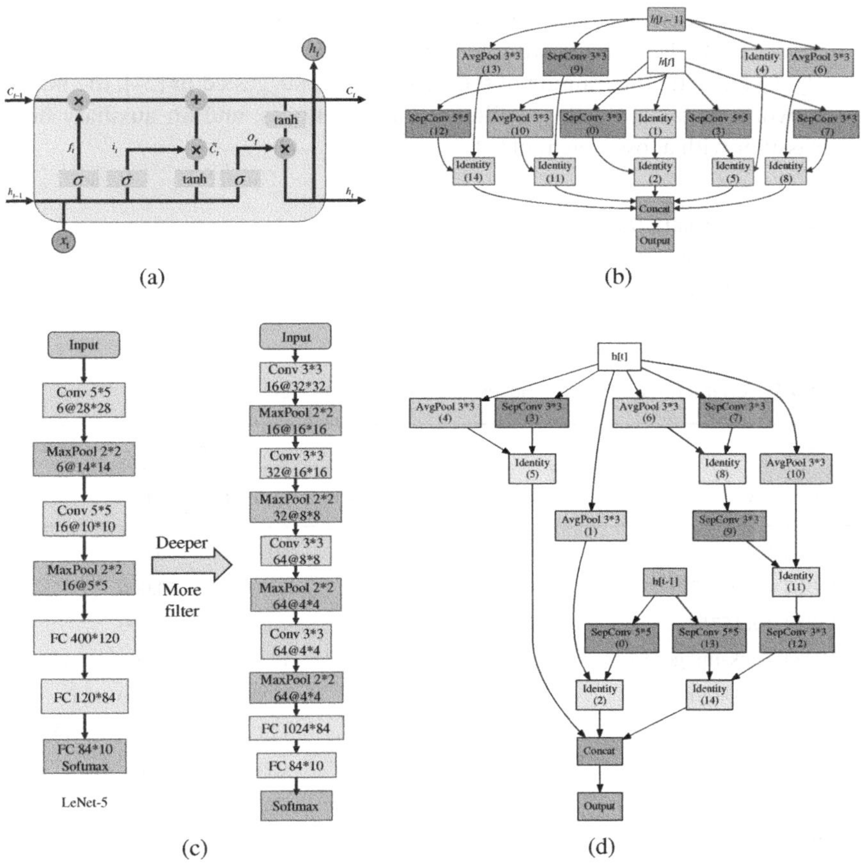

Figure 9.6 The structure of used NNs, including LSTM, LeNet-5 as well as its variant (deeper and more filters), and cell-based CNN. (a) LSTM. (b) Cell-CNN (normal cell). (c) LeNet-5 and its variant. (d) Cell-CNN (reduction cell).

4. Convolutional neural networks (CNNs). CNNs are also powerful NNs for many deep learning tasks, especially computer vision. To investigate the ability of GEMONN in training CNNs, we consider the LeNet-5 [52] and its variant (termed LeNet-deeper) having more convolutional operations and more filters, as shown in Fig. 9.6(c), where a ReLu [53] activation function follows each convolutional operation. To investigate the results on deeper CNNs, a Vgg16 [54] variant with channel setting [8, 16, 32, 55] is employed, whose last two FC layers are set to 512 and 256. Besides, an effective cell-based CNN [56]

is also considered, termed Cell-CNN, whose normal cell and reduction cell are set to the best architecture founded in [18], as plotted in Fig. 9.6(b) and (d). To facilitate the training of the Cell-CNN, as suggested in [24], the neural network contains 6 cells, 16 filters in all the nodes, and an auxiliary head classifier with a loss weight of 0.4.

9.2.4.2 Benchmark Datasets

As listed in Table 9.1, 17 widely used datasets are adopted in the experiments. The first 10 datasets including MNIST [52] as well as its variants [57] and Cifar10-bw [58] are used in the experiments in training AE and SAE. The next

Table 9.1 Characteristics of 17 Chosen Benchmark Datasets.

Benchmark	Dimension	Classes	Size of training	Size of test	Network
MNIST [52][a)]	28∗28	10	60,000	10,000	
MNIST-basic [57][b)]	28∗28	10	12,000	50,000	
MNIST-rot [57][b)]	28∗28	10	12,000	50,000	
MNIST-back-image [57][b)]	28∗28	10	12,000	50,000	AE
MNIST-back-rand [57][b)]	28∗28	10	12,000	50,000	/
MNIST-rot-back-image [57][b)]	28∗28	10	12,000	50,000	SAE
Rectangles [57][c)]	28∗28	2	1,200	50,000	
Rectangles-image [57][c)]	28∗28	2	12,000	50,000	
Convex [57][c)]	28∗28	2	8,000	50,000	
Cifar10-bw [58][d)]	32∗32	10	50,000	10,000	
FordA [59][e)]	500	2	3,601	1,320	
TwoPatterns [59][e)]	128	4	1,000	4,000	
ECG5000 [59][e)]	140	5	500	4,500	RNN
ElectricDevices [59][e)]	96	7	8,926	7,711	
UWaveGestureLibraryX [59][e)]	315	8	896	3,582	
InsectWingbeatSound [59][e)]	256	11	220	1,980	
Cifar10 [58][d)]	3∗32∗32	10	50,000	10,000	CNN

a) http://yann.lecun.com/exdb/mnist/.
b) URL of (b) to the same as (c), or directly combine (b)(c) with the URL of (c).
c) http://www-labs.iro.umontreal.ca/~lisa/icml2007data/.
d) http://www.cs.toronto.edu/~kriz/cifar.html.
e) https://www.cs.ucr.edu/~eamonn/time_series_data_2018/.

six datasets are collected from the UCR time series classification archive [59] and used in the experiments of training LSTMs. Besides, the last dataset Cifar10 [58] is used in the experiments of CNNs.

9.2.4.3 Compared Algorithms

Since the experiments are conducted on AE, SAE, LSTM, and CNNs, different comparison approaches have to be selected for different NNs.

Algorithms for Training AE The following comparison approaches are chosen for training the AE:

- SGD [4]. Stochastic GD;
- Adam [6]. A method for stochastic optimization;
- NSGA-II [44]. NSGA-II with SBX operator;
- NSGA-II (+gradient). Based on NSGA-II, update all offspring solutions by SGD for one step at each generation;
- NSGA-II (+gradient_0). Based on NSGA-II, update all offspring solutions by SSGD for one step at each generation;
- GEMONN (−SSGD). GEMONN without updating the Pareto optimal solutions by SSGD;
- GEMONN (single). GEMONN without optimizing sparsity by Eq. (9.16);
- GEMONN (−S). GEMONN without considering sparsity by Eqs. (9.13) and (9.14);
- GEMONN (−G). GEMONN without using gradient information by Eqs. (9.8), (9.9), (9.11), and (9.12).

SGD and Adam are two most popular gradient-based approaches for training DNNs with millions of weights, which can be used as baselines to validate whether the performance of GEMONN is promising or not. NSGA-II, NSGA-II (+gradient), and NSGA-II (+gradient_0) are used to verify the effectiveness of the proposed genetic operator gSBX, since the other components of NSGA-II and GEMONN, i.e., solution evaluation, mating selection, and environmental selection, are completely the same. The remaining GEMONN (−SSGD), GEMONN(single), GEMONN (−S), and GEMONN (−G) are designed for ablation studies to validate the effectiveness of the novel strategies proposed in GEMONN.

Approaches for Training SAE The following state-of-the-art approaches are used for training SAE:

- SDAE [60]. Stacked denoising AEs;
- SCAE [61]. Stacked contractive AEs;
- SSAE [62]. Stacked sparse AEs;
- EUDNN [11]. An evolutionary approach for training unsupervised DNNs.

As introduced in [11], SDAE, SCAE, and SSAE are three state-of-the-art gradient-based approaches for training SAE, which incorporate various problem independent regularization terms to decrease the network complexity for alleviating overfitting. Besides, the evolutionary approach EUDNN is considered as the state-of-the-art, which can effectively train large-scale DNNs with the help of dimensionality reduction. As a result, the comparison between GEMONN and the chosen algorithms is meaningful and convincing.

Algorithms for Training LSTM and CNNs Only SGD and Adam are considered for training LSTM and CNNs in the comparative studies, since these two approaches are very popular and effective for optimizing the large number of weights together. Besides, existing EAs, including EUDNN, are not well suited for training them due to their poor performance resulted from the curse of dimensionality.

9.2.4.4 Parameter Settings

The parameter settings for all the compared approaches are given below.

Parameters in Training AE The number of hidden units of the AE is set to 500, which means that the total number of weights to be optimized is $(28 * 28 + 1) * 500$. Besides, the following settings are used in the compared approaches:

- SGD and Adam. The learning rate varies in $\{0.1, 0.01, 0.001\}$; the batch size is set to 64; and the maximal number of training steps is set to $50 * 500$, to be consistent with the evolutionary approaches.
- NSGA-II and its variants. The population size is set to 50; the maximum number of generations is set to 500.
- GEMONN and its variants. The population size is set to 50 and the maximum generation is set to 500.

As a result, all the compared approaches have the same number of function evaluations that calculates the output and the gradient over all the training samples for fairness.

Parameters in training SAE The number of units in each layer of SAE is empirically set to $[input, 500, 400, 300, 200, 100, output]$, where *input* is the dimension of features and *output* is the number of classes of the dataset. The following parameter settings are used in the comparative studies, which also ensures that all the compared approaches consume the same number of function evaluations:

- SDAE, SCAE, and SSAE. Following the settings in [11], the SGD is chosen as the training approach, where the learning rate varies in $\{0.1, 0.01, 0.001\}$, and the batch size is set to 64; the maximum number of training epochs is set to $50 * 500$;

the binary corrupted level of SDAE, the sparsity of SSAE, the coefficient of the contractive term in SCAE is set to $\{0.1, 0.3, 0.5, 0.7\}$, respectively.
- EUDNN. The population size is set to 50; the maximum iteration is set to 500; other parameters follow the original settings in the literature.
- GEMONN. The population size is set to 50; maximum generation is set to 500.

Parameters in Training LSTM and CNNs The sizes of the input layer, hidden layer, and output layer of a 2-layer LSTM are set to *input*, 100, and *output*, respectively, and the parameters in CNNs are presented in Section 9.2.4.1. The following settings are used in the compared approaches:

- SGD and Adam. The learning rate varies in $\{0.1, 0.01, 0.001\}$, and the batch size is set to 64 for LSTM and CNNs; The maximum number of training epochs is set to $50 * 100$ for LSTM and $50 * 500$ for CNNs.
- GEMONN. The population size is set to 50; and the maximum number of generations is set to 100 for LSTM and 500 for CNNs.

All models and the gradient-based approaches are implemented in Pytorch [63], and GEMONN as well as variants of NSGA-II are implemented in Numpy [64]. These experiments were performed on a Windows 10 computer with an Intel i7-8700k CPU and an NVIDIA 1080 GPU. The source code of the proposed approach is made available from https://github.com/smartdonckey/GEMONN.

9.2.4.5 Experimental Results on AE

Figure 9.7 presents the solutions obtained by all algorithms under comparison in the objective space. It can also be found that the performance of SGD and Adam highly relies on the learning rate, as the solutions obtained by SGD (0.1) outperform SGD (0.01) and SGD (0.001), while Adam (0.1) is also much worse than Adam (0.01) and Adam (0.001). Moreover, the solutions obtained by SGD and Adam are not sparse at all, which may affect the generalization of the model and result in overfitting.

On the other hand, the solutions obtained by NSGA-II have poor convergence in terms of the MSE and we cannot find any valuable solution due to the poor search ability of the existing genetic operators. With the assistance of the gradient information, the solutions obtained by NSGA-II (+gradient_0) are better than those obtained by NSGA-II, which verifies the importance of introducing gradient information into EAs. Nevertheless, the performance of NSGA-II and its variants is much worse than the gradient-based approaches and the proposed GEMONN.

The proposed GEMONN exhibits quite competitive performance compared to SGD, Adam, and NSGA-II. On the one hand, it can be seen that the solutions acquired by GEMONN are very promising, whose MSE is just worse than those obtained by Adam (0.001) and Adam (0.01) but much better than those obtained

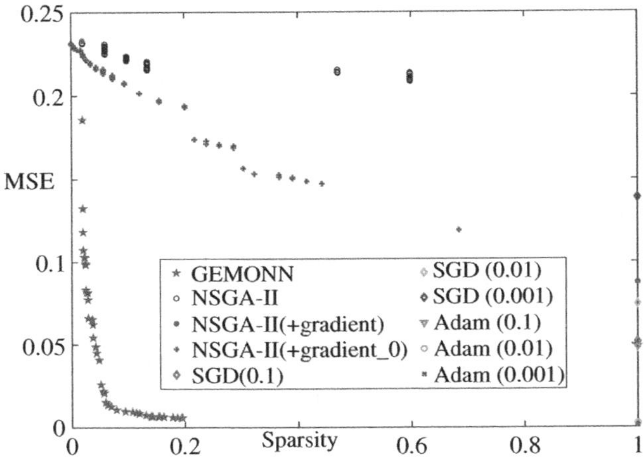

Figure 9.7 The final solutions obtained by the compared approaches on the MNIST dataset, where "SGD (0.1)" denotes the SGD with a learning rate of 0.1.

by the other approaches. On the other hand, the sparsity of the solutions acquired by GEMONN is similar to those obtained by NSGA-II but much better than those obtained by SGD and Adam. In short, the AE trained by the proposed GEMONN can have a relatively low MSE with satisfactory sparsity.

To investigate the convergence speed of the compared approaches, Fig. 9.8 depicts the decline in MSE of all the compared approaches, with the increase of the number of function evaluations, where the observation is consistent with Fig. 9.7. More specifically, GEMONN outperforms NSGA-II and most gradient-based approaches, while it converges slower than Adam (0.01) and Adam (0.001).

As a result, solutions obtained by the proposed GEMONN have better sparsity but worse training loss than those obtained by Adam (0.01) and Adam (0.001). To further investigate which solutions are better for classification, an FC layer is tailored for the AE and trained in a supervised manner. Figure 9.9 presents the mean correct classification rate (CCR) [11] on test set obtained by SGD, Adam, and the proposed GEMONN, averaged over 30 runs. Besides, a three-layer FC network with the same size is directly trained for the classification task and used as the baseline. By contrast, the results of NSGA-II are ignored since the obtained CCR values are quite poor. It can be observed from Fig. 9.9 that the mean CCR values obtained by GEMONN are better than all the other compared approaches including Adam (0.01) and Adam (0.001), though the solutions found by Adam (0.01) and Adam (0.001) are with lower MSE than GEMONN as shown in Fig. 9.8. Therefore, it can be inferred that the AEs trained by Adam overfit the training set to some extent, whereas the sparse AEs obtained by GEMONN can alleviate overfitting.

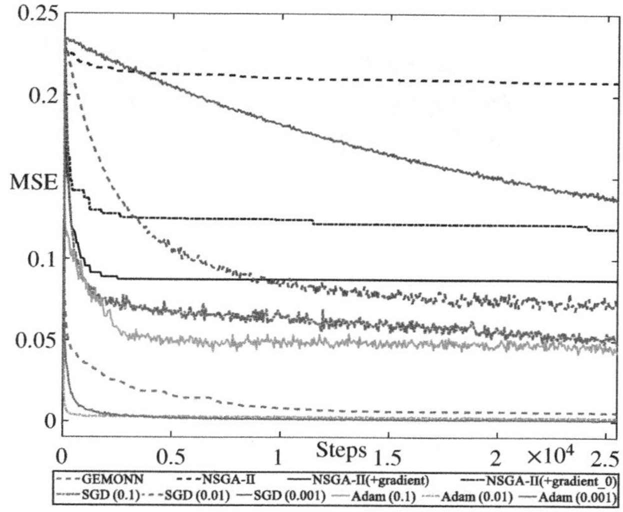

Figure 9.8 Performance changes in MSE of the compared approaches on the MNIST dataset.

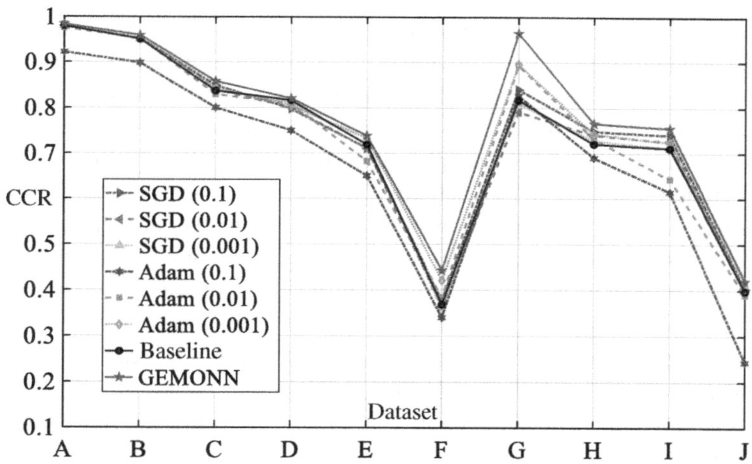

Figure 9.9 The mean CCR values of the proposed GEMONN against SGD, Adam, and the baseline for training the AE tailed by a classification layer. The tick labels "A–J" denote the datasets of MNIST, MNIST-basic, MNIST-rot, MNIST-back-rand, MNIST-back-image, MNIST-rot-back-image, rectangles, rectangles-image, convex, and Cifar10-bw, respectively.

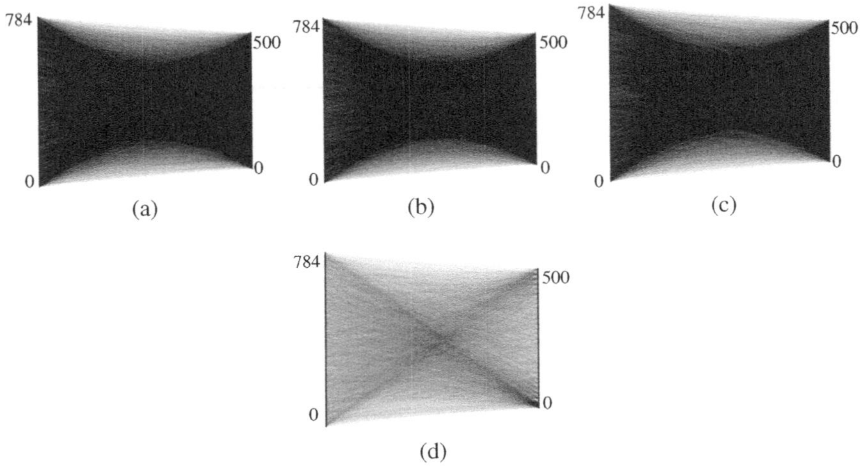

Figure 9.10 Connections between the input and hidden layers of the AEs obtained by Adam (0.01), Adam (0.001), SGD (0.1), and GEMONN on the MNIST dataset. A darker color indicates a denser connection between weights. (a) Adam (0.01). (b) Adam (0.001). (c) SGD (0.1). (d) GEMONN.

For visual observation of the difference between the sparsity of the AEs obtained by the compared approaches, Fig. 9.10 draws the connections between the input layer and hidden layer of the AEs obtained by Adam (0.01), Adam (0.001), SGD (0.1), and GEMONN on the MNIST dataset, where a darker color indicates a denser connection between weights. It is obvious that the AE obtained by GEMONN is much sparser than those obtained by the gradient-based approaches. Moreover, Fig. 9.11 plots the connections between the input layer and the hidden layer of the AE obtained by GEMONN at different generations on the MNIST dataset. For better visualization, each graphic in Fig. 9.11 consists of 500 binary images, where each image has 28×28 pixels and each pixel indicates whether the corresponding input unit is connected to a hidden unit or not. Besides, a gray pixel indicates a connection between the input unit and a hidden unit, while a white dot indicates there is no connection. As can be seen from Fig. 9.11, the AE is not sparse at the beginning, while it gradually becomes sparse as the population evolves. As a consequence, the proposed GEMONN can effectively control the sparsity of the obtained NN, and such a sparse nature is crucial for preventing the model from overfitting.

Afterward, the effectiveness of the novel strategies proposed in GEMONN is verified by ablation studies. First, GEMONN is compared to GEMONN (–SSGD) on the MNIST dataset to verify the effectiveness of the SSGD based the update strategy for Pareto optimal solutions. According to the solutions obtained by

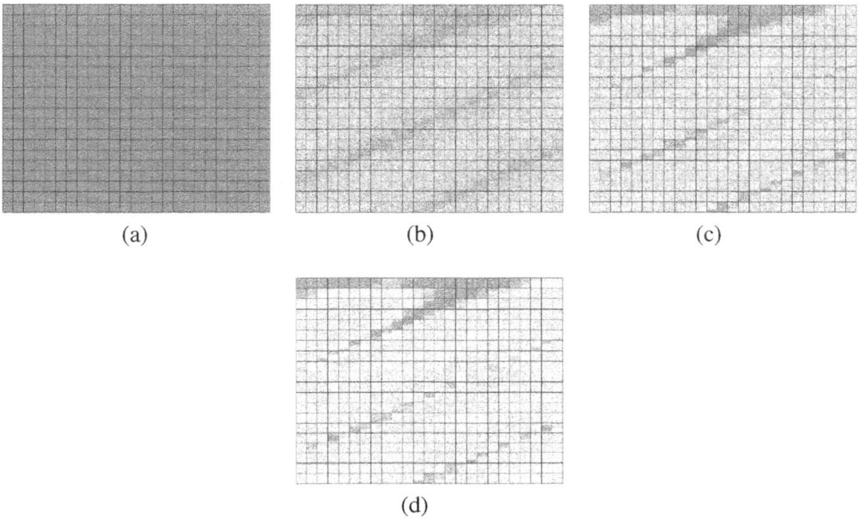

Figure 9.11 Connections between the input and hidden layers of the AE obtained by GEMONN at different generations on the MNIST dataset. A gray pixel indicates an input unit is connected to a hidden unit, while a white pixel indicates there is no connection. (a) 1st generation. (b) 10th generation. (c) 100th generation. (d) 500th generation.

GEMONN and GEMONN (−SSGD) shown in Fig. 9.12(a), it is clear that the solutions obtained by GEMONN dominate most solutions obtained by GEMONN (−SSGD), hence the effectiveness of SSGD is confirmed. Second, Fig. 9.12(b) depicts the effectiveness of considering the optimization of sparsity within GEMONN. As can be seen that the final solutions of GEMONN (single) have higher training CCR values than that of GEMONN on all 10 datasets, however, GEMONN can obtain better test CCR values than GEMONN (single). The crucial reason is that GEMONN (single) encounters the overfitting problem to some extent due to the optimization not considering the sparsity restriction. According to the above comparisons in Figs. 9.8, 9.9, and 9.12(b), it is indeed an effective way to mitigate the overfitting by considering sparsity in optimization. Third, GEMONN is compared to GEMONN (−G) and GEMONN (−S) on the MNIST dataset to verify the effectiveness of the gradient information in the proposed genetic operator. As can be seen from Fig. 9.13(a), the solutions obtained by GEMONN are superior over those obtained by GEMONN (−G) and GEMONN (−S) in terms of both the sparsity and MSE. Besides, GEMONN converges significantly faster than GEMONN (−G) and GEMONN (−S) in terms of the hypervolume indicator (HV) [55], as shown in Fig. 9.13(b). Therefore, the effectiveness of the novel strategies proposed in GEMONN can be verified.

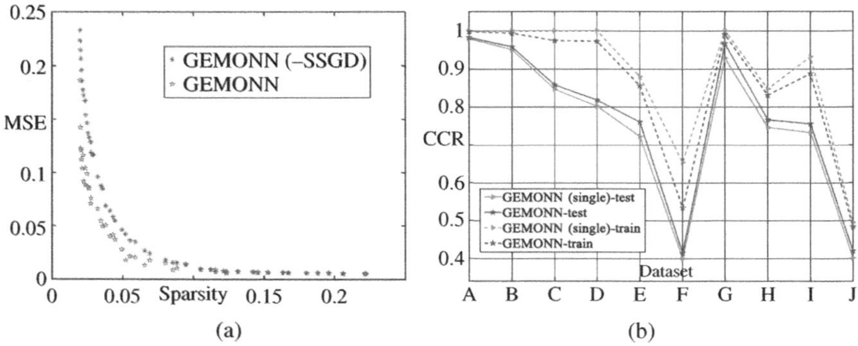

Figure 9.12 The validation of proposed strategies by comparing the performance of GEMONN and its variants. (a) Solutions obtained by GEMONN and GEMONN (–SSGD) on the MNIST dataset. (b) Comparison between GEMONN and GEMONN (single) in terms of training and test CCR values.

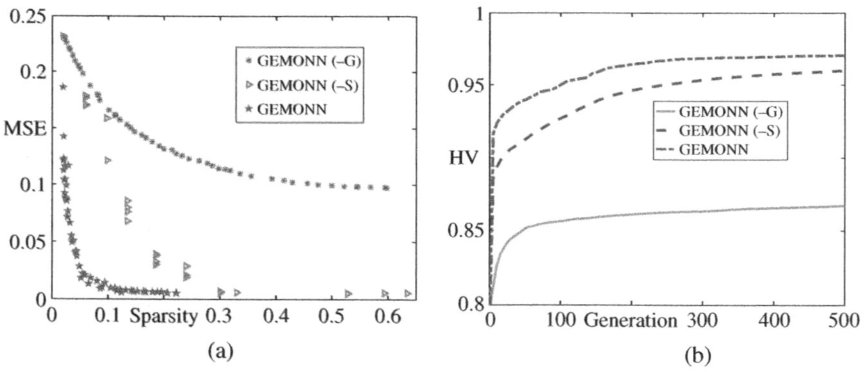

Figure 9.13 Comparison between the performance of GEMONN and its variants on the MNIST dataset. (a) Solutions obtained by GEMONN and its variants. (b) Convergence profiles of HV of GEMONN and its variants.

Lastly, a parameter sensitivity analysis of the population size *Pop* and distribution index η in GEMONN is conducted. Figure 9.14(a) shows the HV values obtained by GEMONN on four datasets MNIST, MNIST-rot, rectangles, and convex with different population sizes, where it can be seen that GEMONN can obtain a relatively good performance when the population size is larger than 50. It is not necessary to set *Pop* to 100 for a slightly higher HV value at the expense of a doubled number of fitness evaluations. The results of sensitivity analysis of the distribution index η in GEMONN on four datasets TwoPatterns, ECG5000, UWaveGestureLibraryX, and InsectWingbeatSound are plotted in Fig. 9.14(b).

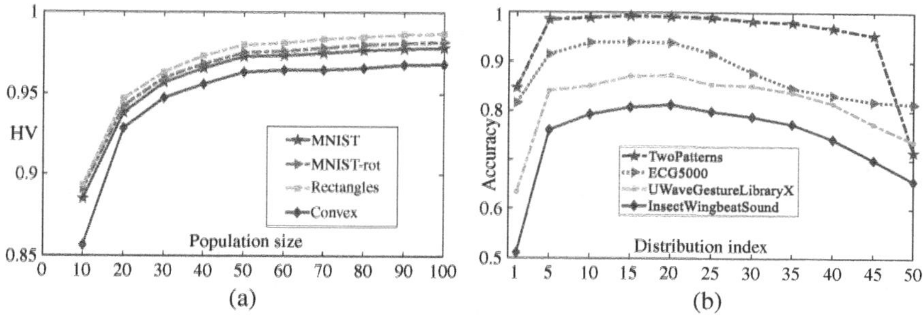

Figure 9.14 Parameter sensitivity analysis to population size and distribution index. (a) Sensitivity to population size. (b) Sensitivity to distribution index.

From this figure, we can see that the proposed GEMONN can obtain the best performance with the distribution index being equal to 15 or 20. Thus, the recommended $\eta = 20$ in [35] is also suitable and effective for GEMONN.

9.2.4.6 Experimental Results on SAE

Table 9.2 lists the CCR values of the SAEs obtained by GEMONN, EUDNN, SSAE, SCAE, and SDAE on the nine Datasets, where each cell in the table records the mean and standard deviation of the CCR values obtained in 30 independent runs. In addition, the Wilcoxon rank sum test [65] at a significance level of 0.05 is employed to perform statistical analysis on the experimental results, where the symbols "+," "−," and "≈" indicate that the result obtained by GEMONN is significantly better than, significantly worse than, and statistically similar to that of the corresponding peer competitors, respectively.

It is clearly presented in Table 9.2 that the proposed GEMONN acquires the best results on seven out of nine datasets, which is followed by EUDNN and SSAE obtaining one best result. In terms of the Wilcoxon rank sum test, the number of datasets where GEMONN performs statistically better than EUDNN, SSAE, SCAE, and SDAE are 6, 7, 8, and 9, respectively. Furthermore, Fig. 9.15 plots the connections between the layers of the SAEs obtained by SDAE, SCAE, SSAE, EUDNN, and GEMONN on the Cifar10-bw dataset. Clearly, the SAE obtained by GEMONN is much sparser than those obtained by the other approaches. Therefore, it can be concluded that the SAE obtained by GEMONN has not only better classification performance but also much lower complexity than those obtained by the other compared approaches. In short, the proposed GEMONN shows significant superiority over the state-of-the-art approaches in training SAE.

Next, Table 9.3 lists the runtimes of the proposed GEMONN, EUDNN, and three approaches trained by SGD on the ten datasets. It can be found that the efficiency of EUDNN is the worst due to its expensive evaluations, and the efficiency of

Table 9.2 The Correct Classification Rate (CCR) Values of the SAEs Obtained by GEMONN, EUDNN, SSAE, SCAE, SDAE on 10 Datasets. The Best Mean Values Are Highlighted in Boldface. The Symbols "+," "−," and "≈" Denote GEMONN Is Statistically Better Than, Worse Than, and Similar to That of the Comparison Approach, Respectively.

Benchmark	GEMONN	EUDNN	SSAE	SCAE	SDAE
MNIST	**0.9879** **(0.0017)**	0.9812^{+} (0.0018)	0.9828^{+} (0.0079)	0.9827^{+} (0.0009)	0.9822^{+} (0.0049)
MNIST-basic	**0.9775** **(0.0012)**	0.9634^{+} (0.0017)	0.9725^{+} (0.0046)	0.9613^{+} (0.0019)	0.9565^{+} (0.0032)
MNIST-rot	**0.9173** **(0.0013)**	0.8871^{+} (0.0025)	0.8857^{+} (0.0041)	0.8975^{+} (0.0043)	0.8765^{+} (0.0081)
MNIST-back-image	0.8771 (0.0064)	$\mathbf{0.8932}^{-}$ **(0.0074)**	$0.8721^{\approx}$ (0.0073)	0.7984^{+} (0.0065)	0.8637^{+} (0.0033)
MNIST-back-rand	0.8609 (0.0024)	0.8857^{-} (0.0012)	$\mathbf{0.8862}^{-}$ **(0.0087)**	$0.8612^{\approx}$ (0.0073)	0.7974^{+} (0.0076)
MNIST-rot-back-image	**0.4975** **(0.0033)**	0.4858^{+} (0.0035)	0.4712^{+} (0.0013)	0.4683^{+} (0.0015)	0.4513^{+} (0.0025)
Rectangles	**0.9757** **(0.0011)**	0.9692^{+} (0.0021)	0.9457^{+} (0.0023)	0.9662^{+} (0.0021)	0.9453^{+} (0.0024)
Rectangles-image	**0.7886** **(0.0027)**	0.7678^{+} (0.0034)	0.7764^{+} (0.0017)	0.7677^{+} (0.0016)	0.7613^{+} (0.0026)
Convex	**0.8186** **(0.0049)**	$0.8135^{\approx}$ (0.0051)	0.8087^{+} (0.0057)	0.7891^{+} (0.0016)	0.8017^{+} (0.0032)
Cifar10-bw	0.4817 (0.0052)	$0.4832^{\approx}$ (0.0013)	0.4753^{+} (0.0039)	$\mathbf{0.4911}^{-}$ **(0.0107)**	0.4694^{+} (0.0152)
$+/-/\approx$		6/2/2	8/1/1	8/1/1	10/0/0

GEMONN is better than EUDNN but worse than SGD, though it has the same time complexity to SGD as analyzed in Section 9.2.3.4. This is mainly because SGD is implemented by Pytorch on GPU, while the proposed GEMONN is implemented on CPU since some complex steps in EA cannot be easily performed on GPU, which is a common difficulty of most EA-based approaches including EUDNN.

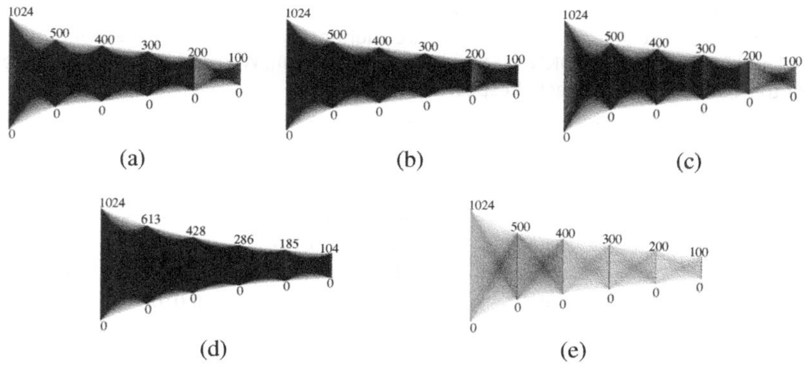

Figure 9.15 Connections between the layers of the SAEs obtained by SDAE, SCAE, SSAE, EUDNN, and GEMONN on the Cifar10-bw dataset. A darker color indicates a denser connection between weights. (a) SDAE. (b) SCAE. (c) SSAE. (d) EUDNN. (e) GEMONN.

Table 9.3 The Average Runtime of the Proposed GEMONN, EUDNN, and Three Approaches Trained by SGD (Including SSAE, SCAE, and SDAE) on 10 Datasets. Note That ">5 h" Means that the Runtime is Significantly Larger Than Five Hours for One Run.

			SGD		
Dataset	**GEMONN**	**EUDNN**	**SSAE**	**SCAE**	**SDAE**
MNIST	2158.4 s	>5 h	1226.8 s	1235.6 s	1270.8 s
MNIST-basic	2163.4 s	>5 h	1232.4 s	1233.7 s	1266.8 s
MNIST-rot	2179.3 s	>5 h	1229.9 s	1239.8 s	1264.3 s
MNIST-back-image	2166.3 s	>5 h	1210.9 s	1229.4 s	1241.6 s
MNIST-back-rand	2211.7 s	>5 h	1219.1 s	1240.4 s	1262.8 s
MNIST-rot-back-image	2138.3 s	>5 h	1226.2 s	1231.9 s	1235.9 s
Rectangles	2158.8 s	>5 h	1212.4 s	1227.6 s	1243.9 s
Rectangles-image	2162.4 s	>5 h	1231.8 s	1235.2 s	1262.8 s
Convex	2201.6 s	>5 h	1227.7 s	1229.2 s	1258.7 s
Cifar10-bw	2681.1 s	>5 h	1338.5 s	1356.4 s	1384.4 s

9.2.4.7 Experimental Results on LSTM and CNNs

The comparison between the proposed GEMONN and gradient-based approaches in training LSTM is presented in Table 9.4. The results are calculated based on 30 independent runs, and the Wilcoxon rank sum test is also employed. Note that due to the poor performance of Adam (0.1) and SGD (0.001), only the results of Adam (0.001), Adam (0.01), SGD (0.01), and SGD (0.1) are listed

Table 9.4 Correct Classification Rate (CCR) Values of the LSTMs Obtained by GEMONN, Adam, and SGD on Six Datasets. Best Values Are Highlighted in Boldface. The Symbols "+," "−," and "≈" Denote GEMONN Is Statistically Better Than, Worse Than, and Similar to That of the Comparison Approach, Respectively.

Benchmark	CCR	GEMONN	Adam (0.001)	Adam (0.01)	SGD (0.01)	SGD (0.1)
FordA	*best*	**1.0000**	0.9848	0.7977	0.7364	**1.0000**
	avg	**0.9994**	0.9702^{+}	0.7737^{+}	0.7192^{+} s	0.9973^{+}
	Std	0.0004	0.0065	0.0120	0.0108	0.0011
TwoPatterns	*best*	**0.9975**	0.9748	0.9765	0.7480	0.9828
	avg	**0.9914**	0.9701^{+}	0.9744^{+}	0.7152^{+}	0.9799^{+}
	Std	0.0112	0.0028	0.0011	0.0265	0.0021
ECG5000	*best*	0.9409	0.9409	**0.9607**	0.9153	0.9389
	avg	0.9387	$0.9372^{\approx}$	$\mathbf{0.9531}^{-}$	0.9108^{+}	0.9334^{+}
	Std	0.0017	0.0089	0.0150	0.0032	0.0042
Electric- -Devices	*best*	**0.9662**	0.8771	0.9598	0.4970	0.8605
	avg	**0.9547**	0.8701^{+}	0.9501^{+}	0.4672^{+}	0.8505^{+}
	Std	0.0075	0.0094	0.0091	0.0195	0.0086
UWaveGesture- -LibraryX	*best*	0.8816	0.8451	0.8068	0.7672	**0.8833**
	avg	0.8724	0.8414^{+}	0.7939^{+}	0.7495^{+}	$\mathbf{0.8772}^{-}$
	Std	0.0083	0.0029	0.0105	0.0084	0.0043
InsectWing- -beatSound	*best*	**0.8197**	0.7990	0.7884	0.8020	0.8172
	avg	**0.8121**	0.7899^{+}	0.7732^{+}	0.7881^{+}	$0.8087^{\approx}$
	Std	0.0103	0.0142	0.0081	0.0082	0.0055
	$+/-/\approx$		5/0/1	5/1/0	6/0/0	4/1/1

in the table. As can be seen, the overall performance of GEMONN is better than the gradient-based approaches, having achieved the largest average CCR values on four out of six datasets. Besides, Adam (0.01) and SGD (0.1) obtain the largest average CCR values on the remaining two datasets. In short, the performance of GEMONN is significantly better than Adam (0.001) as well as SGD (0.01) and competitive to Adam (0.01) and SGD (0.1). In addition, it is worth to note that the performance of Adam is sensitive to the learning rate since Adam (0.01) outperforms Adam (0.001) on ECG5000 and ElectricDevicess, but Adam (0.001) outperforms Adam (0.01) on FordA and UWaveGestureLibraryX.

By contrast, the GEMONN does not contain the learning rate or any other sensitive parameter, which is also a superiority of evolutionary approaches over gradient-based approaches.

The comparative results between GEMONN and the gradient-based approaches in training CNNs on Cifar10 dataset are shown in Fig. 9.16, where the mean test accuracy over 30 independent runs is plotted. Note that the result of SGD (0.001) is not given due to its poor performance. From this figure, it can be observed that GEMONN outperforms gradient-based approaches on LeNet-5, and the performance of GEMONN is also slightly better than the opponents on LeNet-deeper. However, Adam (0.001) and Adam (0.01) get the best performance on Vgg16-variant as well as Cell-CNN, while GEMONN obtains a competitive performance. It is worth noting that the performance of GEMONN degenerates with the architecture of CNN becoming complex since the proportion of convolutional operation parameters increases from 5% to 98%. Essentially, the weights of a whole CNN can be divided into two parts, including the feature extraction (convolutional) part and the classifier (FC) part. Compared to weights of the FC layers, convolutional operations have much fewer weights due to the intrinsic merit of local connection and weight-sharing, where most weights of convolutional layers should not be optimized to zero during the training process [66]. To gain a deep insight into the weights of each part obtained by GEMONN, Fig. 9.17 depicts the percentage of nonzero weights in four CNNs obtained by GEMONN and Adam. It is clear that the weights of convolutional layers obtained by the proposed approach are much sparser than that of Adam; hence, the final performance of GEMONN may be slightly worse than the best gradient-based approaches. Nevertheless, EMONN still holds a competitive performance compared to SGD and Adam.

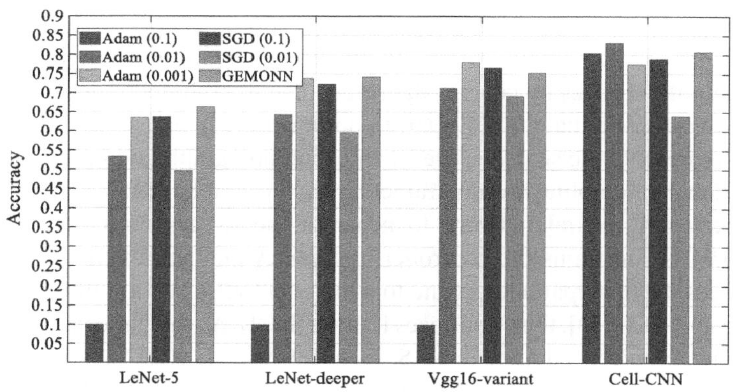

Figure 9.16 The mean Cifar10 test accuracy of the compared approaches on four different CNNs over 30 independent runs.

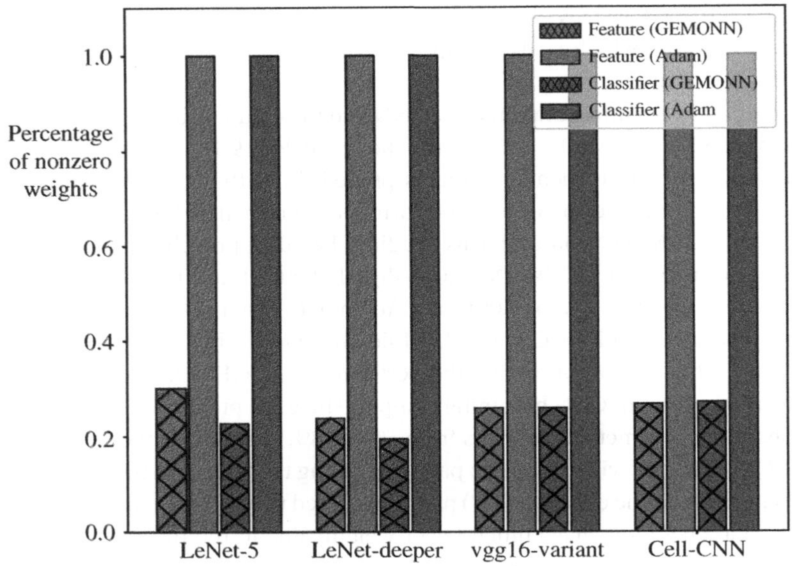

Figure 9.17 The percentage of nonzero weights in feature (extraction) part and classifier part included in four CNNs trained by GEMONN and Adam.

9.3 Action Command Encoding-Based Surrogate-Assisted Evolutionary Algorithm for Neural Architecture Search

NAS appeared in the 1980s [67] and manifested an explosive growth in recent years [23], where various optimizers have been adopted to automatically determine the desired hyperparameters and topologies of neural networks especially CNNs. For example, NEAT [29] optimizes both the topologies and weights of feedforward neural networks by genetic algorithm; The work in [68] adopts the tree-structured Parzen estimator (TPE) [69] to optimize the hyperparameters of CNNs, which represents the search space as an expression graph and optimizes the loss function by sampling hyperparameters in the graph; NAS [17] uses a controller recurrent neural network to generate the hyperparameters and topologies of CNNs and trains the controller by a policy gradient method [70]; NAO [21] embeds the hyperparameters and topologies of CNNs into a continuous space by using an LSTM [50], then optimizes it through GD. As can be seen, there have been many optimizers adopted in NAS, such as EAs (e.g., NEAT), Bayesian optimization (e.g., TPE), RL (e.g., NAS), and GD (e.g., NAO). However, only a few encoding schemes have been designed to provide an effective search space to be explored by these optimizers. In addition, although these search methods can generate better architectures than hand-crafted ones, the performance is improved at the expense of computational resources, where many search methods are criticized for the high computational cost of the search process or the large complexity of the found architectures. In this section, we would like to take a step forward along this line of research, aiming to improve the performance of NAS by presenting a novel encoding scheme, and we also would like to propose an effective EA for NAS with the assistance of the encoding scheme due to the various merits of EAs. However, since EAs are population-based optimizers, it will further aggravate the expensive computational cost in NAS.

To this end, this section introduces an action command encoding-based surrogate assisted EA, termed ACEncoding, for NAS by designing a novel encoding scheme and an effective surrogate model. In the proposed action command encoding scheme, an architecture is constructed by some action commands and each action command is specified by three adjoined integers. By predefining a series of action commands including adding a layer, adding a connection, and cloning a local module, the proposed encoding scheme can use multiple action commands to construct complex architectures in a generative manner. Considering that architectures are represented by a variable-length sequence of integers within a finite range, thus the surrogate model, called Seq2Rank, converts the sequence to a fixed-length vector by a Seq2Seq model [71], then uses a RankNet [72] to predict the rankings of architectures.

In the following, we first present the mathematical formulation of NAS, then the action command encoding method, the performance evaluator-based surrogate model, and the framework of the proposed ACEncoding are subsequently presented. Finally, experimental comparisons between the ACEncoding and state-of-the-art approaches for searching CNNs are presented and discussed.

9.3.1 Problem Formulation of Evolutionary Neural Architecture Search

Given a data set $D = \{D_{train}, D_{val}, D_{test}\}$, the NAS can be formulated as the following single-objective optimization problem.

$$\min_{\mathbf{x}} \ \mathbf{F}(\mathbf{x}) = 1 - f_{val\_accuracy}(\mathbf{x}, D), \tag{9.19}$$

where $\mathbf{x}$ represents the architecture to be optimized and $f_{val\_accuracy}(\mathbf{x}, D)$ denotes the validation accuracy of the architecture $\mathbf{x}$ on D_{val} after being trained for complete epochs on the training dataset D_{train}.

Due to the enormous demand for small-sized networks in real-world applications, the model complexity is also considered in the NAS task [24], and thus NAS can be formulated as a bi-objective optimization problem as follows:

$$\min_{\mathbf{x}} \ \mathbf{F}(\mathbf{x}) = \begin{cases} f_1(\mathbf{x}) &= 1 - f_{val\_accuracy}(\mathbf{x}, D), \\ f_2(\mathbf{x}) &= f_{complexity}(\mathbf{x}), \end{cases} \tag{9.20}$$

where $f_1(\mathbf{x})$ is the classification error rate of $\mathbf{x}$ on validation dataset D_{val} and $f_2(\mathbf{x})$ denotes the model complexity of $\mathbf{x}$. The model complexity can be measured by the number of model parameters, the number of floating-point operations, and inference time [24]. In this chapter, we adopt the bi-objective optimization problem formulation for NAS, where the number of model parameters is used to measure the model complexity.

9.3.2 Action Command Encoding Scheme

9.3.2.1 Action Commands in the Encoding Scheme

The proposed action command encoding scheme is inspired by the process of gene expression, where each architecture is constructed by a series of action commands and each action command is encoded by three integers. Regarding an architecture as a directed acyclic graph $G = (V, E)$, the action command encoding scheme does not directly encode the layer set V or connection set E like existing encoding schemes; instead, the architecture grows from empty according to the action commands in the encoding. In the encoding scheme, the following two basic action commands are defined:

(1) Adding a new layer to the architecture.
(2) Adding a new connection to the architecture.

Obviously, these two action commands are enough to represent arbitrary topologies, which is the same as the graph-based encoding.

However, the encoding will be very long when representing large architectures with dense connections; hence, the following two action commands are also considered in the encoding scheme:

(3) Adding a new layer in parallel with an existing one, i.e., the two layers have the same predecessors and successors.
(4) Adding a new layer after an existing one, i.e., the successors of the new layer are set to the same as the existing layer, then the successor of the existing layer is set to the new layer.

These two action commands can generate some complex topologies like the widely used block-based topology consisting of two parallel layers [56]; hence, we believe that they are helpful to find desired architectures more efficiently. Note that the combination operation is omitted here, which is fixed to the element-wise addition as suggested in many existing encoding schemes [21, 73, 74]. It is obvious that the graph-based encoding should use many operations (i.e., adding a new layer and adding multiple connections) to achieve each of these two action commands, especially when the existing layer is connected to many other layers.

Moreover, it has been well recognized that many successful architectures, including GoogLeNet, ResNet, and those generated by block-based encoding, are composed of repeated modules [75]. Therefore, the following two action commands are designed to repeat the local modules in the architecture:

(5) Cloning an existing local module and locating the two local modules in parallel, i.e., the two modules have the same predecessors and successors.
(6) Cloning an existing local module and locating the two local modules in sequence, i.e., the successors of the new module are set to the same as the existing module, then the successor of the existing module is set to the new module.

Here, a local module is defined as a set of layers connected in a chain-like structure with no branching. Besides, another action command is designed to represent the dense block like those in DenseNet:

(7) Making an existing local module a densely connected block, i.e., each layer is connected to all the others in the module.

By using these seven action commands, the proposed encoding scheme can represent arbitrary topologies and encode some popular blocks in a short sequence. Since these action commands are designed according to the topologies of some popular architectures, the proposed encoding scheme is free to be extended by more action commands for representing the topologies of other effective architectures. Note that each action command does not necessarily appear in the encoding, as an action command will be weeded out during the search process if it always brings a bad performance to the architecture. Besides, it is noteworthy that some basic action commands (e.g., adding a layer or connection) have been adopted in the mutation operator of EA-based search methods [22, 76], but these action commands cannot be achieved by other optimizers. By contrast, the encoding scheme encodes these action commands in a sequence that can be explored by different optimizers, where an EA, RL, and GD are employed in the experiments.

9.3.2.2 Decoding Process of the Proposed Action Command Encoding Scheme

Table 9.5 lists the *codon* of each action command, where each action command is represented by three integers with different meanings. The first integer in each codon varies from 1 to 7 denoting the index of an action command, while the second and third integers can be one of the three variables T, I, and L. T denotes the type of layer to be added, which varies from 1 to 10 denoting one of the following

Table 9.5 Codon of Each Action Command Defined in the Proposed Encoding Scheme, Where Each Action Command Is Represented by Three Integers. The Variable T Denotes the Type of Layer, I Denotes the Index of an Existing Layer, and L Denotes the Length of Local Module.

Codon	Action command
$\{1, T, I\}$	(1) Adding a layer T with predecessor I
$\{2, I, I'\}$	(2) Adding a connection from I to I'
$\{3, T, I\}$	(3) Adding a layer T in parallel with I
$\{4, T, I\}$	(4) Adding a layer T after I
$\{5, I, L\}$	(5) Cloning a module ($L + 1$ layers starting from I) and locating the two modules in parallel
$\{6, I, L\}$	(6) Cloning a module ($L + 1$ layers starting from I) and locating the two modules in sequence
$\{7, I, L\}$	(7) Making a module ($L + 1$ layers starting from I) a densely connected block

10 operations as suggested in [21]: (1) 1×1 convolution; (2) 3×3 convolution; (3) $1 \times 3 + 3 \times 1$ convolution; (4) $1 \times 7 + 7 \times 1$ convolution; (5) 2×2 max pooling; (6) 3×3 max pooling; (7) 5×5 max pooling; (8) 2×2 average pooling; (9) 3×3 average pooling; (10) 5×5 average pooling. Note that all the operations are tailed by batch normalization. I denotes the index of an existing layer; note that the first two indices denote the input layer and output layer rather than the generated convolutional or pooling layers. L denotes the length of a local module with chain-like structure, which consists of $L + 1$ layers starting from an existing layer.

Figure 9.18 depicts an example to illustrate the decoding process of the encoding scheme, where the sequence contains six action commands determined by 18 integers. The initial architecture has two inputs and an output, and it grows according to the following action commands to form an architecture having 9 layers and 12 connections: Adding a convolutional layer, adding a pooling layer, adding a connection, adding a pooling layer, adding a convolutional layer, and cloning a module. It is worth noting that the action commands do not specify the successor when adding a layer, where the successor is set to the output layer by default (e.g., the connection between the convolutional layer and the output layer shown in Fig. 9.18(a)). If this layer is specified as the predecessor of other layers by the subsequent action commands, the default connection between this layer and the output layer is omitted.

In order to provide a finite search space, the variables I and L in the codon are restricted to be less than a maximum value. This maximum value is set to 30 in the experiments, which is enough to represent a relatively large architecture.

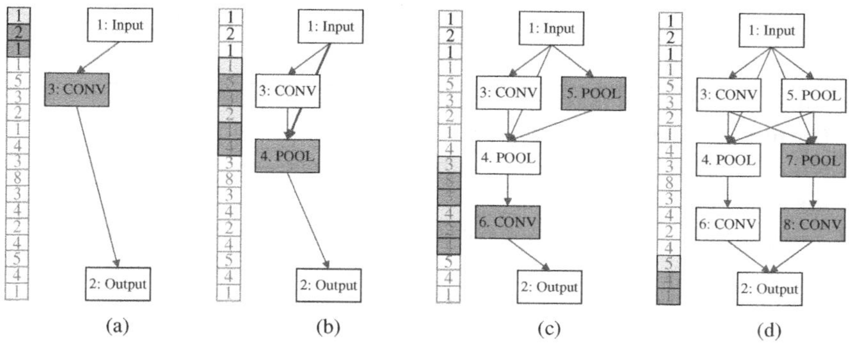

Figure 9.18 An illustration of the decoding process of the proposed action command encoding scheme. (a) Adding a convolutional layer. (b) Adding a pooling layer and a connection. (c) Adding a pooling layer and a convolutional layer. (d) Cloning a module having two layers.

In addition, since some large values of I or L may lead to invalid action commands, they are revised by specific strategies in the decoding process. First, if the layer index I is larger than the number of existing layers N, I is changed to $\text{mod}(I-1, N)+1$. Second, if the module length L exceeds the maximum length of the module (i.e., reaching the output layer or a layer having multiple branches), L is changed to the maximum length of the module. Third, if the architecture becomes cyclic after adding a connection, the connection is reversed. The procedure of the decoding process of ACEnoding is summarized in Algorithm 9.3.

9.3.2.3 Remarks

It can be found from Fig. 9.18 that the proposed action command encoding scheme constructs an architecture with a relatively complex topology by using 18 variables. On the contrary, the adjacent matrix-based encoding requires $[8 \times (7-1)]/2 = 28$ variables to represent the same topology, the graph-based encoding requires 6 variables to store the layers and $12 \times 2 = 24$ variables to store the connections, and the block-based encoding needs to use four blocks determined by $5 \times 4 = 20$ variables to represent a similar architecture with fewer connections. The shorter encoding is owing to the delicate action commands that inherit the advantages of both graph-based encoding and block-based encoding, which provide a more compact search space that facilitates the optimization of architectures. The trade-off between the length of encoding and the complexity of the represented architecture is automatically balanced by optimizers in the search process, where a bias toward the first four action commands in Table 9.5 leads to more complex architectures and a bias toward the last three action commands makes the encoding shorter. The main difference between the proposed action command encoding scheme and existing encoding schemes are summarized in the following.

Algorithm 9.3: Decoding Process of the Proposed Action Command Encoding Scheme

Input: S (a string in the proposed action command encoding scheme)
Output: *Net* (a neural architecture)

Net ← Initialize the architecture with an input node and an output node;
for *each codon* $\in S$ **do**
  [$action, p_1, p_2$] ← Determine the action command of *codon* according to Table 9.5;
  if p_1 *denotes the type T of layer* **then**
    $p_1 \leftarrow \text{mod}(p_1 - 1, 10) + 1$;
  else if p_1 *denotes the index I of a layer* **then**
    $p_1 \leftarrow \text{mod}(p_1 - 1, N) + 1$; //$N$ denotes the number of layers in *Net*
  if p_2 *denotes the index I of a layer* **then**
    $p_2 \leftarrow \text{mod}(p_1 - 1, N) + 1$;
  else if p_2 *denotes the length L of a module* **then**
    $p_2 \leftarrow \min\{p_2, M\}$; //$M$ denotes the maximum length of the module starting from p_1 in *Net*
  Net ← Update the architecture according to *action*, p_1, and p_2;
  Net ← Flip the connections in *Net* to remove loops;
return *Net*;

Action command encoding scheme vs. hyperparameter-based encoding and adjacent matrix-based encoding. The hyperparameter-based encoding and adjacent matrix-based encoding can only represent the hyperparameters and topologies of architectures, respectively, while the proposed action command encoding scheme can represent both of them. Besides, the proposed encoding scheme specifies a topology by using a series of action commands stored in a sequence, which is much shorter than the adjacent matrix.

Action command encoding scheme vs. graph-based encoding. The proposed action command encoding scheme is equivalent to the graph-based encoding when using only the first two action commands, which are enough to represent arbitrary topologies. By contrast, the proposed encoding scheme can represent specific complex topologies by the other action commands, which cannot be represented by the graph-based encoding using a short sequence.

Action command encoding scheme vs. block-based encoding. Both the proposed action command encoding scheme and the block-based encoding can represent complex topologies using a short encoding. However, the topologies that can be represented by block-based encoding are limited by the topology of a single block, whereas arbitrary topologies can be represented by the proposed encoding scheme.

9.3.3 Performance Evaluator-Based Surrogate Model

9.3.3.1 Architecture Embedding

The general architecture of Seq2Rank is illustrated in Fig. 9.19. As can be seen, the model for architecture embedding consists of a FC layer for word embedding

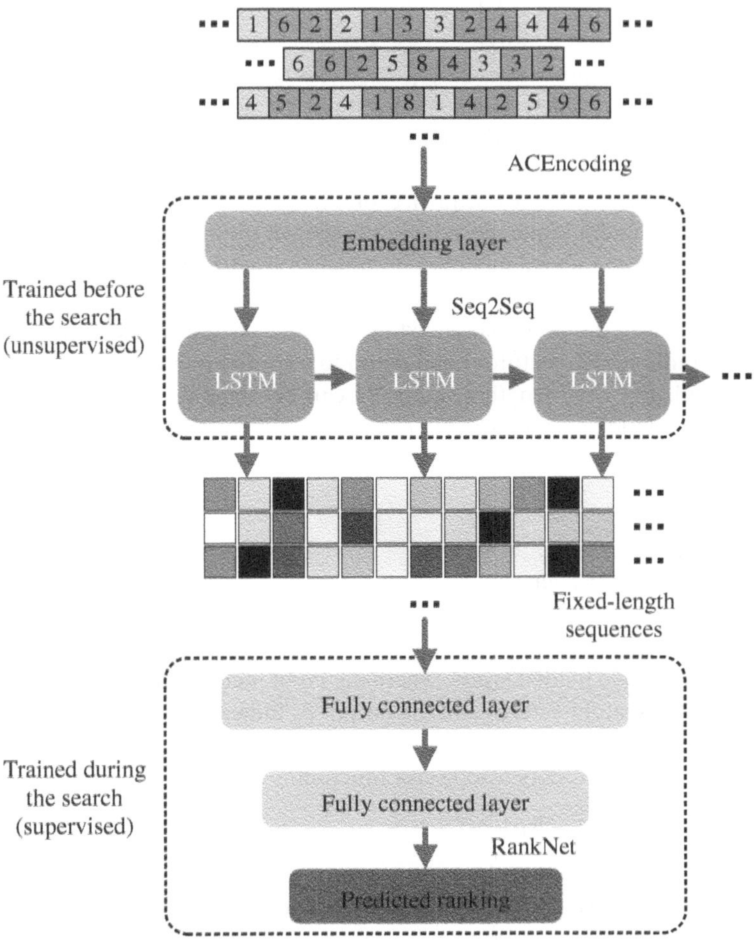

Figure 9.19 An illustration of the Seq2Rank architecture.

and an LSTM layer used in the Seq2Seq model. To begin with, an architecture represented by the proposed action command encoding scheme is tokenized by regarding the codon of each action command as a token; besides, an <SOS> token and an <EOS> token are attached to the start and end of each sequence, respectively, and an <MOS> token is also required to split the action commands for normal cell and reduction cell if the cell structure is adopted. Then, each token is converted to one-hot representation and fed to the embedding layer to obtain a word vector. Afterward, a Seq2Seq model is trained by maximizing the following conditional probability:

$$p(y_1, \dots, y_{T'} \mid x_1, \dots, x_T) = \prod_{t=1}^{T'} p(y_t \mid h_1, \dots, h_T, y_1, \dots, y_{t-1}), \tag{9.21}$$

where $(x_1, \dots, x_T)$ is an input sequence, $(y_1, \dots, y_{T'})$ is the target output sequence, and $(h_1, \dots, h_T)$ is the hidden states of the input sequence. The conditional probability depends on all the hidden states rather than the last one h_T since the attention mechanism is adopted.

To convert the variable-length sequence of the proposed action command encoding to a fixed-length sequence, a large number of training samples should be generated to train the above model in an unsupervised manner. That is, the input sequence and target output sequence of each sample are the same architecture represented by the proposed encoding scheme. Therefore, we randomly generate three million architectures beforehand to train the model until convergence. Afterward, the fixed-length sequence s can be obtained by conducting mean pooling on all hidden states $(h_1, \dots, h_T)$ of an input sequence, i.e.,

$$s = \frac{1}{T} \sum_{i=1}^{T} h_i. \tag{9.22}$$

9.3.3.2 Performance Prediction

The performance of each architecture is estimated by predicting its ranking among multiple architectures via RankNet. The RankNet adopted here is a feedforward neural network with two hidden layers, where the input includes the fixed-length sequences s_i, s_j of two architectures and the output p_{ij} indicates which one of the two architectures has a better performance. Then, the RankNet can be trained by minimizing the following cross-entropy cost function:

$$C_{ij} = -\overline{p}_{ij} \log p_{ij} - (1 - \overline{p}_{ij}) \log(1 - p_{ij}), \tag{9.23}$$

where $\overline{p}_{ij} \in \{0, 1\}$ is the target label.

In contrast to the Seq2Seq model that can be trained in advance, the RankNet should be trained during the search process in a supervised manner, as the target label has to be obtained by comparing the validation accuracies of two architectures. Given the fixed-length sequences $s_1, \dots, s_N$ of N architectures and their validation accuracies $r_1, \dots, r_N$, the training set for RankNet is formed as $\{([s_1, s_2], \overline{p}_{12}), \dots, ([s_i, s_j], \overline{p}_{ij}), \dots, ([s_{N-1}, s_N], \overline{p}_{(N-1)N})\}$, where $i < j \leq N$ and there are $N(N-1)/2$ training samples in total. In practice, a number of architectures should be evaluated via training and validation at the beginning of the search process, then they are used to train the RankNet. During the search process, the performance of most candidate architectures is evaluated via RankNet, while some promising ones can be evaluated via training and validation and used to update the RankNet to enhance its ranking accuracy in the desired search regions.

9.3.4 Framework of ACEncoding

To simultaneously optimize the accuracy and complexity of neural architecture in NAS, the proposed ACEncoding employs NSGA-II as the optimizer due to its high competitiveness in solving bi-objective optimization problems [77]. Compared with existing work on NSGA-II for NAS, the proposed ACEncoding is characterized with two key aspects: an effective action command encoding scheme for NAS, and an efficient performance evaluator-based strategy for accelerating the search. Specifically, Algorithm 9.4 presents the main steps of the proposed ACEncoding with the surrogate model Seq2Rank under NSGA-II, which mainly consists of seven steps.

First, train a Seq2Seq model based on an amount of sequential data in advance. Second, randomly generate a population P consisting of *Pop* solutions which represent corresponding neural architectures, and decode and evaluate the solutions in P after training their corresponding architectures to compute their objective values in Eq. (9.3.1). Then, the surrogate model, i.e., the Seq2Rank performance evaluator, is initially trained with the initial population's information (including all architecture' encoding and performance) based on the pretrained Seq2Seq model. Fourth, the binary tournament selection [24] is applied to the population P to mate individuals in the population and generate the parent population P'. Fifth, an offspring population P_{new} is produced from the parent population P' by implementing the crossover and mutation operators proposed in NSGA-Net [24]. Sixth, determine whether the solutions in P_{new} at the current generation need to train their architecture for evaluation: if yes, decode and evaluate the solutions in P_{new} after training their corresponding architectures to compute their objective values in Eq. (9.3.1); otherwise, directly obtain the architecture performance of all solutions in P_{new} by the surrogate model Seq2Rank. Finally, the main population P and the offspring population P_{new} will be concatenated together, and the standard environmental selection operation in NSGA-II [44] is used to select *Pop* better solutions as the new main population P. The second to the seventh step will repeat

---

Algorithm 9.4: Procedure of ACEncoding

Input: *pop* (population size), *Gen* (maximum number of generations), *GenList* (indexes of generations for real evaluation)

Output: *P* (Set of optimal neural architectures)

Seq2Seq $\leftarrow$ Train a Seq2Seq model in advance;

$P \leftarrow$ Randomly initialize *pop* solutions; `//Each solution is an architecture`

$P \leftarrow$ Decode and evaluate the solutions in P via training and validation;

Seq2Rank $\leftarrow$ Train a Seq2Rank model based on *Seq2Seq* and P;

for $g = 1$ *to Gen* **do**

    $P' \leftarrow$ Select parents from P via the mating selection in [24];

    $P_{new} \leftarrow$ Generate an offspring population based on P' via the proposed crossover and mutation;

    if $g \in GenList$ **then**

        $P_{new} \leftarrow$ Decode and evaluate the solutions in P_{new} via training and validation;

        Seq2Rank $\leftarrow$ Fine-tune *Seq2Rank* based on P_{new};

    else

        $P_{new} \leftarrow$ Evaluate the solutions in P_{new} via *Seq2Rank*;

    $P \leftarrow P \cup P_{new}$;

    $P \leftarrow$ Select half the solutions from P via the environmental selection in [24];

return P;

---

until the evolution termination criteria is satisfied, after which the non-dominated individuals will be output.

9.3.5 Experimental Studies

9.3.5.1 Performance Verification of Action Command Encoding

Actually, the action command encoding scheme devised in the proposed ACEncoding can be used for any other optimizers, including RL and gradient optimization. To validate the effectiveness of the devised action command encoding scheme in encoding, we compared the devised encoding scheme with NSGA-Net, MetaQNN [82] and NAO [21], where NSGA-Net is based on an EA and the adjacent matrix-based encoding, MetaQNN is based on RL and the hyperparameter-based encoding, and NAO is based on GD and the block-based encoding. Hence, it is reasonable to adopt these three methods to compare the proposed action command encoding scheme with other encoding schemes on different optimizers. The parameter settings mainly follow those in the original papers of the three methods, the details of which are given in the following.

Structure of CNN. The widely used cell structure is adopted in ACEncoding to decrease the search space. Specifically, a normal cell and a reduction cell are encoded by the proposed encoding scheme independently, and the whole CNN consists of three normal cells and two reduction cells stacked alternately. For the hyperparameters in the convolutional and pooling layers, the filter size is determined by the encoding, the number of filters is always set to 16, and the stride size is set to 2 for the layers directly connected to the reduction cell's inputs and 1 for all the other layers.

ACEncoding and NSGA-Net. The only difference between NSGA-Net and NSGA-Net is the adopted encoding scheme, where NSGA-Net adopts the adjacent matrix-based encoding. When integrating with the proposed action command encoding, the optimizer randomly initializes a population with 40 architectures having 10–20 action commands for each cell, and then evolves the population for 25 generations to generate 1000 architectures in total. Since the crossover and mutation operators in NSGA-Net are tailored for the binary adjacent matrix, new crossover and mutation operators are designed for the proposed action command encoding scheme. To be specific, the crossover operator randomly selects an anchor point from the encodings of two parent architectures and exchanges the action commands after the anchor point of the two encodings to generate two offspring architectures. The offspring architectures are then mutated by one of the three operations, including adding a random action command, randomly deleting an action command, and randomly changing the variables of an action command. The crossover probability is 1 and the mutation probability is 0.1.

ACEncoding with RL as the optimizer, termed ACEncoding under RL. MetaQNN iteratively generates architectures according to a Q-table and updates the Q-table according to the performance of the generated architectures via Q-learning. When integrating with the proposed action command encoding, each action command is regarded as a state and the Q-table stores the value $Q(s_i, s_j)$ denoting the reward of selecting action command s_j after action command s_i. To generate an architecture, the agent sequentially selects action commands via the ϵ-greedy strategy according to the Q-table, until the termination state is selected or 20 action commands have been selected. This procedure repeats twice to generate the normal cell and reduction cell of an architecture, and the generated architecture is evaluated and stored in a replay dictionary for periodically updating the Q-values between action commands in the Q-table. The Q-learning rate is set to 0.1, the discount factor is set to 1 (i.e., there are no over-prioritize short-term rewards), and the parameter ϵ is set to 1, 0.9, 0.8, 0.7,

0.6, 0.5, 0.4, 0.3, 0.2, and 0.1 when 200, 400, 500, 600, 700, 800, 850, 900, 950, and 1000 architectures have been evaluated.

ACEncoding with GD as the optimizer, termed ACEncoding under GD. NAO adopts an encoder LSTM to embed the architecture into a continuous space and takes the continuous representation as input to predict its performance by a feedforward neural network. Thus, better continuous representation can be obtained by moving the input along the gradient direction induced by the performance, and the updated continuous representation is fed into a decoder LSTM to generate a new architecture. When integrating with the proposed action command encoding, the optimizer randomly initializes 600 architectures having 20 action commands for each cell, as the length of the decoder is fixed in NAO. Then, these architectures are evaluated and used to train the two LSTMs and the feedforward neural network jointly. Afterward, 200 new architectures are generated by performing GD on the best 200 architectures. The last step repeats twice to form 1000 architectures generated in total. The sizes of the embedding layer, the hidden layer of the LSTMs, and the hidden layer of the feedforward neural network are set to 64, 128, and 200, respectively, and the models are trained using Adam [6] for 1000 epochs with a learning rate of 0.001.

Performance evaluation strategy. All the above three methods generate 1000 architectures during the whole search process. Each architecture is evaluated via training and validation without using parameter sharing or surrogate model, where the 50,000 images in CIFAR-10 are randomly split into a training set of 45,000 samples and a validation set of 5000 samples. Each architecture is trained by SGD with a batch size of 128, a momentum of 0.9, and a learning rate ranging from 0.1 to 0.001 following a single-period cosine schedule. The training samples are normalized and randomly augmented by several data pre-processing techniques, including random crop (i.e., padding each image by 4 pixels and cropping a random 32×32 patch), horizontal flipping [13], and cutout (with a mask length of 16) [84]. Besides, the scheduled path dropout (with a probability of 0.2) [56] and an auxiliary classifier (appended to the architecture at 2/3 depth with a constant factor of 0.4) [16] are adopted in the training process.

Figure 9.20 plots the minimum validation error during the search process of the original NSGA-Net, MetaQNN, and NAO and those integrated with the proposed action command encoding scheme, where NSGA-Net with the devised encoding scheme is actually ACEncoding. It can be found that the proposed action command encoding scheme is competitive to the adjacent matrix-based encoding on NSGA-Net, while the devised encoding scheme converges obviously faster than the hyperparameter-based encoding on MetaQNN and the block-based encoding on NAO. Therefore, it can be concluded that the proposed action command encoding scheme is suitable for different optimizers including EAs, RL, and GD, and it is competitive to or more effective than existing encoding schemes.

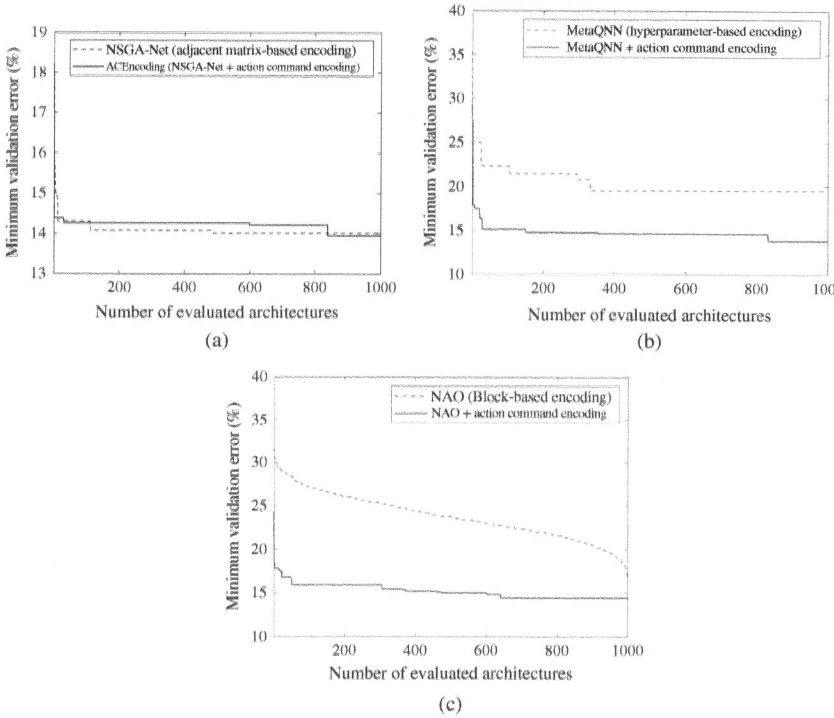

Figure 9.20 Minimum validation error during the search process of the original NSGA-Net (a), MetaQNN (b), and NAO (c) and those integrated with the proposed action command encoding scheme on CIFAR-10.

9.3.5.2 Performance Comparison on CIFAR-10

Table 9.6 presents the test error of 31 CNNs found by human experts and 17 NAS methods. It can be observed that the CNNs obtained by ACEncoding have 2.80% to 3.16% test errors and 2.4M to 3.2M parameters, which are quite competitive with the other CNNs listed in the table. As for ACEncoding, the test error of the obtained CNN is slightly higher than the CNN obtained by ACEncoding without Seq2Rank; by contrast, the search cost can decrease to 1.5 GPU days from 7 GPU days. Furthermore, the CNN obtained by the proposed ACEncoding is enlarged by setting the number of filters to 96, denoted as ACEncoding (more filters), which can lead to a test error of 2.32%, better than all the other CNNs in the table.

Figure 9.21 depicts the test error and number of parameters of the CNNs listed in Table 9.6. It can be found that there are four CNNs having the best trade-offs between test error and complexity, where two of them are obtained by the proposed ACEncoding, one of them is obtained by NASNet, and the remaining

Table 9.6 Comparison of Performance Between the CNNs Found by Human Experts and Neural Architecture Search Methods on CIFAR-10.

Architecture	Test error (%)	Number of parameters (M)	Search cost (GPU days)	Optimizer
Wide ResNet ($k = 2$) [78]	5.33	2.2	—	Manual
Wide ResNet ($k = 4$)	4.97	8.9	—	Manual
Wide ResNet ($k = 10$)	4.17	36.5	—	Manual
DenseNet ($k = 12$) [14]	4.10	7.0	—	Manual
DenseNet ($k = 24$)	3.74	27.2	—	Manual
DenseNet-BC ($k = 40$)	3.46	25.6	—	Manual
AmoeabaNet-A [79]	3.34	3.2	3150	EA
AmoeabaNet-B	3.37	2.8	3150	EA
Hier-EA [80]	3.75	15.7	300	EA
Hier-EA (more filters)	3.63	—	300	EA
Large-scale evolution [76]	5.40	5.4	2666	EA
NSGA-Net [24]	2.75	3.3	4	EA
NSGA-Net (more filters)	2.50	26.8	4	EA
BlockQNN-S [81]	4.38	6.1	96	RL
BlockQNN-S (more filters)	3.54	39.8	96	RL
ENAS [18]	2.89	4.6	0.45	RL
MetaQNN [82]	6.92	11.2	80	RL
NAS v3 [17]	4.47	7.1	3150	RL
NAS v3 (more filters)	3.65	37.4	3150	RL
NASNet-A [56]	2.65	3.3	2000	RL
NASNet-A (more filters)	2.40	27.6	2000	RL
NASNet-B	3.73	2.6	2000	RL
NASNet-C	3.59	3.1	2000	RL
DARTS (1st order) [20]	3.00	3.3	1.5	GD
DARTS (2nd order)	2.76	3.3	4	GD
NAONet [21]	3.18	10.6	200	GD
NAONet (more filters)	2.98	28.6	200	GD
NAONet + WS	3.53	2.5	0.3	GD
BOHB [83]	3.18	27.6	32	BO
NASBOT [22]	12.09	—	40	BO
PNASNet [73]	3.41	3.2	225	HS

Table 9.6 (Continued)

Architecture	Test error (%)	Number of parameters (M)	Search cost (GPU days)	Optimizer
ACEncoding without Seq2Rank	2.80	3.0	7	EA
ACEncoding under RL	3.16	2.9	8.3	RL
ACEncoding under GD	3.03	3.2	12.4	GD
ACEncoding	2.90	2.4	1.5	EA
ACEncoding (more filters)	2.32	9.8	1.5	EA

"BO" refers to Bayesian optimization; "HS" refers to Heuristic search.

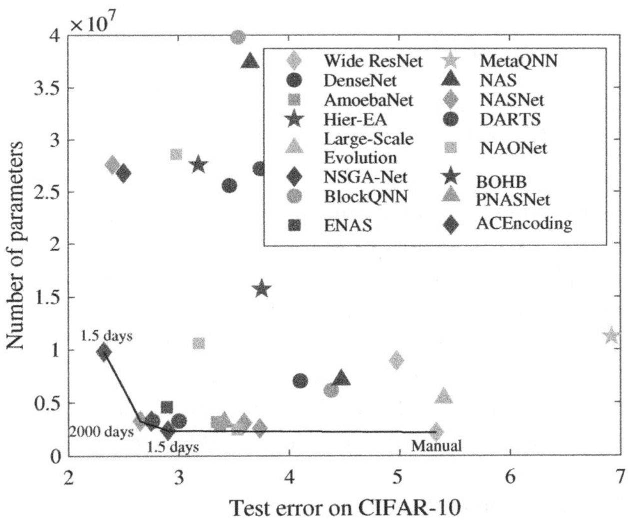

Figure 9.21 Test error and number of parameters of the CNNs found by human experts and neural architecture search methods on CIFAR-10, where the CNNs having the best trade-offs between test error and complexity are connected.

one is wide ResNet. It should be noted that NASNet obtains the CNN by searching for 2000 GPU days, whereas the proposed ACEncoding consumes only 1.5 GPU days. Although the wide ResNet is manually designed with no search cost, its test error is much higher than the other three CNNs. Besides, Fig. 9.22 plots the test error and search cost of the CNNs, where the best trade-off CNN found by the proposed ACEncoding has larger search cost but lower test error than the other trade-off CNNs. In short, the proposed ACEncoding can strike a balance among

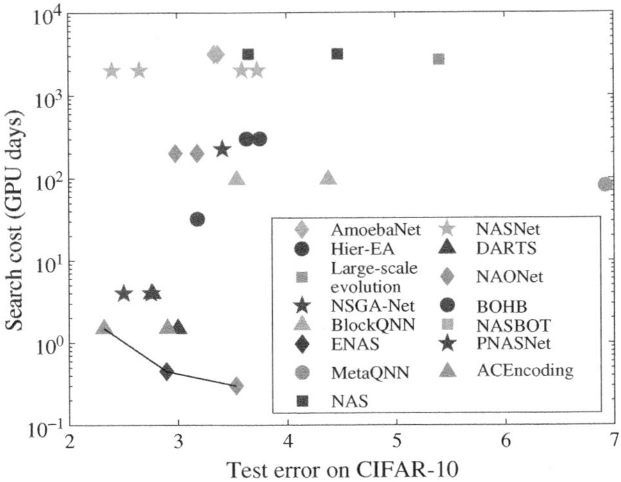

Figure 9.22 Test error and search cost of the CNNs found by neural architecture search methods on CIFAR-10, where the CNNs having the best trade-offs between test error and search cost are connected.

the search performance, the complexity of architectures, and the computational cost of search process.

For a visual observation, Fig. 9.23 depicts the architectures of the normal cell and reduction cell found by the proposed ACEncoding. It can be seen that the architectures have more than 10 layers with many connections. For adjacent matrix-based encoding and graph-based encoding, they require a long sequence of variables to represent such complex architectures. Besides, the block-based encoding is almost impossible to represent these architectures by the block-based topology.

9.3.5.3 Performance Comparison on CIFAR-100

The transferability of the architectures found by the proposed ACEncoding and its variants is then verified on CIFAR-100 [58], where the same CNNs are trained by the 50,000 training samples of CIFAR-100 and tested on the 10,000 test samples. According to the results listed in Table 9.7, the CNNs obtained by the proposed ACEncoding and its variants also have competitive test errors and complexities to the others. Besides, the CNN generated based on the second best architecture found by the proposed ACEncoding is evaluated, which has higher complexity but lower test error than the other CNNs obtained by its variants.

Figure 9.24 depicts the test error and number of parameters of the CNNs listed in Table 9.6. Obviously, the proposed ACEncoding can also obtain two trade-off

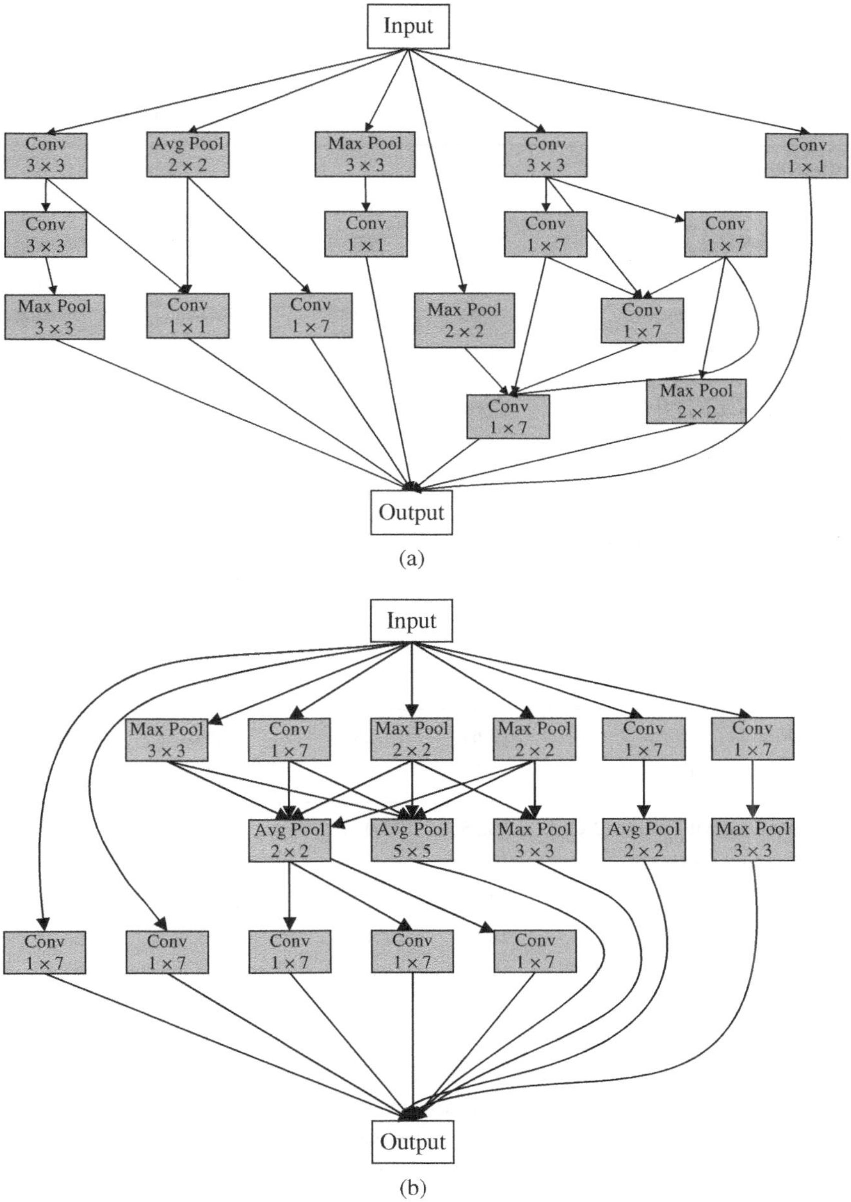

Figure 9.23 Architectures of the normal cell (a) and reduction cell (b) found by the evolutionary algorithm with ACEncoding and Seq2Rank.

Table 9.7 Comparison of Performance Between the Architectures Found by Human Experts and Neural Architecture Search Methods on CIFAR-100.

Architecture	Test error (%)	Number of parameters (M)	Search cost (GPU days)	Optimizer
Wide ResNet ($k = 2$) [78]	26.04	2.2	—	Manual
Wide ResNet ($k = 4$)	22.89	8.9	—	Manual
Wide ResNet ($k = 10$)	20.50	36.5	—	Manual
DenseNet ($k = 12$) [14]	20.20	7.0	—	Manual
DenseNet ($k = 24$)	19.25	27.2	—	Manual
DenseNet-BC ($k = 40$)	17.18	25.6	—	Manual
AmoeabaNet-B [79]	17.66	34.9	3150	EA
Large-scale evolution [76]	23.00	40.4	2666	EA
NSGA-Net [24]	20.74	3.3	8	EA
NSGA-Net (more filters)	19.83	11.6	8	EA
BlockQNN-S [81]	20.65	6.1	96	RL
BlockQNN-S (more filters)	18.06	39.8	96	RL
ENAS [18]	19.43	4.6	0.45	RL
MetaQNN [82]	27.14	11.2	80	RL
NAONet [21]	15.67	10.8	200	GD
NAONet (more filters)	14.75	128.0	200	GD
ACEncoding without Seq2Rank	18.65	3.0	7	EA
ACEncoding under RL	19.65	2.9	8.3	RL
ACEncoding under GD	18.69	3.2	12.4	GD
ACEncoding	18.84	2.4	1.5	EA
ACEncoding (more filters)	17.59	9.8	1.5	EA

CNNs, and NAO can obtain one trade-off CNN. Although the CNN obtained by NAO has the lowest test error, it is found by searching for 200 GPU days, which is much higher than the 1.5 GPU days consumed by the proposed ACEncoding. As can be further observed from Fig. 9.25, the CNN found by the proposed ACEncoding strikes good balance between test error and search cost. In short, the architectures found by the proposed ACEncoding are transferable to more complex datasets.

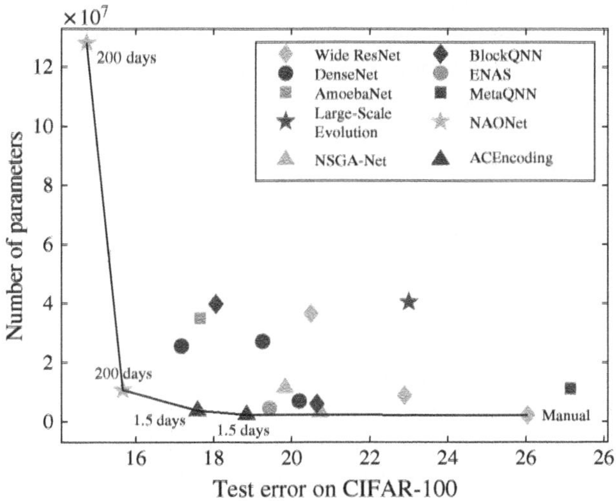

Figure 9.24 Test error and number of parameters of the CNNs found by human experts and neural architecture search methods on CIFAR-100, where the CNNs having the best trade-offs between test error and complexity are connected.

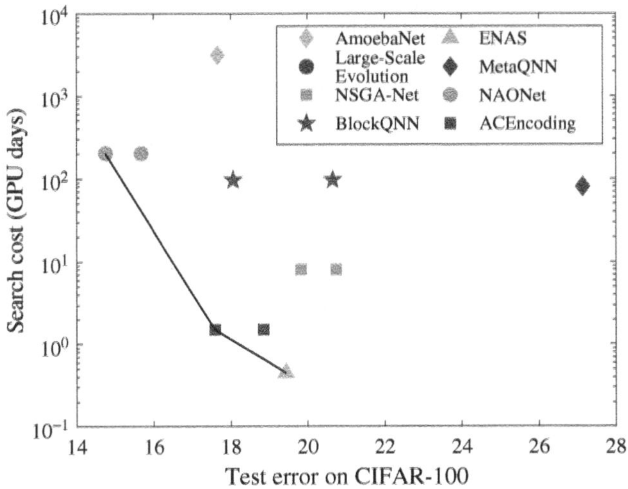

Figure 9.25 Test error and search cost of the CNNs found by neural architecture search methods on CIFAR-100, where the CNNs having the best trade-offs between test error and search cost are connected.

9.3.5.4 Performance Comparison on Fashion-MNIST

Lastly, Table 9.8 shows the test error and number of parameters of the architectures found by human experts and NAS methods on Fashion-MNIST, which is an image classification dataset containing 60,000 training samples and 10,000 test samples [87]. It is obvious that the test error of the CNN obtained by the proposed ACEncoding is lower than all the other CNNs besides the one found by NAONet. Besides, the CNN found by NAONet has a significantly larger complexity than the CNN found by the proposed method. The experimental results further demonstrate that the proposed ACEncoding can hold a good balance between the performance and complexity of architectures.

9.3.5.5 Performance Verification of Seq2Rank

The proposed performance evaluator Seq2Rank is compared to two existing surrogate models for predicting the performance of architectures, namely, E2EPP [88] and Peephole [89]. E2EPP is an end-to-end performance predictor based on random forest [90], which is effective to a limited number of training samples and can directly handle the discrete variables without embedding. Peephole holds a similar pipeline to the proposed Seq2Rank, which embeds the architecture by an

Table 9.8 Comparison of Performance Between the Architectures Found by Human Experts and Neural Architecture Search Methods on Fashion-MNIST.

Architecture (%)	Test error	Number of parameters (M)	Optimizer
VGG16 [54]	6.50	26.0	Manual
GoogLeNet [15]	6.30	101.0	Manual
ResNet ($k = 18$) [13]	5.10	33.2	Manual
Wide ResNet ($k = 10$) [78]	4.01	36.5	Manual
DENSER [85]	4.89	—	EA
EvoCNN [86]	7.28	6.5	EA
NSGANet [24]	4.07	3.4	EA
MetaQNN [82]	10.02	12.5	RL
NAONet [21]	3.50	11.5	GD
DARTS (2nd order) [20]	3.94	3.4	GD
ACEncoding	3.91	2.2	EA

Table 9.9 Kendall's Tau Values of E2EPP, Peephole, and the Proposed Seq2Rank for Predicting the Rankings of the Architectures Represented by the Devised Action Command Encoding.

Number of training samples	E2EPP	Peephole	Seq2Rank
1000	**0.2595**	0.1636	0.1996
3000	0.1168	0.1586	**0.5157**
5000	0.0573	0.0632	**0.6260**
7000	0.0536	0.2253	**0.5366**
9000	0.0304	0.3402	**0.6754**

LSTM with embedding layer and predicts the performance by a feedforward neural network. However, Peephole concatenates all the hidden states but not using mean pooling or attention mechanism, and it predicts the validation accuracy rather than the ranking of each architecture. More importantly, Seq2Rank uses the Seq2Seq model to learn a continuous representation in an unsupervised manner, while Peephole directly trains the whole model in a supervised manner. For the models in Peephole and Seq2Rank, the sizes of the embedding layer, the hidden layer of the LSTM, and the two hidden layers of the feedforward neural network are set to 62, 120, 60, and 30, respectively, and the models are trained using SGD for 25 epochs with a learning rate of 0.1.

A number of 10,000 architectures represented by the devised action command encoding are randomly generated to form the dataset, where different numbers of samples are used as training samples and the rest are used as test samples. Table 9.9 lists the performance of E2EPP, Peephole, and Seq2Rank in terms of the Kendall's Tau metric, which is calculated by [91]

$$KTau = 2 \times \frac{M}{C_N^2} - 1, \tag{9.24}$$

where N is the total number of test samples and M is the concordant pair that the predicted ranking is correct. Obviously, the value of Kendall's Tau ranges from -1 to 1 and a higher value corresponds to a higher accuracy. It can be found from the table that Peephole always has a worse performance than Seq2Rank, and E2EPP outperforms Seq2Rank only with 1000 training samples due to the superiority of random forest on small training sets. To summarize, it can be verified that Seq2Rank is more suitable for predicting the performance of architectures represented by the devised encoding than some existing surrogate-assisted performance evaluators.

9.4 Summary

This chapter presents a gradient guide multi-objective evolutionary algorithm (GEMONN) and an action command encoding-based surrogate-assisted EA (ACEncoding) for the training of DNNs and NAS, respectively. Specifically, the proposed GEMONN focuses on designing a powerful genetic operator gSBX to directly optimize the weights of DNNs, by means of taking advantage of the gradient information. Moreover, the network sparsity is also optimized together with the training loss, which can reduce the network complexity and alleviate overfitting. Experimental results on single-layer NNs, deep-layer NNs, recurrent NNs, and CNNs demonstrate the effectiveness of the proposed GEMONN. In short, this work not only introduces a novel approach for training deep NNs but also enhances the performance of EAs in solving large-scale optimization problems.

For the proposed ACEncoding, a novel encoding scheme and a performance evaluator are presented for NAS. The proposed encoding scheme defines a set of action commands and encodes each architecture in a sequence consisting of the codons of multiple action commands. The proposed encoding scheme inherits the advantages of existing encoding schemes that can represent complex architectures as a relatively short sequence and has shown better effectiveness than some popular encoding schemes in EAs, RL, and GD. The proposed performance evaluator Seq2Rank adopts a Seq2Seq model to convert the variable-length encoding of the action command into a fixed-length sequence and then predicts the ranking of the architecture determined by the encoding via RankNet. The proposed performance evaluator has been empirically verified to be better than some existing surrogate models in predicting the performance of architectures represented by the proposed ACEncoding.

References

1 M. Niepert, M. Ahmed, and K. Kutzkov, "Learning convolutional neural networks for graphs," in *Proceedings of the 2016 International Conference on Machine Learning*, 2016, pp. 2014–2023.

2 T. Young, D. Hazarika, S. Poria, and E. Cambria, "Recent trends in deep learning based natural language processing [review article]," *IEEE Computational Intelligence Magazine*, vol. 13, no. 3, pp. 55–75, 2018.

3 W. Liu, D. Anguelov, D. Erhan, C. Szegedy, S. Reed, C.-Y. Fu, and A. C. Berg, "SSD: Single shot multibox detector," in *Proceedings of the 2016 European Conference on Computer Vision*, 2016, pp. 21–37.

4 H. Robbins and S. Monro, "A stochastic approximation method," *The Annals of Mathematical Statistics*, vol. 22, no. 3, pp. 400–407, 1951.

5 T. Tieleman and G. Hinton, "Lecture 6.5-rmsprop: divide the gradient by a running average of its recent magnitude," *COURSERA: Neural Networks for Machine Learning*, vol. 4, no. 2, pp. 26–31, 2012.

6 D. P. Kingma and J. Ba, "Adam: A method for stochastic optimization," *arXiv preprint arXiv:1412.6980*, 2014.

7 X. Yao, "Evolving artificial neural networks," *Proceedings of the 1999 IEEE*, vol. 87, no. 9, pp. 1423–1447, 1999.

8 X. Cui, W. Zhang, Z. Tüske, and M. Picheny, "Evolutionary stochastic gradient descent for optimization of deep neural networks," in *Proceedings of the 2018 Advances in Neural Information Processing Systems*, vol. 31, 2018, pp. 6048–6058.

9 Y. Jin and B. Sendhoff, "Pareto-based multiobjective machine learning: an overview and case studies," *IEEE Transactions on Systems, Man, and Cybernetics, Part C (Applications and Reviews)*, vol. 38, no. 3, pp. 397–415, 2008.

10 J. Liu, M. Gong, Q. Miao, X. Wang, and H. Li, "Structure learning for deep neural networks based on multiobjective optimization," *IEEE Transactions on Neural Networks and Learning Systems*, vol. 29, no. 6, pp. 2450–2463, 2017.

11 Y. Sun, G. G. Yen, and Z. Yi, "Evolving unsupervised deep neural networks for learning meaningful representations," *IEEE Transactions on Evolutionary Computation*, vol. 23, no. 1, pp. 89–103, 2018.

12 F. Girosi, M. Jones, and T. Poggio, "Regularization theory and neural networks architectures," *Neural Computation*, vol. 7, no. 2, pp. 219–269, 1995.

13 K. He, X. Zhang, S. Ren, and J. Sun, "Deep residual learning for image recognition," in *Proceedings of the 2016 IEEE Conference on Computer Vision and Pattern Recognition*, 2016, pp. 770–778.

14 G. Huang, Z. Liu, L. Van Der Maaten, and K. Q. Weinberger, "Densely connected convolutional networks," in *Proceedings of the 2017 IEEE Conference on Computer Vision and Pattern Recognition*, 2017, pp. 4700–4708.

15 C. Szegedy, W. Liu, Y. Jia, P. Sermanet, S. Reed, D. Anguelov, D. Erhan, V. Vanhoucke, and A. Rabinovich, "Going deeper with convolutions," in *Proceedings of the 2015 IEEE Conference on Computer Vision and Pattern Recognition*, 2015, pp. 1–9.

16 C. Szegedy, V. Vanhoucke, S. Ioffe, J. Shlens, and Z. Wojna, "Rethinking the inception architecture for computer vision," in *Proceedings of the 2016 IEEE Conference on Computer Vision and Pattern Recognition*, 2016, pp. 2818–2826.

17 B. Zoph and Q. V. Le, "Neural architecture search with reinforcement learning," *arXiv preprint arXiv:1611.01578*, 2016.

18 H. Pham, M. Guan, B. Zoph, Q. Le, and J. Dean, "Efficient neural architecture search via parameters sharing," in *Proceedings of the International Conference on Machine Learning*, 2018, pp. 4095–4104.

19 Y. Chen, G. Meng, Q. Zhang, S. Xiang, C. Huang, L. Mu, and X. Wang, "RE-NAS: Reinforced evolutionary neural architecture search," in *Proceedings of the 2019 IEEE Conference on Computer Vision and Pattern Recognition*, 2019, pp. 4787–4796.

20 H. Liu, K. Simonyan, and Y. Yang, "Darts: differentiable architecture search," *arXiv preprint arXiv:1806.09055*, 2018.

21 R. Luo, F. Tian, T. Qin, E. Chen, and T.-Y. Liu, "Neural architecture optimization," in *Proceedings of the 2018 Advances in Neural Information Processing Systems*, 2018, pp. 7827–7838.

22 K. Kandasamy, W. Neiswanger, J. Schneider, B. Poczos, and E. P. Xing, "Neural architecture search with Bayesian optimisation and optimal transport," in *Proceedings of the 2018 Advances in Neural Information Processing Systems*, 2018, pp. 2020–2029.

23 T. Elsken, J. H. Metzen, and F. Hutter, "Neural architecture search: a survey," *The Journal of Machine Learning Research*, vol. 20, no. 1, pp. 1997–2017, 2019.

24 Z. Lu, I. Whalen, V. Boddeti, Y. Dhebar, K. Deb, E. Goodman, and W. Banzhaf, "NSGA-Net: neural architecture search using multi-objective genetic algorithm," in *Proceedings of the 2019 Genetic and Evolutionary Computation Conference*, 2019, pp. 419–427.

25 L. D. Whitley, *The GENITOR algorithm and selection pressure: why rank-based allocation of reproductive trials is best*. Colorado State University, Department of Computer Science, 1989.

26 D. J. Montana and L. Davis, "Training feedforward neural networks using genetic algorithms," in *Proceedings of the 11th International Joint Conference on Artificial Intelligence*, 1989, pp. 762–767.

27 D. Whitley, T. Starkweather, and C. Bogart, "Genetic algorithms and neural networks: optimizing connections and connectivity," *Parallel Computing*, vol. 14, no. 3, pp. 347–361, 1990.

28 E. Ronald and M. Schoenauer, "Genetic lander: an experiment in accurate neuro-genetic control," in *Proceedings of the 1994 International Conference on Parallel Problem Solving from Nature*, 1994, pp. 452–461.

29 K. O. Stanley and R. Miikkulainen, "Evolving neural networks through augmenting topologies," *Evolutionary Computation*, vol. 10, no. 2, pp. 99–127, 2002.

30 K. O. Stanley, "Compositional pattern producing networks: a novel abstraction of development," *Genetic Programming and Evolvable Machines*, vol. 8, no. 2, pp. 131–162, 2007.

31 M. Gong, J. Liu, H. Li, Q. Cai, and L. Su, "A multiobjective sparse feature learning model for deep neural networks," *IEEE Transactions on Neural Networks and Learning Systems*, vol. 26, no. 12, pp. 3263–3277, 2015.

32 Y. Tian, X. Zheng, X. Zhang, and Y. Jin, "Efficient large-scale multiobjective optimization based on a competitive swarm optimizer," *IEEE Transactions on Cybernetics*, vol. 50, no. 8, pp. 3696–3708, 2019.

33 F. P. Such, V. Madhavan, E. Conti, J. Lehman, K. O. Stanley, and J. Clune, "Deep neuroevolution: genetic algorithms are a competitive alternative for training deep neural networks for reinforcement learning," *arXiv preprint arXiv:1712.06567*, 2017.

34 C. He, L. Li, Y. Tian, X. Zhang, R. Cheng, Y. Jin, and X. Yao, "Accelerating large-scale multiobjective optimization via problem reformulation," *IEEE Transactions on Evolutionary Computation*, vol. 23, no. 6, pp. 949–961, 2019.

35 K. Deb and R. B. Agrawal, "Simulated binary crossover for continuous search space," *Complex Systems*, vol. 9, no. 2, pp. 115–148, 1995.

36 H. Zhu, H. Zhang, and Y. Jin, "From federated learning to federated neural architecture search: a survey," *Complex & Intelligent Systems*, vol. 7, pp. 639–657, 2021.

37 S. Liu, H. Zhang, and Y. Jin, "A survey on computationally efficient neural architecture search," *Journal of Automation and Intelligence*, vol. 1, no. 1, p. 100002, 2022.

38 H. Zhu and Y. Jin, "Real-time federated evolutionary neural architecture search," *IEEE Transactions on Evolutionary Computation*, vol. 26, no. 2, pp. 364–378, 2021.

39 Y. Jin, T. Okabe, and B. Sendhoff, "Neural network regularization and ensembling using multi-objective evolutionary algorithms," in *Proceedings of the 2004 IEEE Congress on Evolutionary Computation*, 2004, pp. 1–8.

40 H. Kita, I. Ono, and S. Kobayashi, "Multi-parental extension of the unimodal normal distribution crossover for real-coded genetic algorithms," *Transactions of the Society of Instrument and Control Engineers*, vol. 36, no. 10, pp. 875–883, 2000.

41 R. Storn and K. Price, "Differential evolution–a simple and efficient heuristic for global optimization over continuous spaces," *Journal of Global Optimization*, vol. 11, no. 4, pp. 341–359, 1997.

42 N. Hitomi and D. Selva, "A classification and comparison of credit assignment strategies in multiobjective adaptive operator selection," *IEEE Transactions on Evolutionary Computation*, vol. 21, no. 2, pp. 294–314, 2016.

43 Y. Tian, X. Zhang, C. Wang, and Y. Jin, "An evolutionary algorithm for large-scale sparse multi-objective optimization problems," *IEEE Transactions on Evolutionary Computation*, vol. 24, no. 2, pp. 380–393, 2019.

44 K. Deb, A. Pratap, S. Agarwal, and T. Meyarivan, "A fast and elitist multiobjective genetic algorithm: NSGA-II," *IEEE Transactions on Evolutionary Computation*, vol. 6, no. 2, pp. 182–197, 2002.

45 P. Vincent, H. Larochelle, Y. Bengio, and P.-A. Manzagol, "Extracting and composing robust features with denoising autoencoders," in *Proceedings of the 25th International Conference on Machine Learning*, 2008, pp. 1096–1103.

46 L. Bottou, F. E. Curtis, and J. Nocedal, "Optimization methods for large-scale machine learning," *Siam Review*, vol. 60, no. 2, pp. 223–311, 2018.

47 J. Branke, K. Deb, H. Dierolf, and M. Osswald, "Finding knees in multi-objective optimization," in *Proceedings of the 2004 International Conference on Parallel Problem Solving from Nature*, 2004, pp. 722–731.

48 Y. Mu, W. Liu, X. Liu, and W. Fan, "Stochastic gradient made stable: a manifold propagation approach for large-scale optimization," *IEEE Transactions on Knowledge and Data Engineering*, vol. 29, no. 2, pp. 458–471, 2016.

49 J. Zhai, S. Zhang, J. Chen, and Q. He, "Autoencoder and its various variants," in *Proceedings of the 2018 IEEE International Conference on Systems, Man, and Cybernetics*, 2018, pp. 415–419.

50 S. Hochreiter and J. Schmidhuber, "Long short-term memory," *Neural Computation*, vol. 9, no. 8, pp. 1735–1780, 1997.

51 G. E. Hinton, S. Osindero, and Y.-W. Teh, "A fast learning algorithm for deep belief nets," *Neural Computation*, vol. 18, no. 7, pp. 1527–1554, 2006.

52 Y. LeCun, L. Bottou, Y. Bengio, and P. Haffner, "Gradient-based learning applied to document recognition," *Proceedings of the 1998 IEEE*, vol. 86, no. 11, pp. 2278–2324, 1998.

53 G. E. Dahl, T. N. Sainath, and G. E. Hinton, "Improving deep neural networks for LVCSR using rectified linear units and dropout," in *Proceedings of the 2013 IEEE International Conference on Acoustics, Speech and Signal Processing*, 2013, pp. 8609–8613.

54 K. Simonyan and A. Zisserman, "Very deep convolutional networks for large-scale image recognition," *arXiv preprint arXiv:1409.1556*, 2014.

55 L. While, P. Hingston, L. Barone, and S. Huband, "A faster algorithm for calculating hypervolume," *IEEE Transactions on Evolutionary Computation*, vol. 10, no. 1, pp. 29–38, 2006.

56 B. Zoph, V. Vasudevan, J. Shlens, and Q. V. Le, "Learning transferable architectures for scalable image recognition," in *Proceedings of the 2018 IEEE Conference on Computer Vision and Pattern Recognition*, 2018, pp. 8697–8710.

57 H. Larochelle, D. Erhan, A. Courville, J. Bergstra, and Y. Bengio, "An empirical evaluation of deep architectures on problems with many factors of variation," in *Proceedings of the 24th International Conference on Machine Learning*, 2007, pp. 473–480.

58 A. Krizhevsky and G. Hinton, *Learning multiple layers of features from tiny images*. University of Toronto, 2009.

59 H. A. Dau, A. Bagnall, K. Kamgar, C.-C. M. Yeh, Y. Zhu, S. Gharghabi, C. A. Ratanamahatana, and E. Keogh, "The UCR time series archive," *IEEE/CAA Journal of Automatica Sinica*, vol. 6, no. 6, pp. 1293–1305, 2019.

60 P. Vincent, H. Larochelle, I. Lajoie, Y. Bengio, and P.-A. Manzagol, "Stacked denoising autoencoders: learning useful representations in a deep network with a local denoising criterion," *Journal of Machine Learning Research*, vol. 11, no. 12, pp. 3371–3408, 2010.

61 S. Rifai, P. Vincent, X. Muller, X. Glorot, and Y. Bengio, "Contractive auto-encoders: explicit invariance during feature extraction," in *Proceedings of the 28th International Conference on International Conference on Machine Learning*, 2011, pp. 833–840.

62 B. A. Olshausen and D. J. Field, "Sparse coding with an overcomplete basis set: a strategy employed by V1?" *Vision Research*, vol. 37, no. 23, pp. 3311–3325, 1997.

63 A. Paszke, S. Gross, S. Chintala, G. Chanan, E. Yang, Z. DeVito, Z. Lin, A. Desmaison, L. Antiga, and A. Lerer, "Automatic differentiation in pytorch," in *Proceedings of the 31st Conference on Neural Information Processing Systems*, 2017, pp. 1–4.

64 S. Van Der Walt, S. C. Colbert, and G. Varoquaux, "The NumPY array: a structure for efficient numerical computation," *Computing in Science & Engineering*, vol. 13, no. 2, pp. 22–30, 2011.

65 R. G. D. Steel, J. H. Torrie, and D. A. Dickey, *Principles and procedures of statistics: a biometrical approach*. McGraw-Hill College, 1980.

66 J. Frankle and M. Carbin, "The lottery ticket hypothesis: finding sparse, trainable neural networks," in *Proceedings of the 2018 International Conference on Learning Representations*, 2018, pp. 1–13.

67 X. Yao, "A review of evolutionary artificial neural networks," *International Journal of Intelligent Systems*, vol. 8, no. 4, pp. 539–567, 1993.

68 J. Bergstra, D. Yamins, and D. D. Cox, "Making a science of model search: hyperparameter optimization in hundreds of dimensions for vision architectures," in *Proceedings of the 30th International Conference on Machine Learning*, 2013, pp. 115–123.

69 J. S. Bergstra, R. Bardenet, Y. Bengio, and B. Kégl, "Algorithms for hyper-parameter optimization," in *Proceedings of the 2011 Advances in Neural Information Processing Systems*, 2011, pp. 2546–2554.

70 R. J. Williams, "Simple statistical gradient-following algorithms for connectionist reinforcement learning," *Machine Learning*, vol. 8, no. 3-4, pp. 229–256, 1992.

71 I. Sutskever, O. Vinyals, and Q. V. Le, "Sequence to sequence learning with neural networks," *Proceedings of the 27th International Conference on Neural Information Processing Systems*, pp. 3104–3112, 2014.

72 C. Burges, T. Shaked, E. Renshaw, A. Lazier, M. Deeds, N. Hamilton, and G. Hullender, "Learning to rank using gradient descent," in *Proceedings of the 22nd International Conference on Machine Learning*, 2005, pp. 89–96.

73 C. Liu, B. Zoph, M. Neumann, J. Shlens, W. Hua, L.-J. Li, L. Fei-Fei, A. Yuille, J. Huang, and K. Murphy, "Progressive neural architecture search," in *Proceedings of the 2018 European Conference on Computer Vision*, 2018, pp. 19–34.

74 X. Dong and Y. Yang, "One-shot neural architecture search via self-evaluated template network," in *Proceedings of the 2019 IEEE International Conference on Computer Vision*, 2019, pp. 3681–3690.

75 R. Kozma, C. Alippi, Y. Choe, and F. C. Morabito, *Artificial intelligence in the age of neural networks and brain computing*. Academic Press, 2018.

76 E. Real, S. Moore, A. Selle, S. Saxena, Y. L. Suematsu, J. Tan, Q. V. Le, and A. Kurakin, "Large-scale evolution of image classifiers," in *Proceedings of the 2017 International Conference on Machine Learning*, 2017, pp. 2902–2911.

77 C. A. Coello Coello, *Evolutionary algorithms for solving multi-objective problems*. Springer, 2007.

78 S. Zagoruyko and N. Komodakis, "Wide residual networks," *arXiv preprint arXiv:1605.07146*, 2016.

79 E. Real, A. Aggarwal, Y. Huang, and Q. V. Le, "Regularized evolution for image classifier architecture search," in *Proceedings of the 2019 AAAI Conference on Artificial Intelligence*, 2019, pp. 4780–4789.

80 H. Liu, K. Simonyan, O. Vinyals, C. Fernando, and K. Kavukcuoglu, "Hierarchical representations for efficient architecture search," *arXiv preprint arXiv:1711.00436*, 2017.

81 Z. Zhong, J. Yan, W. Wu, J. Shao, and C. L. Liu, "Practical block-wise neural network architecture generation," in *Proceedings of the 2018 IEEE Computer Society Conference on Computer Vision and Pattern Recognition*, 2018, pp. 2423–2432.

82 B. Baker, O. Gupta, N. Naik, and R. Raskar, "Designing neural network architectures using reinforcement learning," in *Proceedings of the 5th International Conference on Learning Representations*, 2017, pp. 1–11.

83 A. Zela, A. Klein, S. Falkner, and F. Hutter, "Towards automated deep learning: efficient joint neural architecture and hyperparameter search," *arXiv preprint arXiv:1807.06906*, 2018.

84 T. DeVries and G. W. Taylor, "Improved regularization of convolutional neural networks with cutout," *arXiv preprint arXiv:1708.04552*, 2017.

85 F. A. ao, N. Lourenço, P. Machado, and B. Ribeiro, "DENSER: Deep evolutionary network structured representation," *Genetic Programming and Evolvable Machines*, vol. 20, pp. 5–35, 2019.

86 Y. Sun, B. Xue, M. Zhang, and G. G. Yen, "Evolving deep convolutional neural networks for image classification," *IEEE Transactions on Evolutionary Computation*, vol. 24, no. 2, pp. 394–407, 2019.

87 H. Xiao, K. Rasul, and R. Vollgraf, "Fashion-MNIST: A novel image dataset for benchmarking machine learning algorithms,"*arXiv preprint arXiv:1708.07747*, 2017.

88 Y. Sun, H. Wang, B. Xue, Y. Jin, G. G. Yen, and M. Zhang, "Surrogate-assisted evolutionary deep learning using an end-to-end random forest-based performance predictor," *IEEE Transactions on Evolutionary Computation*, vol. 24, no. 2, pp. 350–364, 2019.

89 B. Deng, J. Yan, and D. Lin, "Peephole: Predicting network performance before training," *arXiv preprint arXiv:1712.03351*, 2017.

90 T. K. Ho, "Random decision forests," in *Proceedings of the 3rd International Conference on Document Analysis and Recognition*, 1995, pp. 278–282.

91 P. K. Sen, "Estimates of the regression coefficient based on Kendall's Tau," *Journal of the American Statistical Association*, vol. 63, no. 324, pp. 1379–1389, 1968.

Index

Evolutionary Large-Scale Multi-Objective Optimization and Applications: Integrating Evolutionary Computation, Machine Learning, and Data Science, First Edition. Xingyi Zhang, Ran Cheng, Ye Tian, and Yaochu Jin.

d

k

l

m

q

r

s

t

Printed and bound by CPI Group (UK) Ltd, Croydon, CR0 4YY

17/09/2026

14964308-0003